Chemical Evolution and the Origin of Life

Horst Rauchfuss

Chemical Evolution and the Origin of Life

Translated by

Terence N. Mitchell

 Springer

Author
Prof. Dr. Horst Rauchfuss
Sandåkergatan 5
432 37 Varberg
Sweden
horst.rauchfuss@tele2.se

Translator
Prof. Dr. Terence N. Mitchell
Universität Dortmund
Fachbereich Chemie
44221 Dortmund
Germany

ISBN: 978-3-540-78822-5   e-ISBN: 978-3-540-78823-2

Library of Congress Control Number: 2008929511

© 2008 Springer-Verlag Berlin Heidelberg

Cover design: J.A. Piliero

Printed on acid-free paper

9 8 7 6 5 4 3 2 1

springer.com

Foreword

How did life begin on the early Earth? We know that life today is driven by the universal laws of chemistry and physics. By applying these laws over the past fifty years, enormous progress has been made in understanding the molecular mechanisms that are the foundations of the living state. For instance, just a decade ago, the first human genome was published, all three billion base pairs. Using X-ray diffraction data from crystals, we can see how an enzyme molecule or a photosynthetic reaction center steps through its catalytic function. We can even visualize a ribosome, central to all life, translate genetic information into a protein. And we are just beginning to understand how molecular interactions regulate thousands of simultaneous reactions that continuously occur even in the simplest forms of life. New words have appeared that give a sense of this wealth of knowledge: The genome, the proteome, the metabolome, the interactome.

But we can't be too smug. We must avoid the mistake of the physicist who, as the twentieth century began, stated confidently that we knew all there was to know about physics, that science just needed to clean up a few dusty corners. Then came relativity, quantum theory, the Big Bang, and now dark matter, dark energy and string theory. Similarly in the life sciences, the more we learn, the better we understand how little we really know. There remains a vast landscape to explore, with great questions remaining.

One such question is the focus of this book. The problem of the origin of life can be a black hole for researchers: If you get too close, you can disappear from sight. Only a few pioneering scientists, perhaps a hundred or so in the international community, have been brave enough to explore around its edges. The question of life's origin is daunting because the breadth of knowledge required to address it spans astronomy, planetary science, geology, paleontology, chemistry, biochemistry, bioenergetics and molecular biology. Furthermore, there will never be a real answer. We can never know the exact process by which life did begin on the Earth, but at best we will only know how it could have begun. But if we do understand this much, we should be able to reproduce the process in the laboratory. This is the gold that draws the prospectors into the hills. We know the prize is there, but we must explore a vast wilderness of unknowns in order to find it.

Perhaps most exciting is that we are now living in a time when enough knowledge has accumulated so that there are initial attempts to fabricate versions of living cells in the laboratory. Entire genomes have been transferred from one bacterial species to another, and it is now possible to reconstitute a system of membranes, DNA, RNA and ribosomes that can synthesize a specific protein in an artificial cell.

Other investigators have shown that the informational molecules of Life – RNA and DNA – themselves can be synthesized within lipid vesicles.

We are getting ever closer to the goal of synthetic life, and when that is achieved we will see more clearly the kinds of molecular systems that were likely to have assembled in the prebiotic environment to produce the first forms of life.

We now think about the beginning of life not as a process restricted to the early Earth, but instead as a narrative that takes into account the origin of the biogenic elements in exploding stars, the gathering of the ashes into vast molecular clouds light years in diameter, the origin of new stars and solar systems by gravitational accretion within such clouds, and finally delivery of organic compounds to planetary surfaces like that of the Earth during late accretion. Only then can the chemical reactions and self-organization begin that leads to the origin of life.

This is the scope covered in this book, hinted at by the images on the cover that range from galaxies to planets to a DNA molecule. Horst Rauchfuss is among those rare few individuals who understand the greater evolutionary narrative, and his book is an account of the conceptual map he has drawn to help others find their own path through the wilderness.

The book begins with a brief history of biogenesis, a word that Rauchfuss prefers to use rather than phrases like "origin of life" or "emergence of life." The first chapter brings the reader from the ancient Greeks up to the present when we are seeing a near-exponential growth of our knowledge. Here he makes an effort to define life, always a difficult task, but succeeds as well as any. The book then steps through nine basic concepts that must be taken into account to understand biogenesis, with a chapter given to each. For instance, Chapters 2 and 3 describe the origin of galaxies, stars and planets, and Chapter 4 discusses chemical evolution, which is central to our ideas about life's beginnings. The material is presented at a level that can be understood by students in an introductory chemistry course. The next six chapters present facts and concepts underlying protein and nucleic acid functions in modern cells, with constant references to how these relate to biogenesis. In Chapter 10 Rauchfuss brings it all together to describe the evidence for the first forms of cellular life. This chapter is a nice example of how Rauchfuss tries to present information in a clear and interesting manner. For instance, there is considerable controversy about the evidence related to the first life on the Earth, which is based on isotopic analysis and microfossils, and the controversy is presented along with the scientists on both sides of the argument. In the last chapter and epilogue, Rauchfuss gives an overview of astrobiology, which in fact is the unifying theme of the book, and raises a series of unanswered questions that are a guide to the major gaps that still remain to be filled by experiments, observations and theory.

Chemical Evolution and the Origin of Life is well worth reading by young investigators who seek an overview of biogenesis. It is also enjoyable reading for scientists like myself who will discover that the book fills in blank spaces in their own knowledge of the field. We owe a "danke sehr!" to Horst Rauchfuss for putting it all together.

July 2008   Professor David W. Deamer

                                                        Department of Chemistry and Biochemistry
                                                        University of California
                                                        Santa Cruz, CA
                                                        USA

Preface to the English Edition

The first edition of this book was published in German, a language which is now not so widely read as it was even a generation ago. So I am very happy that Springer decided to publish an English edition. Naturally, I have tried to bring the book up to date, as the last years have seen considerable progress in some areas, which this book tries to cover.

It was unfortunately impossible to mention all the many new results in the extremely broad area of the "origin of life". Selections often depend on the particular interests of the writer, but I have tried to act as a neutral observer and to take account of the many opinions which have been expressed.

I thank my colleagues Günter von Kiedrowski (Ruhr-Universität Bochum), Wolfram Thiemann (Universität Bremen) and Uwe Meierhenrich (Université de Nice, Sophia Antipolis). Particular thanks go to my colleague Terry Mitchell from the Technische Universität Dortmund for providing the translation and for accommodating all my changes and additions.

This year has sadly seen the deaths of two of the pioneers of research on the origin of life: Stanley L. Miller and Leslie Orgel. They provided us with vital insights and advances, and they will be greatly missed. Their approach to scientific research should serve as a model for the coming generation.

Varberg, July 2008   Horst Rauchfuss

Preface

The decision to write a book on the origin (or origins) of life presupposes a fascination with this "great problem" of science; although my first involvement with the subject took place more than 30 years ago, the fascination is still there. Experimental work on protein model substances under simulated conditions, which may perhaps have been present on the primeval Earth, led to one of the first books in German on "Chemical and Molecular Evolution"; Klaus Dose (Mainz) had the idea of writing the book and was my co-author.

In recent years, the huge enlargement and differentiation of this research area has led to the formation of a new, interdisciplinary branch of science, "Exo/Astrobiology", the ambitious goal of which is the study of the phenomenon of "life" in our universe.

The following chapters provide a review of the manifold attempts of scientists to find answers to the question of "where" life comes from. Successes will be reported, but also failures, discussions and sometimes passionate controversies. It will also be made clear that very many open questions and unsolved riddles are still awaiting answers: there are more such questions than is often admitted! The vast amount of relevant scientific publications unfortunately makes it impossible to report in detail on all the components of this interdisciplinary area of natural science.

The description of scientific facts and issues is generally dealt with by two different types of author: either by scientists working on the particular problem under discussion and developing hypotheses and theories, or by "outsiders". In each case there are advantages and disadvantages: the researcher brings all his or her expertise to bear, but there is a danger that his or her own contributions and related theories may to some extent be judged one-sidedly. The "outsider", however, should be able to provide a neutral appraisal and evaluation of the scientific contributions in question. In an article in the "Frankfurter Allgemeine Zeitung" (July 9th, 2001) entitled "Warum sich Wissenschaft erklären muß", the neurophysiologist Prof. Singer refers to this problem: "on the other hand, researchers tend to overvalue their own fields, and the intermediary must be able to confront this problem with his own critical ability".

The intermediary is often forced to present complex material in a simple manner, i.e., to carry out a "didactic reduction". Such processes naturally cause problems, resembling a walk on a jagged mountain ridge. On the one side is the abyss of an inordinate simplification of the scientific conclusions (and the resulting condemnation by the experts), on the other that of the complexity of scientific thought, which is only really understood by the specialist.

Presentation of the biogenesis problem is difficult, because there is still not one single detailed theory of the emergence of life which is accepted by all the experts working in this area. There has been important progress in recent years, but the single decisive theory, which unites all the experimental results, has still not emerged. In other words, important pieces in the jigsaw puzzle are still missing, so that the complete picture is not yet visible.

This book is organised as follows: first, a historical introduction, followed by a survey of the origin of the universe, the solar system and the Earth. Planets, meteorites and comets are discussed in the third chapter, while the next deals with experiments and theories on chemical evolution. Proteins, peptides and their possible protoforms are characterized in Chaps. 5 and 6, as well as the "RNA world". Further chapters deal with important hypotheses and theories on biogenesis, for example, inorganic systems, hydrothermal vents and the models proposed by Günter Wächtershäuser, Manfred Eigen, Hans Kuhn, Christian de Duve and Freeman Dyson, as well as the problem of the origin of the genetic code. Chapter 9 provides a discussion of basic theoretical questions and the chirality problem. The search for the first traces of life and the formation of protocells are dealt with in the tenth chapter, while the last covers the question of extraterrestrial life forms, both within and outside our solar system.

Looking back, I must thank my academic teachers, Gerhard Pfleiderer and Theodor Wieland, for introducing me to biochemistry and natural product chemistry, and thus to the phenomenon of "life", the origins of which are still hidden in the darkness of the unknown.

I thank Dr. Gerda Horneck (DLR, Cologne) and my colleagues Clas Blomberg (Royal Institute of Technology, Stockholm), Johannes Feizinger (Ruhr University, Bochum), Niels G. Holm (University of Stockholm), Günter von Kiedrowski (Ruhr University, Bochum), Wolfram Thiemann (University of Bremen) and Roland Winter (University of Dortmund).

Thanks are also due to many colleagues across the world for allowing me to make use of images and information and for encouraging me to continue the work on this book.

I also thank the members of the planning office for chemistry in the Springer Verlag, Peter W. Enders, senior editor chemistry and food sciences, Pamela Frank and Birgit Kollmar-Thoni for their patience and helpfulness.

To Dr. Angelika Schulz go thanks for her exemplary editorial support in the preparation of the book, and to Heidi Zimmermann for preparing most of the illustrations.

Maj-Lis Berggren (Varberg) provided invaluable help in avoiding all the pitfalls which computers can generate. Special thanks go to my wife, who showed great patience during the time of preparing the manuscript.

Finally, a quote from Georg Christoph Lichtenberg, to whom we owe thanks for so many apposite, polished aphorisms. Lichtenberg (1742–1799) was a scientist, satirist and Anglophile. He was the first professor of experimental physics in Germany. I hope that, with respect to most of his points, Lichtenberg made gigantic mistakes in the following lines!

Eine seltsamere Ware
als Bücher gibt es wohl schwerlich
in der Welt. Von Leuten gedruckt
die sie nicht verstehen; von Leuten
verkauft, die sie nicht verstehen;
gebunden, rezensiert und gelesen,
von Leuten, die sie nicht verstehen,
und nun gar geschrieben von
Leuten, die sie nicht verstehen.

Here is one possible translation:

There could hardly be
stranger things in the world than books.
Printed by people who do not understand them;
sold by people who do not understand them;
bound, reviewed and read by people who do not understand them,
and now even written by
people who do not understand them.

Varberg, 2004   Horst Rauchfuß

Author's note: Some figures in this book are published additionally in colour in order to make them clearer.

Contents

Color Figures

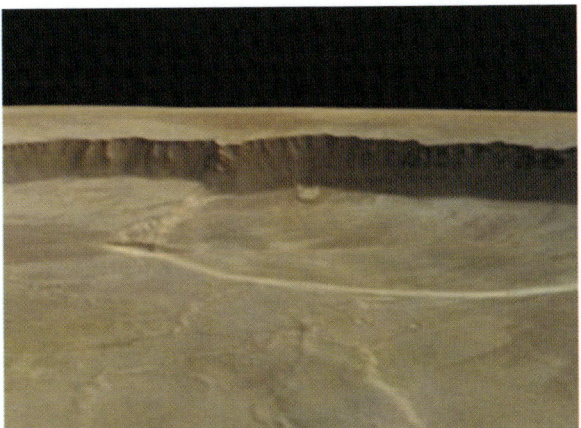

Fig. 3.1 Perspective view of part of the caldera of Olympus Mons on Mars. This view was obtained from the digital altitude model derived from the stereo channels, from the nadir channel (vertical perspective) and the colour channels on the Mars Express Orbiter. The photograph was taken on 21 January 2004 from a height of 273 km. The vertical face is about 2.5 km high, i.e., about 700 m higher than the north face of the Eiger mountain (Switzerland). With permission of the DLR

Fig. 3.3 An artist's impression of the planned "hydrobot" mission to Europa. The robot has bored through the ice layer in the moon's intermediate aqueous layer and is investigating the ocean floor. From NASA

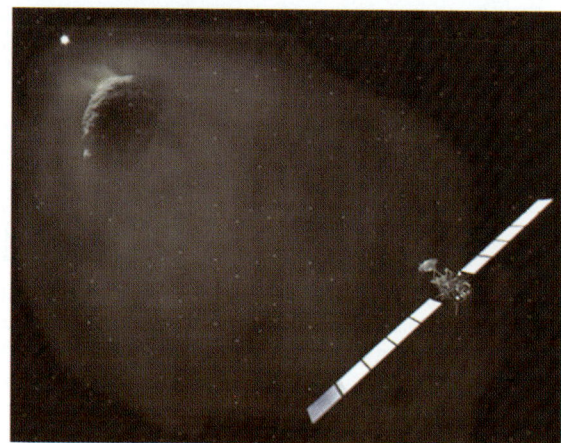

Fig. 3.6 Artist's impression of the planned approach of "Rosetta" to the comet 67P/Churyumov/Gerasimenko in the year 2014. ESA picture

Fig. 3.12 Model of an agglomerate consisting of many small interstellar dust particles. Each of the rod-shaped particles consists of a silicate nucleus surrounded by yellowish organic material. A further coating consists of ice formed from condensed gases, such as water, ammonia, methanol, carbon dioxide and carbon monoxide. Photograph: Gisela Krüger, University of Bremen

Fig. 7.5 Pyrite (FeS_2) crystals, with quartz

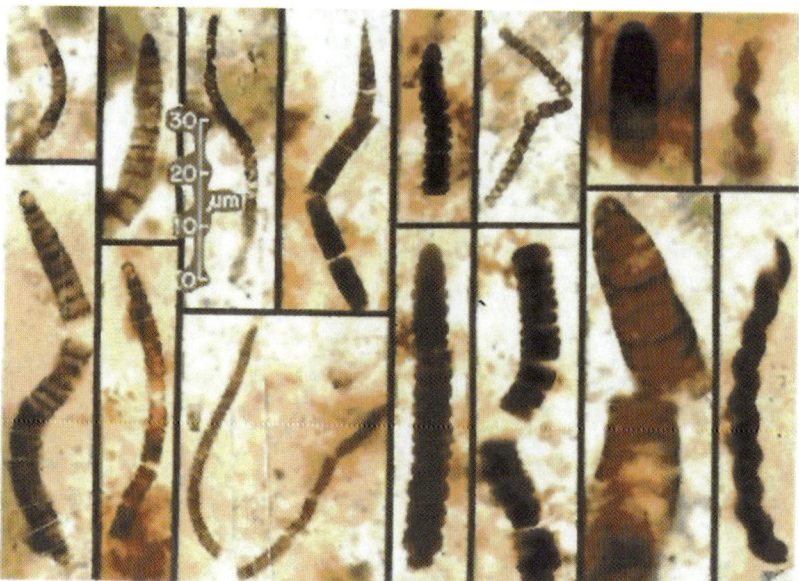

Fig. 10.1 Cellular, petrified, filamentous microfossils (cyanobacteria) from the Bitter Springs geological formation in central Australia; they are about 850 million years old. With kind permission of J. W. Schopf

MEDIUM DIAMETER (2-5μm) FILAMENTS, CYLINDRICAL CELLS

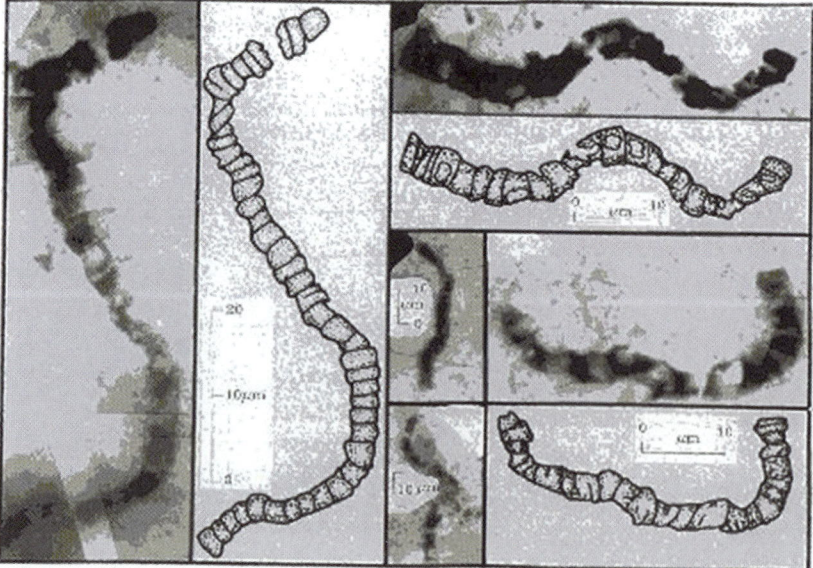

Primaevifilum amoenum

Fig. 10.2 Cyanobacteria-like, filamentous carbonaceous fossils from the 3.456-billion-year-old Apex chert in northwestern Australia; their origin and formation are still under discussion. The photographs are accompanied by the corresponding drawings. With kind permission of J. W. Schopf

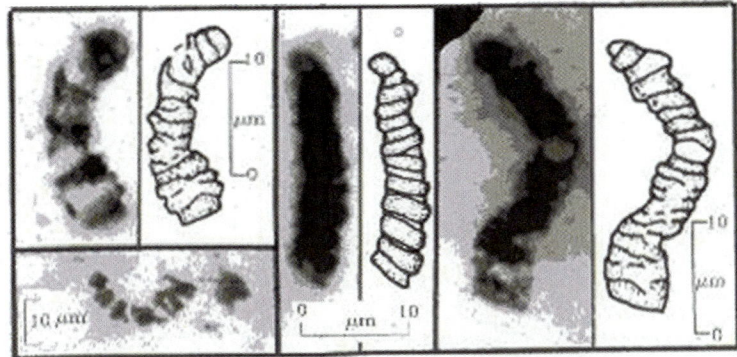

ROUNDED END CELLS - *Archaeoscillatoriopsis disciformis*

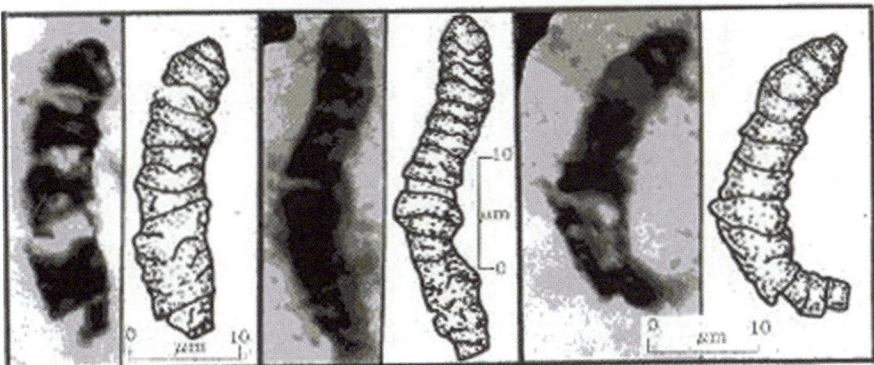

CONICAL END CELLS - *Primaevifilum conicoterminatum*

Fig. 10.3 Microfossils with differently formed end cells, from the same source as in Fig. 10.2 and thus of the same age. Again, the corresponding drawings are shown to make the structures clearer. With kind permission of J. W. Schopf

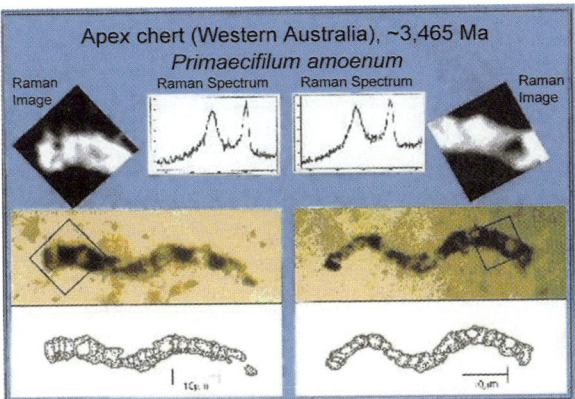

Fig. 10.4 Fossilized cellular filamentous microorganisms (two examples of *Primaevifilum amoenum*). They are 3.456 billion years old and come from the Apex chert region in northwestern Australia. As well as the original images, drawings and the Raman spectra and Raman images, which indicate that the fossils have a carbonaceous (organic) composition, are shown. With kind permission of J. W. Schopf

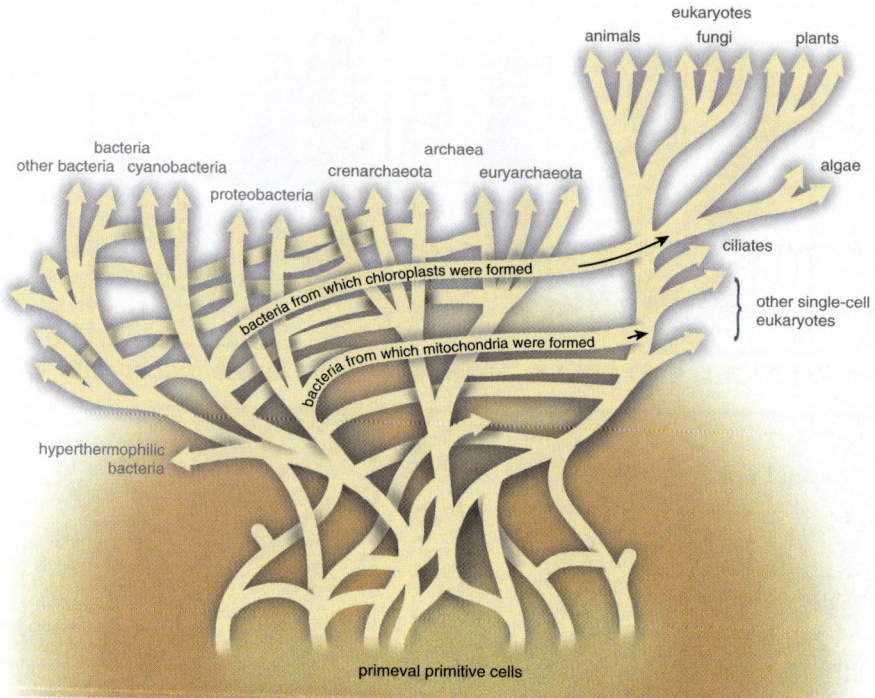

Fig. 10.11 The "modified tree of life" still has the usual tree-like structure and also confirms that the eukaryotes originally took over mitochondria and chloroplasts from bacteria. It does, however, also show a network of links between the branches. The many interconnections indicate a frequent transfer of genes between unicellular organisms. The modified tree of life is not derived, as had previously been assumed, from a single cell (the hypothetical "primeval cell"). Instead, the three main kingdoms are more likely to have developed from a community of primitive cells with different genomes (Doolittle, 2000)

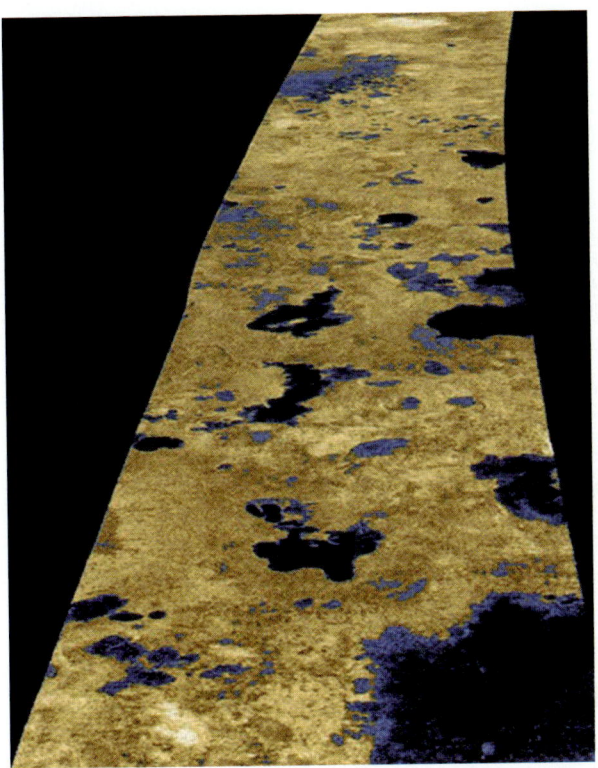

Fig. 11.1 Pseudo-colour radar picture of the north polar region of Titan (NASA/JPL, 2007)

Fig. 11.5 One of the telescopes in the Darwin flotilla. With kind permission of ESA

Introduction

For more than 50 years, scientists have been working diligently towards finding a solution to the "biogenesis" problem. We have chosen to use this word rather than the expression "origin of life" or "emergence of life". Biogenesis research has involved many individual disciplines—more than normally participate in work on other scientific challenges—from astrophysics, cosmochemistry and planetology to evolutionary biology and paleobiochemistry. Biogenetic questions also have their roots in the humanities. Thus Wolfgang Stegmüller, a philosopher who taught at the University of Munich, stated in the introduction to the second volume of his "Hauptströmungen der Gegenwartsphilosophie" ("Important Trends in Modern Philosophy") that science was presently trying to "... answer questions about the construction of the universe, the basic laws of reality and the formation of life. Such questions form the basis of the oldest philosophical problems; the key difference is only that the vast arsenal of modern science was not available to the Greek thinkers when they were trying to devise their solutions." This arsenal has been greatly increased in the last years and decades.

The problem in its entirety can be characterised by means of analogies. Thus the chemist Leslie Orgel, who carried out successful experiments on chemical evolution for many years, compared the struggle to solve the biogenesis problem with a crime novel: the researchers are the detectives looking for clues to solve the "case". But there are hardly any clues left, since no relics remain from processes which took place on Earth more than four billion years ago.

Research into the biogenesis puzzle is special and differs from that carried out in many other disciplines. The philosophy of science divides scientific disciplines into two groups:

Operational science: a group including those disciplines which explain processes which are repeatable or repetitive, such as the movements of the planets, the laws of gravity, the isolation of plant ingredients, etc.

Origin science: a group which deals with processes which are non-recurring, such as the formation of the universe, historical events, the composition of a symphony, or the emergence of life.

H. Rauchfuss, *Chemical Evolution and the Origin of Life*,
© Springer-Verlag Berlin Heidelberg 2008

Origin science cannot be explained using normal traditional scientific theories, since the processes with which it deals cannot be checked by experiment and are thus also not capable of falsification.

So is the work done on the biogenesis problem in fact not scientific in nature at all? Surely it is! There is a way out of this dilemma: according to John Casti, if enough thoroughly thought-out experiments are carried out, the unique event will become one which can be repeated. The hundreds of simulation experiments which will be described in Chaps. 4–8 represent only tiny steps towards the final answer to the problem. However, modern computer simulations can lead to new general strategies for problem solving.

In recent years, the number of scientists working on the biogenesis problem has increased considerably, which of course means an increase in the number of publications.

Unfortunately, biogenesis research cannot command the same financial support as some other disciplines, so international cooperation is vital. The biogenesis community is still relatively small, and most of its members have known each other for many years. The International Society for the Study of the Origin of Life, ISSOL, has been in existence for around 40 years and has just added the tagline "The International Astrobiology Society" to its name; it organises international conferences every three years. The atmosphere at these conferences is very pleasant, even though there is complete unity on only a few points in biogenesis. Opponents of the evolution and biogenesis theories naturally use such uncertainties for their own arguments. The most radical of these opponents are the creationists, a group based in the USA which takes the biblical account of creation literally; they consider the beauty and complexity of life forms to be evidence for their notions.

The chapters which now follow will provide a survey of the multifarious aspects of the question of "where" life on our Earth came from.

Chapter 1
Historical Survey

1.1 The Age of Myths

When we are debating the sense of our existence, the question as to "where" all living things come from keeps coming back to plague us. Human beings have been seeking answers to this question for hundreds, or even thousands, of years. But only since the middle of the last century have attempts been made to solve the problem of biogenesis with the help of scientific methods.

In the mists of time, myths dominated the thoughts, emotions and deeds of our ancestors. The Greek thinkers used myths as a possibility of structuring the knowledge obtained from mankind's encounters with Nature; the myths mirrored people's primeval experiences. The forces of nature dominated the lives of our ancestors in a much more direct and comprehensive manner than they do today. Life was greatly influenced by numerous myths, and in particular by creation myths. These often dealt with the origin of both the Earth and the universe and with the creation of man (or of life in general). In ancient Egypt, the god Ptah, the god of the craftsmen, was originally worshipped in Memphis, the capital of the Old Kingdom. Ptah was one of the most important gods. Each of the most important religious centres had its own version of the origin of the Earth. In Memphis, the priests answered the question as to how creation had taken place by stating that Ptah had created the world "with heart and tongue". By this they meant that Ptah had created the world only through the "word"; in other words, the principle of *will* dominated creation. Jahweh, the god of the Bible, and Allah (in the Koran) created the world by the power of the word: "There shall be...."

There is no doubt that in those times, all civilisations considered that there was a connection between natural events and their myths of the Earth's creation. Thus most of the Egyptians—whichever gods they worshipped—shared the common belief that the creation of the Earth could be compared with the appearance of a mound of land from the primeval ocean, just as every year they experienced the re-emergence of the land from the receding Nile floods.

A similar connection between the world around us and cosmology can be found in the land between the Tigris and Euphrates. The Earth was regarded as a flat disc, surrounded by a vast hollow space which was in turn surrounded by the firmament of heaven. In the Sumerian creation myth, heaven and Earth formed

An-Ki, the universe ("heaven–Earth"). An infinite sea surrounded heaven and Earth. In Mesopotamia, water was regarded as the origin of all things, and from it had sprung both the Earth's disc and the firmament, i.e., the whole universe. The Babylonian Enuma-Elish legend describes the birth of the first generation of gods, which included Anu (the god of the heavens) and Ea (the god of the Earth) from the primordial elements: Apsu (fresh water), Tiamat (the sea) and Mummu (the clouds).

In the Nordic creation myth, which can be found at the beginning of the Edda, we encounter Ginnungagap, a timeless, yawning void. It contains a type of supreme god, Fimbultyr, who willed the formation of Niflheim in the north, a cold, inhospitable land of fog, ice and darkness, and in the south Muspelheim (with light and fire). Sparks from Muspelheim flew onto the ice of Niflheim. This caused life to emerge, and the ice giant Ymir and the huge cow Audhumbla were formed.

From "The Seeress's Prophecy" (3, 57):
Young were the years when Ymir made his settlement,
There was no sand nor sea nor cool waves;
Earth was nowhere nor the sky above,
Chaos yawned, grass was there nowhere.
(Larrington, 1999)

Under Ymir's left arm were formed a giant and a giantess. Since the cow Audhumbla found no grass on which to feed, she licked salty ice blocks, and from under her tongue emerged Buri the Strong, who had a son, Bör. He in turn had three children with Bestla: Odin (Wotan), the most important of the gods, Vili and Vé. The Earth itself was formed only at this stage. The frost giant Ymir was vanquished, and from his corpse came Midgard, the land of men, from his blood the oceans, from his bones and teeth the mountains and cliffs, from his hair the trees and from his

Fig. 1.1 Rune singer with his instrument, the kantele

skull the heavens. His brain was thrown into the air by the gods, and from it were formed the clouds. Flowers and animals just appeared. One day, the three sons of Bör were walking on the beach and came upon Ask, the ash, and Embla, the elm. Man and woman were formed from the two trees, and Odin breathed life and spirit into their bodies. Vili gave them intelligence and emotions, and from Vé they got their faces and their language. We know neither when these myths first appeared, nor the history of their emergence.

Several hundred kilometres further east, in Finnish Karelia, the nineteenth century saw the birth of legends which were passed down by word of mouth from generation to generation. Elias Lönnrot, a doctor, collected these fables and used them to create the Finnish national epic, the "Kalevala", which starts with a creation myth. In the first rune, the daughter of the air lets herself fall into the sea. She is made pregnant by the wind and the waves. The duck, as water mother, comes to her, builds a nest on her knee, and lays her eggs. These roll into the sea and break, giving rise to the Earth, the heavens, the sun, the moon and the stars:

> From one half the egg, the lower,
> Grows the nether vault of Terra:
> From the upper half remaining,
> Grows the upper vault of Heaven;
> From the white part come the moonbeams,
> From the yellow part the sunshine,
> From the motley part the starlight,
> From the dark part grows the cloudage.
> (Kalevala, Rune I, translated by John Martin Crawford, 1888)

At the beginning of the orchestral prelude to his opera "Rheingold", Richard Wagner brilliantly shaped the myth of creation in music, which describes nature in its primordial state, at the absolute beginning of all things. For many bars there is no modulation, no chordal variation. Then a chord in E flat minor appears; first the tonic can be heard in unfathomable depths, followed by the addition of a fifth, which finally becomes a triad. The "nature motive" develops as the leitmotif of all creation (Donington, 1976).

 The nature leitmotif

But now let us go back again, many centuries: the Greek philosophers tried to explain the formation of living systems by compounding matter (which is by nature lifeless) with the active principle of "gestalt". The "gestalt" principle is so powerful that it can breathe life into inert matter.

For Aristotle (384–322 BC) there was only one type of matter; this could, however, exist in four basic forms: earth, air, fire and water, all of which could be converted one into the other. Observations of natural phenomena only came second in ancient Greece, though. Biological processes were considered to be very important, and attempts were made to explain the behaviour of, for example, water, air, rain, snow and heat. The Greeks did this by relating their observations to cause and effect. For Aristotle, experiments (in the sense of questions posed to nature) were not suitable ways of getting information, as they involved menial operations which were only carried out by slaves. Aristotle's teachings actually represent a cognition theory, in which general observations are used to make decisions on individual cases.

The atomists, for example, Leucippus, Democritus and Epicurus, thought that a phenomenon could be explained when its individual elements were known; in contrast, Aristotle was of the opinion that that was not enough, since such information refers only to the material basis. In order to be able to understand things and processes, three further "origins", "principles" and "reasons" must be known.

The "four reasons why", which Aristotle attributed to all things which were subject to change, are: causa materialis, the material cause; causa efficiens, the efficient cause; causa formalis, the formal cause, and causa finalis, the final cause. The first three causes exist for the last one, as it is the whole reason that the other three causes are implemented; they are to the final cause what the means are to the end, and form the process of which the final cause is the goal.

The final cause was the most important for Aristotle, just because it was what was actually reached at the end of the process. Aristotle's teaching dominated the way people thought well into the Middle Ages. Thus, the "four reasons why" were of great importance for western philosophy.

Interestingly, the teachings of Democritus (460–371 BC) did not become so important, although in the sense of natural science (as we now know it), they were much more relevant. Leucippus was Democritus's teacher, and thus the scholar took over the basic ideas of atomic theory from his teacher: atoms as tiny particles, too tiny to be visible, which were everlasting and could not be destroyed. They were supposedly made from the same material, but were of different sizes and weights. According to Democritus, life arises from a process in which the small particles of the moist earth combine with the atoms of fire.

Empedocles, born around 490 BC in Agrigent (Sicily), was a member of the group known as the eclectists (the selectors), because they selected ideas from systems which already existed and put them together to form new theories. According to Empedocles, the lower forms of life were formed first, and then the higher organisms; first plants and animals, then human beings. Initially both sexes were united in one organism; the separation into male and female took place later. These ideas appear to contain elements of modern scientific theory.

1.2 The Middle Ages

Many centuries passed between the hypotheses of the Greek philosophers and the development of new ideas, and of vague models of how life on Earth might have developed. However, a completely new methodology was now used: while the Greeks had merely reflected on how things might have happened, their successors used experiments.

The often luckless alchemists were looking for the "transmutatio metallorum", the transmutation of non-noble metals into gold. Here, of course, they remained unsuccessful. Attempts to create a "homunculus", a human being in a test tube, also failed completely. The work "De generatione rerum naturalium" (On the generation of natural things) by Paracelsus did the most to spread the idea of tiny creatures in a test tube. Three hundred years later, the "homunculus" found its way into world literature in Goethe's "Faust".

The idea of "spontaneous generation", the emergence of life from dead matter, dominated medieval ideas of biogenesis. It was supported and confirmed by experiments. Thus, mice, frogs, worms and other animals could apparently appear from decaying, but formerly living, material. The famous Doctor van Helmont demonstrated an experiment for the "original procreation" of mice: a jug (with no lid) was filled with wheat and dirty underclothes, and after 21 days, changes occurred—particularly in the smell! A certain "ferment" from the underclothes permeated the wheat and turned it into mice! There were, however, critical observers: while at the court of Ferdinand II of Tuscany, the Italian doctor and poet Francesco Redi (1626–1698) showed that the white maggots found in decaying meat came from eggs laid by flies: no maggots are formed if the decaying meat is stored in a vessel covered with gauze. In spite of such proofs, the theory of spontaneous emergence of life remained attractive.

L. Joblot also showed that it is not possible for life to occur spontaneously: he prepared an extract of hay, which he poured into two vessels, one of which was immediately sealed with parchment. As expected, microorganisms grew only in the open vessel. Regrettably, Joblot's results were not taken seriously by his contemporaries.

In the middle of the eighteenth century, there was a violent scientific argument about the spontaneous generation of life between the Englishman J. T. Needham (1713–1781) and the Frenchman G. de Buffon on the one side, and the Italian L. Spallanzani (1729–1799), who taught natural history at the University of Pavia, on the other. Both parties carried out experiments similar to those of Joblot, but came to opposite results. Needham filled vessels with mutton broth or other organic materials and sealed them. Because he did not work in a sterile manner, microorganisms grew in the vessels. He and Buffon interpreted this result as a proof of spontaneous generation. Spallanzani, however, carried out his experiments very carefully and under sterile conditions—and obtained completely different results. Both sides then carried out many other experiments; however, they could not convince each other, and so the question of the spontaneous emergence of life remained open.

The learning process with respect to the problem of the origin of life took place in a manner similar to the three stages described by the French philosopher Auguste Comte (1798–1857) for the linear history of progress in human culture. These three stages are:

Stage 1: The theological and mythological period. Reality is described as the result of supernatural forces (polytheism, monotheism, animism).
Stage 2: The age of metaphysics. The supernatural beings (gods) are replaced by abstract terms, powers or entities.

Stage 3: The scientific or positive age. The unification of theory and practice, which is the result of a combination of rational thinking and observation, allows us to recognize relationships and similarities. Ideally it is possible to describe many single phenomena on the basis of one unified postulate, i.e., to formulate a scientific law.

Comte's three-stage principle can be applied not only to the intellectual development of all mankind, but also to the individual development of a single human being. It can also be applied to the development of an individual science: at first there is a dominance of theological and mythical concepts, followed by the phase of metaphysical speculation, and finally the advanced stage of positive knowledge.

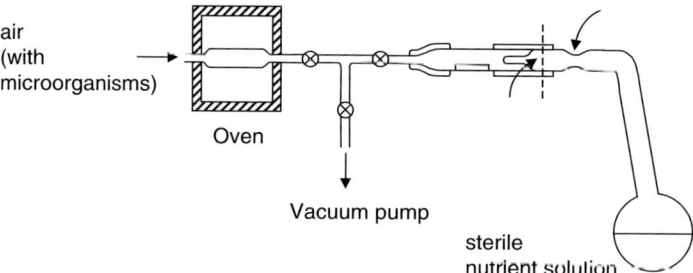

Fig. 1.2 Pasteur's apparatus: if the oven is not switched on, the microorganisms in the air enter the sterile culture solution and multiply. If the oven is switched on, they are killed by the heat. After Conaut (1953)

Around 1860, the French Academy of Science decided to award a prize to the scientist who could unambiguously settle the question of the spontaneous emergence of life. Louis Pasteur (1822–1895) used some elegant experiments to show that a *de novo* synthesis of microorganisms from various materials of organic origin was not possible. He demonstrated that all microbes are descended from existing microorganisms. Pasteur showed that air itself contained various types of microorganism; if air is filtered through guncotton, the latter retains the microorganisms. If the guncotton is then dissolved in a mixture of ethanol and ether, the cells can readily be identified under the microscope in the solution, and they multiply if the latter is transferred to a sterile culture medium. If, however, the air is heated before being passed into boiled culture broth, the cells are killed by the heat. Pasteur's opponents argued that by heating the stream of air, he had destroyed the vital force.

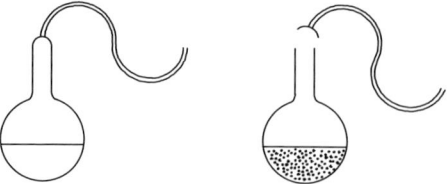

Fig. 1.3 Pasteur's swan-necked flasks: in the first flask, the unbroken neck hinders contamination; if the neck is broken off as in the second flask, the sterile culture medium is invaded by microorganisms. After Pasteur (1862)

In order to disprove this theory, Pasteur used swan-necked flasks; unheated air could now enter the sterile culture solution. But in this case, the microorganisms in the air were deposited in the long S-shaped neck and did not enter the culture medium. If, however, the neck of a flask was broken off, they could enter the solution and multiply.

In 1864, Louis Pasteur received the well-deserved prize of the Academy in recognition of his achievements. However, Pasteur's experiments provided no information on *how* life was formed.

At around this time, there was much scientific debate about the theory of the origin of species proposed by Charles Darwin (1809–1882), a theory which was to change the world. Darwin himself was very cautious about making statements on biogenesis. It was still too early to answer such questions, because neither results from the science of cell biology nor an extensive knowledge of our planet, the solar system and the cosmos were available.

1.3 Recent Times

The huge disquiet which had been caused by Darwin's principles also led to new ideas on the origin of life. According to H. Kamminga from the University of Cambridge (1991), there are two approaches (from about 1860 and 1870), which differ greatly in their profound metaphysical assumptions on the nature of life and of living organisms. The first assumed that life is an aspiring property of nature. Living things are a product of lifeless matter and evolved in the course of the history of the universe. The other approach postulated that life is a fundamental property of the cosmos and that living things have always existed somewhere in the universe. This second approach, considered scientifically, cannot provide an answer to the question as to the origin of life; it reappeared in the form of the panspermia hypothesis.

The ideas of the well-known physiologist from Bonn, Eduard Pflüger (1829–1910), seem to predate modern theories: he assumed that, under the specific conditions of the primordial Earth, fundamental constituents of protoplasma could have developed from cyanide-type compounds or polymers derived from them (Pflüger, 1875).

The idea that microbes could migrate across the universe was supported by scientists with a worldwide reputation, such as H. von Helmholtz, W. Thomson (later Lord Kelvin) and Svante Arrhenius. This hypothesis was still accepted by Arrhenius in the year 1927, when he reported in the "Zeitschrift für Physikalische Chemie" on his assumption that thermophilic bacteria could be transported within a few days from Venus (with a calculated surface temperature of 320 K) to the Earth by the radiation pressure of the sun (Arrhenius, 1927). The panspermia hypothesis, which seemed to have disappeared in the intervening decades, was reintroduced in the ideas of Francis Crick (Crick and Orgel, 1973). It still exists in a modified form (see Sect. 11.1.2.4).

Fig. 1.4 The Swedish
physical chemist Svante
Arrhenius (1859–1927), who
received the Nobel Prize for
chemistry in 1903 for his
work on electrolytic
dissociation

The deciding impulse which introduced biogenesis into scientific discussion came from Russia. After the upheavals of the civil war, that country was the subject of worried observation by the rest of the world. It was assumed that no great scientific achievements would be possible there. Then, in 1924, a book on the material basis of the origin of life on Earth appeared in "Red Russia". Its author was Alexandr Ivanovich Oparin (1894–1980) from the Bakh Institute of Biochemistry in Moscow (Oparin, 1924). Basically, the Oparin hypothesis makes the following assumptions:

> The prebiotic atmosphere had reducing properties, so that the bioelements C, O, N and S were present in reduced form as CH_4, H_2O, NH_3 and traces of H_2S.
>
> This primeval atmosphere was subjected to various energy sources, such as electrical discharge, solar radiation and heat from volcanoes; these led to the formation of small organic molecules.
>
> These chemical substances accumulated in the hydrosphere, which thus became a "dilute soup" from which the first forms of life evolved spontaneously.

Not all points of this hypothesis are now accepted. Some of the assumptions on the physicochemical state of the primeval Earth have undergone considerable revision in the light of more recent results. Oparin answered the question as to how he came to think that organic molecules could be formed from methane, ammonia, water and hydrogen by referring to ideas he obtained from Mendeleev's hypothesis on the inorganic origin of oil (Oparin, 1965). The concept of a reducing primeval atmosphere was supported by the idea that free oxygen would have immediately destroyed organic molecules by oxidation. In addition, it was already known in 1924 that the sun consisted mainly of hydrogen.

Only four years after Oparin's book was published in Russia, the English scientist J. B. S. Haldane (1928) published an article whose ideas strongly resembled those of Oparin. We now know that Haldane had no knowledge of Oparin's publication, and when the two first met, many years later, they immediately agreed that Oparin had priority. Haldane's assumption of a reducing primeval atmosphere was based on completely different observations: he concluded from anaerobic glycolysis, which is used by many contemporary living organisms as their primary source of energy, that life must have originated in a reducing environment. The ideas described above have gone down in scientific history as the "Oparin–Haldane Hypothesis". Unlike Haldane, Oparin continued to study the biogenesis problem until his death and, in particular, published articles on the formation of protocells. A recent short but detailed survey and assessment of Oparin's life's work was provided by Miller et al. (1997) in their article "Oparin's 'Origin of Life': 60 Years After".

Other scientists took up Oparin's ideas, used them for their own concepts, and tried to form organic molecules from inorganic starting materials. The Mexican scientist A. L. Herrera reported in 1942 in an article entitled "A New Theory of the Origin and Nature of Life" on his investigations with "sulphobes" (Herrera, 1942). These are morphological units ("lifelike forms") which he obtained from reactions between thiocyanates and formalin. Sulphobes are spherical in form, with a diameter between 1 and 100 μm, and can interact with their surroundings; thus they can adsorb dyestuffs. In some ways, they resemble the coacervates studied by Oparin and his school (Sect. 10.2.2).

Another type of experiment on chemical evolution was due first to Groth and Suess and later to Garrison. They studied the type of energy which must be applied to a simulated primeval atmosphere in order to form organic building blocks for biomolecules, starting from inorganic materials. Groth and Suess (1938) studied the influence of UV light on simple molecules, while Garrison (1951) carried out similar experiments using ionising radiation.

Then came the year 1953, and with it important events, both political and scientific in nature: the death of Stalin and the determination of the structure of DNA; in addition, a scientific article was published in "Science" by a previously unknown author, Stanley L. Miller. Its title was "A Production of Amino Acids under Possible Primitive Earth Conditions" (Miller, 1953).

In a footnote, Miller thanked the Nobel Prize winner Harold C. Urey for supervising his Ph.D. thesis work. Thus, this experiment became known as the "Miller–Urey experiment" (Sect. 4.1). Not only was the broader public impressed by these results, but also the small group of scientists who were more or less closely involved with the question of the evolution of life. The successful synthesis of protein building blocks from a simulated primeval Earth atmosphere generated activity in several laboratories, leading in the next few years to important new results. The great importance of the Miller–Urey experiment is due particularly to the fact that it showed for the first time that the problem of the origin of life can be approached by means of scientific method, i.e., *experimentally*.

1.4 The Problem of Defining "Life"

Scientific theory states that one of the most important tasks of science, and scientists, is the task of definition. Thus it becomes absolutely necessary to define the phenomenon known as "life". Very few terms which are used so frequently have been defined in such an unsatisfactory manner. The paradox is that the more we know about life, the more difficult it becomes to define it satisfactorily. There is still *no* clear definition of the term "life" which is accepted by all the scientists studying this phenomenon (Cleland and Chyba, 2002).

Various definitions have been proposed, and, depending on one's scientific standpoint, a suitable one may be available. Several of these definitions will be presented below. A completely satisfactory answer will, however, probably only be found when more detailed results on the origin of life become available.

Sixty years ago, Erwin Schrödinger asked the question, What is life? His English-language book with that title, which appeared in 1944 (Schrödinger, 1944), is based on a series of lectures which he had given at the University of Dublin. He was seeking an answer to the question, How can the processes in time and space, which take place within the limits of a living organism, be explained by physics and chemistry? There is no doubt that his book had an important influence on the development of modern biology, and it already hinted at certain lines of development in molecular biology.

As stated above, biologists and scientists from other related areas have so far not been able to agree on a single definition of the term "life" (Barrow, 1991). This is in no way surprising, since more than 100 attributes and properties have been found to characterize life (Clark, 2002). There is a certain amount of agreement on the distinguishing features of a living system. In his lecture given at a conference held in Trinity College Dublin in September 1993 to celebrate the 50th anniversary of the Schrödinger lectures on the subject "What is Life?", Manfred Eigen defined three basic characteristics which have so far been found in all living systems:

> Self-reproduction: without this process, information would be lost after every generation.
> Mutation: without it, information would be invariant—and thus no development of the species would be possible.
> Metabolism: without this, a living system would reach an equilibrium state, from which, again, no development would be possible.

The physical chemist Luigi Luisi, ETH Zürich (1998), made clear the vital importance of an agreed definition for future progress in biogenesis research. He proposed five definitions for the term "life" and suggested that a definition agreed on by as many scientists as possible would make it possible to define the goals for future research projects, on the basis of that general definition.

When life is to be defined, it is necessary for the purposes of biogenesis research to limit the discussion to the simplest life forms. This type of reduction is necessary in order to be able to make a clear division between inanimate and animate objects. Even for "reduced systems", the boundaries between the two become unclear, as shown by the example of viruses. A definition of minimal life makes it possible to ignore the complex properties of higher living organisms, such as consciousness, intelligence or ethics.

According to Luisi, a definition of life must satisfy the following criteria:

> It should be possible to make the distinction between animate and inanimate as clearly and as simply as possible, by means of experiments.
> The criteria for making the distinction should be verifiable across a wide range.
> The definition should include both forms of life which are already known and hypothetical pre-life forms. It should be logically self-consistent.

The definitions of "life" which have been formulated in the NASA Exobiology Program as general working definitions are as follows:

1. "Life is a self-sustained chemical system capable of undergoing Darwinian evolution."

This definition was previously used by Horowitz and Miller (1962). An undefined external energy source was included in this definition. The growing influence of the "RNA world" can be seen in the second NASA definition:

2. "Life is a population of RNA molecules (a quasispecies) which is able to self-replicate and to evolve in the process."

The following definitions proposed by L. Luisi go further than the NASA definitions:

3. "Life is a system which is self-sustaining by utilizing external energy/nutrients owing to its internal process of component production."

Instead of "reproduction" or "replication", the more general term "production" was used. The third definition includes the first definition. However, because it contains neither Darwinian nor genetic specification, this definition takes both coded and uncoded life into account. Since the term "population" is not included, the definition can be applied to single objects such as robots.

In the next definition, there is a limitation of the smallest life forms:

4. "Life is a system which is spatially defined by a semipermeable compartment of its own making and which is self-sustaining by transforming external energy/nutrients by its own process of components production."

This definition excludes all systems which do not have a spatial boundary to their synthetic machinery, for example pure RNA replication. The walls of a test tube or the banks of a "warm, little pond"[1] cannot be included as boundaries in the sense of definition four.

Taking these limitations into account, Luisi suggests a fifth and last definition:

5. "Life is a system which is self-sustaining by utilising external energy/nutrients owing to its internal process of component production and coupled to the medium via adaptive changes which persist during the time history of the system."

Here there is no limitation, as some scientists consider one to be unnecessary. The order of the definitions is not arranged with respect to their quality.

These attempted definitions are extremely useful, since they force biogenesis researchers to define their own standpoints. They make it possible to develop new working hypotheses for future research projects. According to Luisi, "Once you have the intellectual clarification in front of you, you have the challenge to realize it in the laboratory." However, the definitions presented above are not good enough for all the scientists working in this area.

Other characteristics of life have been formulated by Daniel E. Koshland Jr. (University of California at Berkeley) as the "Seven Pillars of Life". They are as follows:

1. A program
2. Improvisation
3. Compartmentalisation
4. Energy
5. Regeneration
6. Adaptability
7. Seclusion

This list contains life characteristics which are contained in most of the definitions we have seen. However, two or three of the "pillars" are unusual:

Point 2 describes the possibility that a system can change its program in order to adapt to new environmental conditions.
Point 5 takes into account that thermodynamic losses must be compensated for.

The last pillar can perhaps be compared with "privacy" in the social world. This property of life makes it possible for many biochemical processes to take place independently in a cell without disturbing one another (Koshland, 2002).

The search for life in the cosmos requires a generalised, universal definition of life. This must take into account the properties of systems ranging from viruses, prions, denucleated cells or endospores to life in a test tube, computer viruses or even to robots which are capable of self-replication.

---

[1] This phrase is taken from a letter written by Charles Darwin (1871) that contains vague references to chemical evolution: "...if we could conceive in some warm, little pond with all sorts of ammonia and phosphoric salts, light, heat, electricity etc. present that a proteine compound was chemically formed...".

Results from philosophical considerations on language show that attempts to define life lead to a dilemma, similar to that which occurred when trying to define water before molecular theory existed. Since no analogous theory of the nature of living systems exists, an infinite controversy as to the definition of life is unavoidable (Cleland and Chyba, 2002).

"The definitions of life are extremely controversial". So begins a publication on the problem of the definition of life which appeared as late as 2004. This publication is written by three Spanish scientists from the Centre for Astrobiology (INTA/CSIC) in Madrid, the University of València and the University of the Basque Country in San Sebastian (Ruiz-Mirazo et al., 2004). Their "general definition" of life introduces two new terms into the discussion: "autonomy" and "open-ended evolution capacities".

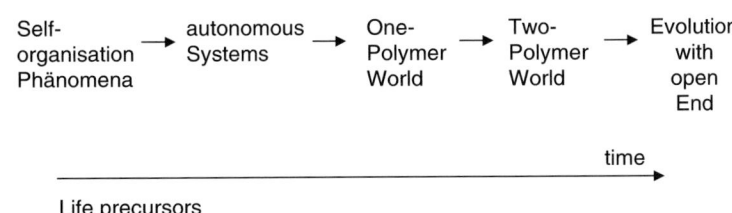

Life precursors

Fig. 1.5 Schematic representation of the evolution of life from its precursors, on the basis of the definition of life given by the authors. If bioenergetic mechanisms have developed via autonomous systems, the thermodynamic basis for the beginning of the archiving of information, and thus for a "one-polymer world" such as the "RNA world", has been set up. Several models for this transition have been discussed. This phase of development is possibly the starting point for the process of Darwinian evolution (with reproduction, variation and heredity), but still without any separation between genotype and phenotype. According to the authors' definition, life begins in exactly that moment when the genetic code comes into play, i.e., in the transition from a "one-polymer world" to a "two-polymer world". The last phase, open-ended evolution, then follows. After Ruiz-Mirazo et al. (2004)

In addition, the authors suggest that all such systems must have a semi-permeable active boundary (membrane), an energy transduction apparatus and (at least) two types of functionally interdependent macromolecular components (catalysts and records). Thus, the phenomenon of life requires not only individual self-replication and self-sustaining systems, but it also requires of such individual systems the ability to develop a characteristic, evolutionary dynamic and a historical collectivist organisation.

A hypothesis put forward by the British physicist James Lovelock, the Gaia hypothesis, is related to the problems just discussed. This hypothesis is supported by several well-known scientists, such as the American biologist Lynn Margulis and the theoretical physicist Freeman Dyson (Dyson, 1992). According to the Gaia hypothesis, the Earth itself can be regarded as a type of living organism. In ancient Greece, Gaia was the Earth goddess, who balanced out inequilibria which developed from interactions between heaven and Earth. There are various arguments in support of

Gaia; on the other hand, it also appears possible that the Earth is a highly resistant system which can deal with changes such as those induced by catastrophes.

In an alternative theory, the results of population dynamics rather than Darwinian natural selection are responsible for the regulation of environmental conditions (Staley, 2002).

It is not yet possible to make a final decision on Gaia, a hypothesis which also requires further studies and experiments to give a clear answer and thus a deeper understanding of our existence.

References

Arrhenius S (1927) Z Phys Chem 130:516

Barrow J (1991) Theories of Everything: The Quest for Ultimate Explanation. New York, Oxford University Press

Clark B (2002) Second Astrobiology Conference, NASA-Ames Research Center http://www. astrobiology.com/asc2002/abstract.html

Cleland C, Chyba C (2002) Orig Life Evol Biosphere 32:387

Conaut JB (1953) Pasteurs and Tyndalls Study of Spontaneous Generation. Harvard University Press, Cambridge, Mass

Crick FHC, Orgel LE (1973) Icarus 19:341

Donington R (1974) Wagner's "Ring" and its symbols. The Music and the Myth. Faber and Faber Limited, London

Dyson F (1992) From Eros to Gaia. Penguin Books

Eigen M (1995) What will endure of 20th century biology? In: Murphy MP, O'Neill LAJ (Eds) What is Life? – The Next Fifty Years. Cambridge University Press, pp 5–23

Garrison WM, Morrison DC, Hamilton JG, Benson AA, Calvin M (1951) Science 114:416

Groth W, Suess H (1938) Naturwissenschaften 26:77

Haldane JBS (1928) The Origin of Life. Rationalist Annual 148: 3. Reprinted in: Science and Human Life, Harper Brothers, New York, 1933

Herrera AA (1942) Science 96:14

Horowitz N, Miller SL (1962) in: Progress in the Chemistry of Natural Products 20:423

Kalevala, translated by John Martin Crawford (1888)

Kamminga H (1991) Uroboros 1:1.95

Koshland Jr DE (2002) Science 295:2215

Larrington C (translator) (1999) The Poetic Edda. Oxford University Press

Luisi PL (1998) Orig Life Evol Biosphere 28:613

Miller SL (1953) Science 117:528

Miller SL, Schopf J, Lazcano A (1997) J Mol Evol 44:351

Oparin AI (1924) The Origin of Life, 1. Edition (Russian: Proiskhozdenic Zhizni). Moskovskiy Rabochii, Moskau

Oparin AI (1965) History of the subject matter of the conference. In: Fox SW (Ed) The Origins of Prebiological Systems. Academic Press, New York London, p 91

Pasteur L (1862) Ann Physik 64:5

Pflüger E (1875) Archiv für gesamte Physiologie 10:251

Ruiz-Mirazo K, Peretó J, Moreno A (2004) Orig Life Evol Biosphere 34:323

Schrödinger E (1944) What is Life? Cambridge University Press

Staley M (2002) J Theor Biol 218:35

Chapter 2
The Cosmos, the Solar System and the Primeval Earth

2.1 Cosmological Theories

The question of the origin of life on Earth leads directly to the question of the formation of our planet, of the solar system and of the universe. The ancient philosophers, as we have seen, attempted to answer such questions, but the models which we discuss and argue about today were proposed by scientists only in the last century.

Since cosmological theories are not a direct concern of this book, only a brief outline of this area will be given. Two developments in the last century were of particular importance and led to huge advances in knowledge:

> Albert Einstein's general relativity theory and
> The discovery of the flight of the galaxies by Edwin Hubble.

The relativity theory, looked at in a very simple manner, is a theory of gravitation which brings together space and time to form one single unified phenomenon. The universe is then no longer a static system, but a dynamic one which is continually expanding. The question then arises as to whether this expansion process will continue infinitely, or whether it can be put into reverse if gravitation forces the system to collapse. This could happen if the density of the matter in the universe were to exceed a certain limiting value.

In 1922, the Russian scientist A. A. Friedman made use of Einstein's equations and concluded that the universe was expanding; the Belgian physicist G. E. Lemaître came to a similar conclusion in 1927. The latter assumed that the universe must have had its origin as an extremely small volume of matter. He invented the idea of the "primeval atom" (l'atome primitif). Only two years later, Erwin Hubble discovered the "flight of the galaxies": he compared the positions of the spectral lines originating from certain galaxies with those obtained in laboratory experiments and found that the lines from the galaxies were shifted slightly towards the red end of the spectrum. He interpreted this effect as being due to the galaxies moving away from the Earth and recognized the phenomenon as a Doppler effect. If this motion is calculated in reverse, the result is a very small volume of space in which some type of primeval explosion must have occurred. This process was described in 1955 by the British astronomer Fred Hoyle as the "big bang"; at that time, Hoyle was a convinced proponent of the "steady state hypothesis", which postulated a type of

H. Rauchfuss, *Chemical Evolution and the Origin of Life*,
© Springer-Verlag Berlin Heidelberg 2008

equilibrium state in which material was continually being formed. Thus there was no "beginning" and no "end": the universe as a whole remained unchanged.

Fig. 2.1 George Gamow (1904–1968) was born in Odessa, studied in Leningrad (St. Petersburg) and emigrated in 1934 to the USA, where he taught in Washington, D.C., until 1965 and at the University of Colorado in Boulder for the last three years of his life

Today the "big bang" theory is favoured by most cosmologists. Apart from "Abbé" Lemaître, the man who did the most to popularize it and to formulate its theoretical background was George Gamow. Gamow, a Russian-born scientist living and working in the USA, had forecast the 3K background radiation of the universe.

This radiation amounts to about 400 photons per cubic centimetre and fills the whole universe. The afterglow of the big bang was discovered in 1964 by A. Penzias and W. Wilson as 3K microwave emissions, and in 1978 the two scientists were rewarded with the Nobel Prize for physics. Apart from 3K radiation and red shift, there is a third point which supports the big bang theory: calculations of the amount of helium which must have been formed since the big bang during the cooling of the expanding universe gave a value of 23–24%, which agrees very well with values determined experimentally.

The big bang theory suggests that the formation of the universe took around 15×10^9 years. The process started with a state called the "singularity", i.e., the beginning of time, space and matter. At the beginning of the big bang, there was an extremely hot blazing ball of matter and radiation. The closer one got to time zero, the higher the temperature of this plasma became. In this state, the four fundamental forces (strong and weak atomic forces, electromagnetic force and gravitation) are united: the normal laws of physics no longer apply. Perhaps this state cannot even be described in words. The laws which apply to the explosion itself are also unknown: the extreme values of pressure, temperature, energy and density are unimaginable for us, and no attempt at simplification should be made!

A fraction of a second after the explosion, however, the first structures emerged. Results from particle physics allow us to calculate and predict cosmic processes; we can expect that, within the first second, groups of three quarks united to form protons or neutrons. The temperature fell to around 10^{10} K. The energy density was such that electrons and the corresponding antiparticles, the positrons, could not be formed from photons. Positrons and electrons annihilate each other, and the result is a small excess of electrons. One minute after the explosion, groups of two neutrons and two protons united to form the atomic nucleus He^{2+}. After three minutes, the temperature had fallen to 10^9 K. At that stage, the expanding universe consisted of about 24% helium and 76% hydrogen nuclei, as well as traces of light elements. Elements with an atomic number higher than helium (known to astronomers as "metals") were formed in later stages of development of the universe. Further cooling led to the formation of hydrogen and helium atoms (by electron capture) as well as of traces of lithium. This process led to a drastic reduction in the number of free electrons, and the universe became "transparent", i.e., photons were now able to pass through space without being scattered by free electrons.

After another few hundred million years (some astrophysicists speak of around a billion years), the temperature was around 18 K and then sank further to a value of 3 K (or to be exact, 2.73 ± 0.01 K) (Unsöld and Baschek, 2001).

In a short interview, Larson and Bronn (2002) reported on the latest models, calculations and computer simulations. According to these, the first stars were formed about 100–250 million years after the big bang. They formed small protogalaxies, which were themselves the result of small density fluctuations in the still young universe. Although the universe was generally homogeneous in its early days, slight density fluctuations led to the formation of filament-like structures, similar to those of a network. At the nodes, the material (only hydrogen and helium, no metals) was denser, and the first stars were formed. To quote from the book of Genesis, "And there was light."

How do these first stars differ from those of today? As we have already mentioned, it is mainly because of their different composition. In addition, calculations show that they must have been much heavier (100–1,000 solar masses) and thus much brighter (up to a million times brighter than our sun).

A further important difference is that the first stars did not live as long, only a few million years. As they consisted only of hydrogen and helium, the energy generation occurred in a different manner than in today's stars, in which certain elements act as catalysts in nuclear fusion; without these catalysts, the nuclear fusion would be much less efficient. Thus the young stars needed to reach higher temperatures and to be more compact. It is assumed that temperatures around 17 times higher than that of our sun were normal. Some of the early stars exploded, forming supernovas. The heavier metals which were formed during the explosions diffused through space and influenced further developments in the universe, for example the formation of planets.

In recent years, the development of new cosmological models has caused frequent rethinking. The well-known book by Stephen Weinberg *The First 3 Minutes* (1977) gives an account of the initial processes.

James E. Peebles, professor emeritus at Princeton (2001), offers his own description. He states that "at present the house of cosmological theories resembles scaffolding which is solidly assembled but still has large gaps. The open questions are those of 'dark matter', 'inflation' and 'quintessence'. We live in exciting times for cosmology."

Table 2.1 Grades for cosmological theories (from Peebles, 2001)

Hypothesis	Grade	Remarks
The universe developed from a hot, dense beginning.	*Very good*	Huge amount of supporting evidence from various areas of biology and physics.
The universe expanded according to the general theory of relativity.	*Good*	Passes all previous tests, but only a few of these were stringent.
Galaxies consist mainly of dark matter built up from exotic particles.	*Satisfactory*	Much indirect evidence, but the particles still have to be discovered and alternative theories disproved.
The mass of the universe is in general evenly distributed; it acts as Einstein's cosmological constant and accelerates expansion.	*Poor*	Agrees well with most of the recent measurements, but the evidence is still thin, and theoretical problems are still unsolved.
The universe initially went through a phase of rapid expansion, the so-called inflation.	*Fail*	Elegant theory, but still no evidence; requires huge extension of the laws of physics.

The "quintessence" hypothesis was proposed by J. P. Ostriker and Steinhardt (2001). The authors use the term quintessence ("fifth substance") to describe a quantum force field which is gravitationally repulsive. It has a certain similarity to an electrical or magnetic field and could lead to an invisible energy field which accelerates cosmic expansion.

The most modern instruments provide ever more exact data on the structure of the cosmos and the possibility of penetrating ever deeper, almost to the boundaries of the universe. Data processing and simulation using high-performance computers increase the possibilities of devising new approaches to the solution of the many still unanswered questions. An attempt to relate the big bang theory to the string theory led American physicists to the model of the "ekpyrotic universe". According to this hypothesis, the universe was formed in a collision of two three-dimensional worlds (branes) in a space with an extra (fourth) spatial dimension, and *not* via the big bang, the favourite model of many astrophysicists; while the big bang can explain many phenomena of cosmophysics, it cannot answer them all. Some of the basic cosmological questions are still unanswered, as is shown by the most recent research results and by models derived from them, which cast doubt on some of the previous assumptions and hypotheses.

An international research team including many French members has used the analysis of data from NASA's Wilkinson Microwave Anisotropy Probe (WMAP) to devise an amazing new model of our universe. According to this, the cosmos is not infinite and expanding because of pressure from dark energy (the cosmological standard model); instead, it is finite and has an extremely rigid topology, possibly in the form of a Poincaré dodecahedral space (Luminet et al., 2003; Ellis, 2003). There is no doubt that we can expect many new results from cosmophysics in the next few years when the results of future missions have been interpreted.

2.2 Formation of the Bioelements

The well-known textbook *General Chemistry* by Atkins and Beran (1992) starts by telling the reader that "the cradle of chemistry lies in the stars." One can hardly think of a better way of emphasising the role of cosmochemistry. The synthesis of the elements, which are now logically ordered in the periodic table, can be divided into three stages, which are separated in both time and space:

  The synthesis of the light elements hydrogen, helium and lithium (including their isotopes), which occurred just after the big bang;
  The synthesis of the intermediate elements, which were formed in various "burning processes" and
  The synthesis of the heavy elements in supernova explosions.

The temperature of the universe about three minutes after the big bang was around a billion degrees. On further cooling, tritium (^{3}H) and the helium isotopes ^{3}He and ^{4}He remained stable. Heavier elements could not be formed because of the low concentration of deuterium: the ^{2}H nuclei decomposed rapidly (Weinberg, 1977). Further expansion, and thus further cooling, led to a change in the behaviour of the deuterium nuclei, and in this phase, they became stable, while their concentration, however, remained low. The universe was composed of about 24% helium at that time. About 300,000 years after the big bang, the temperature was low enough to permit electrons and nuclei to unite to form atoms. Later, concentrations of matter took place at some points in the universe, and the first stars were formed. The complex processes occurring in those stars led to the synthesis of heavier chemical elements. Exactly which elements were formed depended to a large extent on the mass of the stars, which is generally referenced in publications to the mass of our own sun; thus we speak of "solar masses" as the unit. The reactions taking place in the interior of the stars are referred to pictorially as "burning".

Table 2.2 lists the most important syntheses occurring in the stars. The main products include the bioelements C, O, N and S. The synthesis of the elements began in the initial phase after the big bang, with that of the proton and the helium nucleus. These continue to be formed in the further development of the stars. The stable nuclide ^{4}He was the starting material for subsequent nuclear syntheses. Carbon-12 can be formed in a triple α-process, i.e., one in which three helium nuclei collide. However, such processes occur relatively seldom, while E. Salpeter (Cornell University) showed that a two-step reaction should be more easily realisable. A collision of two helium nuclei leads to the formation of a beryllium nucleus, which decomposes very rapidly to the starting materials unless it is hit by a further helium nucleus; the newly-formed nucleus ^{12}C is stabilized by radiation emission. The lifetime of the beryllium nucleus is only about 0.05 s (Hillebrand and Ober, 1982); thus, the density of the helium nuclei must be very high in order to give a high collision probability.

Table 2.2 The pre-supernova burning stages of a star with 25 solar masses. From: Maciá et al. (1997)

Burning process	T (in 10^9 K)	Main products	Time taken
H	0.02	^{4}He, ^{14}N	7×10^6 years
He	0.2	^{12}C, ^{16}O, ^{20}Ne	5×10^5 years
C	0.8	^{20}Ne, ^{23}Na, ^{24}Mg	6×10^2 years
Ne	1.5	^{16}O, ^{24}Mg, ^{28}Si	1 year
O	2.0	^{28}Si, ^{32}S, ^{40}Ca	180 days
Si and e$^-$ process	3.5+	^{54}Fe, ^{56}Ni, ^{52}Cr	1 day

Further capture of α-particles leads to the formation of oxygen and neon. ^{16}O itself forms the basis for the synthesis of sulphur. The only biogenic element missing in Table 2.2 is phosphorus, which is an exception in that it is formed by a complex nuclear synthesis (Maciá et al., 1997). In large stars, the reactions listed in the table take place in the following series, without stopping but over long periods of time.

$$\text{H} \rightarrow \text{He} \rightarrow \text{C}, \text{O} \rightarrow \text{Ne} \rightarrow \text{Mg}, \text{Si} \rightarrow \text{Fe}, \text{Ni} \tag{2.1}$$

The result is a type of onion-like model of the star with an iron–nickel core in the centre. The situation is somewhat different for smaller stars: the path branches at the point where "carbon burning" (^{12}C + ^{12}C) begins. While the heavier stars are not affected by this process, the smaller ones (4–8 solar masses) are completely torn apart by carbon burning.

In the heavier stars, a stage in which ^{20}Ne is destroyed occurs subsequently to the carbon burning, but before the absorption of oxygen. The α-particles formed are used up by the nuclei already present (also from neon itself) in so-called neon burning.

$$^{20}\text{Ne} + \gamma \rightarrow {}^{16}\text{O} + \alpha \quad \text{and} \quad ^{20}\text{Ne} + \alpha \rightarrow {}^{24}\text{Mg} + \alpha \rightarrow {}^{28}\text{Si} + \gamma \tag{2.2}$$

These reactions take place in the inner zone of stars heavier than 15 solar masses. Hydrostatic carbon burning is followed by explosive neon burning at temperatures of around 2.5×10^9 K. Under these conditions, phosphorus (^{31}P) can be formed, although complex side reactions also occur. In comparison with the formation of the other five biogenic elements, the synthetic pathways which lead to phosphorus appear quite involved (Macià et al., 1997). ^{31}P nuclei can be formed only in those classes of stars which, because of their mass, are able to carry out carbon and nickel burning. Some of the nuclear reaction pathways occur in only very low yields (around 2.5%), which explains the relatively low proportion of this important bioelement. The largest amount of the natural ^{31}P nuclide is probably formed via the following reaction pathways:

$$^{12}C + ^{12}C \rightarrow {}^{24}Mg^* \rightarrow {}^{23}Na + p \rightarrow {}^{23}Na(\alpha, p) \rightarrow {}^{26}Mg(\alpha, y) \rightarrow {}^{30}Si(p, \gamma) \rightarrow {}^{31}\mathbf{P}$$

$$(2.3)$$

The reaction of $^{24}Mg^*$ to give ^{23}Na takes place in around 50% yield, with the following reaction only in 5% yield. A large part of the ^{31}P is destroyed by the reaction ^{31}P (p, α) $\rightarrow {}^{28}Si$. More details of phosphorus synthesis and that of its compounds can be found in Maciá et al. (1997) and Maciá (2005).

2.3 The Formation of the Solar System

Two types of theory have been put forward to explain the formation and development of our solar system: catastrophe and evolution. The former assumes a collision or coming together of two stars. As early as 1745, the French scientist Count Buffon postulated that the Earth had been torn out of the sun by a passing comet. He estimated the age of the Earth to be 70,000 years, while theology proclaimed that the Earth was less than 6,000 years old.

It is generally accepted today that our solar system was formed in evolution processes. René Descartes (1596–1650) suggested that the solar system was formed from a gigantic whirlpool within a universal fluid and that eddies in the flow produced planets; his theory tried to explain both the formation of the sun and the motions of the planets. More than a hundred years later, the "Kant–Laplace nebular hypothesis" was put forward; this theory was much closer to modern ideas on the origin of the solar system and was due to the philosopher Immanuel Kant (1724–1804) (who was born in Königsberg/Kaliningrad) and Pierre Simon, Marquis de Laplace (1749–1827). Kant's work "Universal Natural History and Theory of Heaven" appeared in 1755. Kant and Laplace developed their theories independently of each other, Kant describing his ideas about 40 years earlier than Laplace. Both hypotheses share the postulate that slightly denser regions of the gas-filled universe contracted more and more under the influence of gravitation (Neukum, 1987). However, the Laplace hypothesis, formulated as it is in terms of mathematical formulae, has certain weaknesses which led others to propose new catastrophe scenarios. There are indeed basic differences between the two approaches. Kant postulates a rotating primeval nebula, which forms a group of clouds. These in turn become planets as the result of further density increases, while the rest of the nebula condenses to form the sun. Laplace, however, postulates a hot rotating gas disc, which shrinks on cooling. The disc spins very fast and casts off rings of gas, which form the planets, with the remaining matter forming the sun (Struve and Zebergs, 1962).

The process in which the solar system was formed was certainly extremely complex, so there is as yet no generally accepted theory to describe it. The different types of heavenly body (sun, planets, satellites, comets, asteroids) have very different characteristics which need to be explained using mechanisms which are valid for them all.

According to present-day concepts, our solar system was formed from a huge gas–dust cloud several light years across in a side arm of the Milky Way. The particle density of this interstellar material was very low, perhaps 10^8–10^{10} particles or molecules per cubic metre, i.e., it formed a vacuum so extreme that it can still not be achieved in the laboratory. The material consisted mainly of hydrogen and helium with traces of other elements. The temperature of the system has been estimated as 15 K.

An unknown event disturbed the equilibrium of the interstellar cloud, and it collapsed. This process may have been caused by shock waves from a supernova explosion, or by a density wave of a spiral arm of the galaxy. The gas molecules and the particles were compressed, and with increasing compression, both temperature and pressure increased. It is possible that the centrifugal forces due to the rotation of the system prevented a spherical contraction. The result was a relatively flat, rotating disc of matter, in the centre of which was the primeval sun. Analogues of the early solar system, i.e., protoplanetary discs, have been identified from the radiation emitted by T Tauri stars (Koerner, 1997).

More than 99% of the total mass of the whole system was present in the protosun. The formation of this disc is demonstrated by the coplanar movement of the planets and by the fact that they all rotate in the same direction around the sun. The increasingly concentrated matter in the primeval sun influenced the rotating disc of matter so that its diameter decreased and the rate of rotation of the whole system increased.

We can assume that the primeval sun rotated much faster than the present-day one and thus had a very high angular momentum. Today, however, the sun accounts for only around 0.5% of the total angular momentum of our solar system. How can we explain the discrepancy between the mass of the sun (around 99.8% of the total mass of the solar system) and its angular momentum? The "angular momentum problem" can be explained on the basis of magnetic interactions between the sun and the rotating disc of matter, which is made up of charged particles (ions and electrons). Lüst and Schlüter suggest a possible mechanism in the form of coupling between the interplanetary plasma and the sun, as in an eddy-current brake. Since (according to the law of conservation of energy) angular momentum cannot be destroyed, the sun must have given up a large part of its angular momentum to the rotating interplanetary disc, and thus to the planets which were slowly being formed.

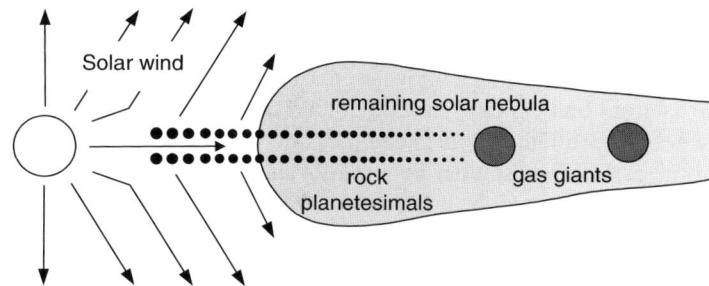

Fig. 2.2 The state of the incipient solar system during the T Tauri phase of the young sun. The central region around the sun was "blown free" from the primeval dust cloud. Behind the shock front is the disc with the remaining solar nebula, which contained the matter formed by the influence of the solar wind on the primeval solar nebula. From Gaffey (1997)

The young sun went through the "T Tauri phase", in which huge streams of hot gas were blown off into space. This is an unstable phase in a star's development and must have lasted for 10^5–10^8 years, depending on the mass of the star. The velocity of the gas streams may have been up to 200–300 km/s. Immense amounts of material were blown out into the outer regions of the gas–dust disc, i.e., into the regions where the larger planets were later formed. In the region of the terrestrial planets, there must have been enough of the heavier elements present to withstand the solar wind, in spite of the higher temperatures and the nearness to the sun. The energy for the T Tauri phase probably came from the fusion reactions (conversion of hydrogen into helium) occurring in the sun's interior. At this point, the sun's atmosphere must have radiated at a temperature of between 10,000 and 100,000 degrees and emitted a vast amount of UV light. The primeval nebular disc was characterized by enormous temperature differences, depending on the distance from the sun. Its density was probably greater in the neighbourhood of the sun than in the distant regions.

Tiny microparticles came together to form microagglomerates, and these in turn formed larger clots, which then formed larger bodies, the diameter of which was initially measured in centimetres but later increased to metres: such planetary building blocks are known as "planetesimals". Computer simulations indicate that these existed around four and a half billion years ago (Wetherhill, 1981). Planetesimals grew to form bodies which were several kilometres across, and there were often collisions in which larger bodies were swallowed up by smaller ones: a process which is not unknown in modern economics!

In the region of the terrestrial planets, there may have been several thousand planetesimals of up to several hundred kilometres in diameter. During about ten million years, these united to form the four planets—Mercury, Venus, Earth and Mars—which are close to the sun. Far outside the orbit of the planet Mars, the heavier planets were formed, in particular Jupiter and Saturn, the huge masses of which attracted all the hydrogen and helium around them. Apart from their cores, these planets have a similar composition to that of the sun. Between the planets Mars and Jupiter, there is a large zone which should really contain another planet. It seems clear that the huge mass of Jupiter prevented the formation of a planet from the planetesimals, which already had diameters measured in kilometres. Thus, in this part of the solar system, we find only asteroids orbiting the sun. It has been estimated that the asteroid belt contains around 50,000 objects, only about 10% of which have so far been identified; asteroids can measure up to 900 km in diameter. The total mass of the asteroids is smaller than that of the Earth's moon. The three largest, Ceres, Pallas and Vesta, with diameters of 933, 523 and 501 kilometres, account for half the total mass of the asteroids.

Binzel et al. (1991) give an account of the origin and the development of the asteroids, while Gehrels (1996) discusses the possibility that they may pose a threat to the Earth. The giant planets, and in particular Jupiter, caused a great proportion of the asteroids to be catapulted out of the solar system: these can be found in a region well outside the solar system, which is named the "Oort cloud" after its discoverer, Jan Hendrik Oort (1900–1992). The diameter of the cloud has been estimated as around 100,000 AU (astronomic units: one AU equals the distance between the Earth and the sun, i.e., 150 million kilometres), and it contains up to 10^{12} comets. Their total mass has been estimated to be around 50 times that of the Earth (Unsöld and Baschek, 2001).

Oort was able to show that the gravitational force of the sun in these regions is so weak that passing stars can cause great changes in the orbits of the Oort comets; they can either be steered into interstellar space, or their elongated ellipsoid orbits can bring them into the solar system (Weissman, 1998). The Oort cloud is regarded as a type of "refrigerator" for active, long-period comets. The short-period comets, however, seem to come from a region of the solar system known as the Kuiper belt, which lies beyond the orbits of Neptune and Pluto. As early as 1951, the Dutch astronomer Gerald Peter Kuiper (1905–1973) proposed that the outermost region of the solar system contained a collection of primeval material; the matter in the Kuiper belt probably derives from the period in which our solar system was formed. More than 30 smaller objects with diameters between 100 and 500 km have so far been discovered (Luu and Jewitt, 1996).

2.4 The Formation of the Earth

The early stages of the formation of the Earth are relatively closely linked to that of the formation of the other three terrestrial planets. Their nearness to the sun meant that light gases such as hydrogen, helium, methane and ammonia could not be held back by the protoplanets but were blown away by the solar wind and the sun's heat. Liquids such as water could not condense and went the same way as the gases. Thus, a type of fractionation occurred in the young solar system: a large proportion of the substances with high vaporisation temperatures, such as metals and silicates, remained close to the sun (Press and Siever, 1994). Elements with higher atomic numbers were not the result of processes occurring in the sun, but were derived from the interstellar cloud from which the solar system had been formed.

Because of their similar history, the four terrestrial planets have similar layer structures. However, their surfaces and atmospheres show enormous physical and chemical differences. The development of the primeval Earth via the agglomeration of planetesimals was accompanied by a vast temperature increase caused by contributions from three different phenomena:

The energy set free by collisions with planetesimals,
The Earth's gravitation and
The radioactivity of the planet's interior.

The kinetic energy set free in collisions with planetesimals was proportional to the square of the velocity of the body which hit the Earth. Thus, if a planetesimal hit the Earth's surface with a velocity of 11 km/s, the amount of energy set free would correspond to the explosion of the corresponding amount of the explosive TNT (trinitrotoluene). The increased compression due to the increase in mass led to pressure increases in the interior of the planet and thus to temperature increases up to around 1,270 K (Press and Siever, 1994).

It has been estimated that the radioactive decay of the various elements provided enough heat to cause temperature increases up to 2,300 K: the long-lived radioactive isotopes ^{235}U, ^{232}Th and ^{40}K still heat up the Earth's interior today. However, this energy alone was not sufficient to melt the primeval Earth. The energy set free when the denser, heavier elements (such as iron and nickel) melted and concentrated at the centre of the Earth provided an additional heat source, and gravitational energy was set free in this process. The time required for the formation of the planets depended to a large extent on their mass. It has been suggested that it took 100–200 million years for the terrestrial planets to accrete, while the giant planets probably required about a billion years.

The melting process and the differentiation of the Earth's matter according to its density caused the lighter crust minerals to migrate to the outer layers of the still young Earth, whose surface temperature at that time was such that it was covered by a sea of melted rock (Wills and Bada, 2000). This separation of materials led to the layer structure of the Earth:

The crust,
The mantle (upper and lower mantle) and
The core (outer and inner core).

The formation of the core, the mantle and the crust can be explained using two basically different accretion models:

Homogeneous accretion and
Heterogeneous (inhomogeneous) accretion.

According to the *homogeneous* model, the metal-containing materials (in particular iron and nickel) and the silicate-containing material of the primeval solar cloud condensed out at about the same time. The proto-Earth thus formed was composed of a mixture of these two types of matter, which differed greatly in their densities. At that time, the Earth's temperature was probably only a few hundred degrees, and its composition corresponded roughly to that of the carbonaceous chondrites (see Sect. 3.3.2). Only later did the metals concentrate at the centre of the proto-Earth as described above.

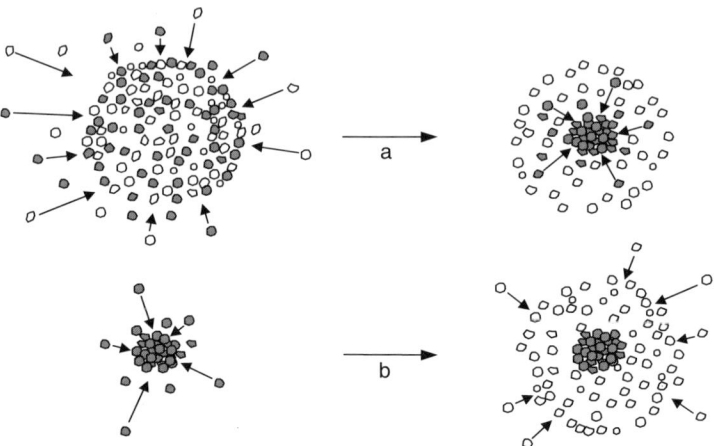

Fig. 2.3 According to the *homogeneous* accretion model (**a**), iron-containing material (black) and silicate-containing material (colorless) condensed out at the same time, i.e., the proto-Earth consisted of a mixture of the two. The concentration of iron in the Earth's core took place later. According to the *heterogeneous* model (**b**), the iron condensed out of the primeval solar nebula first, while the silicates later formed a crust around the heavy core. From Jeanloz (1983)

In the *heterogeneous* model, the metals condensed first and formed the core, while the silicates, which condensed later, formed an outer layer—the mantle.

Of the two models, homogeneous accretion is generally favoured. H. Wancke from the Max Planck Institute in Mainz (1986) described a variant of this model, in which the terrestrial planets were formed from two different components. Component A was highly reduced, containing elements with metallic character (such as Fe, Co, Ni, W) but poor in volatile and partially volatile elements. Component B was completely oxidized and contained elements with metallic character as their oxides, as well as a relatively high proportion of volatile elements and water. For the Earth, the ratio A:B is calculated to be 85:15, while for Mars it is 60:40. According to this model, component B (and thus water) only arrived on Earth towards the end of the accretion phase, i.e., after the formation of the core. This means that only some of the water was able to react with the metallic fraction.

The chemical composition of the Earth's interior determined the character (the oxidation state) of the primeval atmosphere. If metallic iron had collected in the Earth's core in the early phase of the accretion, the exhalations from the interior of the Earth would have consisted mainly of CO_2 and H_2O, since the gas from the interior could only have come into contact with FeO and Fe_2O_3 silicates in the mantle. If, however, metallic iron had been distributed throughout the mantle, the iron and the FeO silicates would have had a reductive influence on the gases: the gas exhaled into the atmosphere would then have consisted of CH_4, H_2 and NH_3 (Whittet, 1997).

The thin, newly formed Earth's crust, consisting of light silicates, swam on the surface of the sea of magma. It was often broken apart by collision with planetesimals of various sizes. The formation of the crust was a complex process, many details of which are as yet not understood. This admission points to the fact that we do not have much geological evidence from this early phase of the Earth's formation.

A vital event in the further development of the Earth was its collision with a smaller planet, possibly as big as Mars. It is assumed that this gigantic collision took place between four and four and a half billion years ago (Sleep et al., 2001), and that it also resulted in the birth of our moon (Luna), which was formed from partially vaporized matter from the Earth. It is likely that not all of the proto-Earth was melted by the energy set free in the collision, but that sections of it remained in their original form. However, more exact information is not yet available.

A corroboration of the theory that the moon was formed mostly from material coming from the Earth is due to researchers from the Max Planck Institute for chemistry in Mainz (Münker et al., 2003). The chemical analysis of material from the surface of the moon shows great similarity with material from the Earth's crust; however, there are certain differences. For example, the concentration of iron on the moon is much lower than that on Earth.

The two rare earth elements niobium (Nb) and tantalum (Ta) were the main subject of study in the investigation referred to. Both elements have very similar properties and almost always occur together in our solar system. However, the silicate crust of the Earth contains around 30% less niobium (compared to its "sister" tantalum). Where are the missing 30% of niobium? They must be in the Earth's FeNi core. It is known that the metallic core can only take up niobium under huge pressures, and the conditions necessary for this may have been present on Earth. Analyses of meteorites from the asteroid belt and from Mars show that these do not have a niobium deficit.

A similar niobium deficit to that on Earth was found on the moon, although the latter's lower mass would preclude the existence of pressures high enough to lead to an absorption of niobium by the FeNi core. It is thus very likely that the moon was formed from material derived from the heavenly body which collided with the Earth and from the proto-Earth's silicate-rich crust around 4.4 billion years ago.

The earlier assumption that Luna was a body which had been captured by the Earth can now be regarded as relatively unlikely. The same is true for the "double planet hypothesis", according to which Luna and the Earth were formed at the same time from condensing primordial matter (Taylor, 1994). There are, however, still disagreements on the point in time at which the collision occurred and on the masses and the physical states of the heavenly bodies involved (Halliday and Drake, 1999).

An evaluation of the number of moon craters per unit area (differentiated according to the diameter of the craters) as a function of the time at which the collisions leading to their formation occurred indicates that the processes involved were similar to those which could have occurred on Earth. It is likely that the bombardment reached a maximum around four billion years ago and dropped after about another billion years to the present rate of collision (Neukum, 1987).

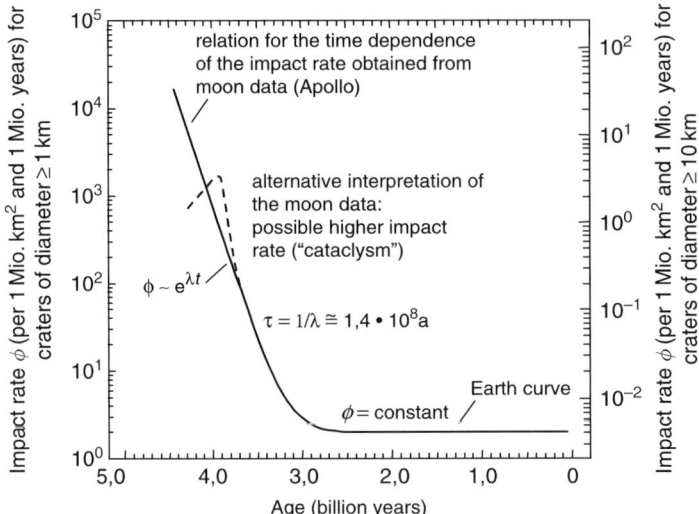

Fig. 2.4 Time dependence of the rate of impact of comets and asteroids on the surface of the Earth and primeval Earth (derived from Apollo moon data). With kind permission of Prof. Neukum (1987)

There are still great uncertainties as to the time frame in which the Earth's cooling occurred, and thus as to the formation of the Earth's crust and the continents. N. H. Sleep et al. from the Department of Geophysics at Stanford University and from NASA's Ames Research Center listed those factors involved in the cooling process, which should be taken into account. They concluded that temperatures at the primeval Earth's surface in the range 333–383K were present for only a relatively short time; in geological terms, "short" means several million years (perhaps as little as one million years). This temperature range is exactly that in which thermophilic microorganisms can exist. Since the composition of the primeval atmosphere, and thus the magnitude of the CO_2 greenhouse effect, is not known, the time available for the formation of the first continents is also unclear. Initial answers to the question of the size and nature of the early continents can be obtained by measurements on isotopes with long half-lives, such as the neodymium isotope ^{143}Nd. This is a product of decomposition of the radioactive isotope samarium-147 (Hofmann, 1997).

Many properties and characteristics of the Earth are determined by plate tectonics, according to the theory of which the lithosphere is not a closed shell; instead it consists of about a dozen large, rigid plates. These are constantly in motion—on a geological timescale. Each of the plates moves as an independent unit and "swims" on the softer, but more dense, asthenosphere (Press and Siever, 1995).

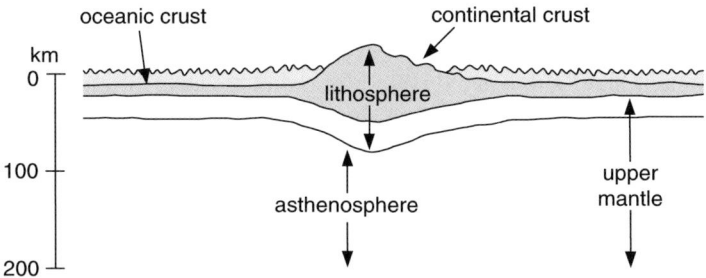

Fig. 2.5 The outer shell of the Earth, the lithosphere, is a solid, rigid layer. It consists of the crust and the outer parts of the mantle. The lithosphere swims on the flexible, partially melted part of the mantle (the asthenosphere). Figure simplified after Press and Siever (1994). With permission of W. H. Freeman and Company, New York

The thickness of the Earth's continental crust is only about five thousandths of its radius. The crust and the oceans together make up only about 0.4% of the total mass of the Earth. The two magmatic minerals, basalt and granite, are present mainly as the material of the ocean floor (basalt) or the continental plates (granite). The latter (with an average density of 2,700 kg/m^3) is the much older material. Thus, some types of granite are up to 3.8 billion years old and have in that time never undergone fusion; the basalts, however, have probably been through several fusion cycles due to subduction.

2.5 The Primeval Earth Atmosphere

All the models of the chemical composition of the atmosphere of primeval Earth are hypothetical. Samples from this period of development of the Earth are not available! And the oldest rocks give us only a limited amount of information.

"The chemical composition of the primeval atmosphere is a central point of argument in the debate on the formation of life." This short remark made by M. Gaffey (1997) from Rensselaer Polytechnic Institute, Troy, New York, hit the nail on the head, and nothing has changed since!

However, we need information on the atmospheric composition in order to plan and carry out simulation experiments. Although the four terrestrial planets originated from the same solar matter, their atmospheres are completely different. This is due to:

The strengths of their gravitational fields,
Their distances from the sun,
Their ability to reflect solar radiation (albedo) and
In a later phase of development, the existence or non-existence of life.

Among the terrestrial planets, the situation of the Earth is special. Its atmosphere (around 21% oxygen and 78% nitrogen by volume) is completely different from those of its neighbours. Venus and Mars have atmospheres consisting almost solely of CO_2 (around 95% by volume) but with very different partial pressures. In the case of nitrogen content, the Earth has only one primordial "relative" in the whole solar system: Saturn's moon Titan, with its thick nitrogen envelope (see Sect. 3.1.6). Since its formation, the atmosphere of the Earth has changed its composition drastically several times. Only traces of the components of the primordial solar nebula have been found. It is likely that cosmic material had already undergone segregation before its aggregation to form the planets (Schidlowski, 1980). The present low concentration of the noble gases on Earth indicates that only between 10^{-7} and 10^{-11} of the primordial noble gases remain. The elements helium, neon and argon are among the most common in the universe. Their rarity on the Earth, and their low chemical reactivity, were the reasons for their late discovery, only about 110 years ago. The noble gases have two very different origins:

Radioactive decomposition of labile elements (such as uranium, thorium or potassium) and
Synthesis during nuclear processes occurring in the interior of the sun.

Only the lightest gases, such as hydrogen and helium, could easily escape the gravitational field of the Earth. In contrast to earlier assumptions, it is now believed that the young Earth probably had either no atmosphere at all or only a very thin one, since the proportion of the primeval solar nebula from which the terrestrial planets were formed consisted mainly of non-volatile substances.

About 50 years ago, it was thought that the primordial Earth must have been surrounded by an envelope with the composition of the primeval solar nebula. The gas masses around the giant planets Jupiter and Saturn with their strongly reducing atmospheres of hydrogen, (helium), methane, ammonia and water were considered to be the models. This idea, of which Oparin and Urey were the main proponents, is still around today, although in a much modified form. It appears certain that the primeval atmosphere contained no oxygen. The thesis of a strongly reducing primeval atmosphere was strongly supported by the sensational experiments carried out by Miller and Urey (1953) (see Sect. 4.1). However, two years earlier, the American geochemist William Rubrey (1951) had suggested that volcanic exhalations, with their high concentration of CO_2, were the main source of the gases of the primeval atmosphere. The Miller/Urey experiments were followed by other successful syntheses under strongly reducing atmospheric conditions, so that Rubrey's postulate was initially ignored. However, doubts soon arose, due to two points:

Because of its low mass, the Earth was (and is) unable to retain large amounts of hydrogen and
The volcanic exhalations of today consist mainly of water and CO_2. There are good geological, geochemical and geophysical grounds for the assumption that today's exhalations are not much different than those produced around four billion years ago. However, we must assume that at that time there was very much more volcanic activity than today.

If the primeval Earth's atmosphere was indeed formed only from volatile components emitted by the primitive, newly formed Earth's crust, its composition must have depended on the time at which it was formed, i.e., whether this was before or after the formation of the iron-rich Earth's core (Joyce, 1989):

Gas emission *before* core formation: contact with metallic iron leads to a strongly reducing atmosphere containing only H_2, H_2O, CH_4 and CO.

Gas emission *after* core formation: the redox state in the iron-containing minerals of the Earth's crust is determined by the ratio of Fe^{2+} to Fe^{3+}.

The result would then be a weakly reducing atmosphere containing H_2O, CO_2, H_2 and CO, but *almost no* CH_4! In addition, strongly reducing molecules such as CH_4 and NH_3 would have been relatively quickly decimated by photodecomposition (Owen, 1979).

According to James F. Kastings (1993) from the Institute of Geosciences at Pennsylvania State University, an expert on this problem, reducing gases could only have been set free if the tendency for oxygen release from the CH_4 and NH_3 dissolved in erupting magma had been several orders of magnitude lower. It has also been suggested that CH_4 and NH_3 could have been transported to the primeval Earth by comets and meteorites. The photochemical reduction of CO_2 in the presence of Fe^{2+} has also been discussed. A tragic natural catastrophe which occurred some years ago shows that CO_2 escapes from the Earth's crust in large amounts even today. Lake Nyos, a lake in Cameroon, occupies the crater of an extinct volcano. A gas cloud which suddenly erupted from the lake (its volume has been estimated as around one cubic kilometre) flowed over the edge of the crater and down the mountain, killing 1,700 people and 3,000 animals (Decker, 1997).

H. D. Holland (1984) estimated the average ratios of the content of volcanic exhalations as follows: $H_2/H_2O = 0.01$ and $CO/CO_2 = 0.03$. Nitrogen is very difficult to detect, and only traces of ammonia are found. In addition, highly variable amounts of the following are present: SO_2, H_2S, elementary sulphur, HCl and B_2O_3. Small amounts of H_2, CH_4, CO and HF have been detected. As early as 1962, Holland suggested that the primeval atmosphere must have gone through two stages:

A highly reduced state, which was characterized by gases which were in equilibrium with metallic iron and

A more oxidized state, in which the gases found today in volcanic exhalations were present.

This initial hypothesis was later revised, since some researchers (such as Walker et al., 1983) were able to show that, according to the model of inhomogeneous accretion, metallic iron was removed from the Earth's crust in a very early phase and accumulated in the core. These results led to the now generally accepted theory that the young Earth was surrounded by a weakly reducing atmosphere.

The CO_2 content of the planetary atmosphere plays a vital role. A relatively high CO_2 partial pressure was certainly an important precondition for solving the problem of the "faint, young sun". It is assumed that the sun was much cooler four billion years ago than it is today, as first suggested by Sagan and Müllen (1972). Theories on the structures and development of the stars show that the radiation intensity of the sun has increased by 25–30% in the course of the history of the solar system. According to Gough (1984), the sun was colder because of the lower He/H ratio in the sun's nucleus. In general, the surface temperature of a planet depends on three factors:

> The radiation energy emitted by the sun.
>
> The fraction of the sun's energy which is reflected back into space (albedo); the non-reflected energy maintains the temperature of the atmosphere and the surface.
>
> The "greenhouse effect" of the atmosphere: a fraction of the infrared radiation is emitted from the surface, absorbed by the atmosphere and reflected back to the surface.

If we assume a radiation loss of the sun of 25–30% in comparison with today's values, the primeval Earth would have had a surface temperature below the freezing point of water (provided that all other factors which influence the surface temperature remained basically unchanged).

Geological proof that liquid water was prevalent on the primeval Earth's surface is provided by sedimentary rocks, whose age has been shown to be greater than 3.8 billion years, as well as by stromatolite-forming bacteria which have been dated to 3.5 million years ago. It appears hardly possible that these could have existed on an ice-covered Earth's surface. Another indication of the presence of liquid water has apparently been found by Stephan Mojzsis and co-workers (University of California at Los Angeles), who found an enrichment of the oxygen isotope ^{18}O in zirconia crystals which are between 3.9 and 4.28 billion years old. This leads to the assumption that the zirconia ($ZrSO_4$) crystallized from molten rock which was in contact with water (Mojzsis et al., 2001). If the cool young sun did not go through an albedo catastrophe, the presence of a larger greenhouse effect than that present on Earth today must be assumed.

Sagan and Mullen (1972) showed that water vapour alone cannot be responsible for the required greenhouse effect. Ammonia, a photochemically unstable compound, cannot have served as an additional component; it is also not found in abiotic sources. Carl Sagan and Christian Chyba (1997) suggested the following: an atmospheric distribution ratio of around $10^{-5\pm1}$ for ammonia could have been enough to compensate for the heat deficit of the weak, young sun. Perhaps organic molecules in aerosols in the higher layers of the atmosphere absorbed the UV irradiation from the sun. According to Owen and Cess (1979), carbon dioxide and water sufficed to solve the problem of the weak, young sun, if it is assumed that the CO_2 concentration in the primeval atmosphere was 100–1,000 times higher than today. Since CO_2 and water are still the major exhalation products of active volcanoes, this assumption appears justified. If the Earth had been tectonically more active, a higher CO_2 output would have been expected. The bioelement nitrogen probably remained in the atmosphere, as an inert element, during the whole history of the Earth.

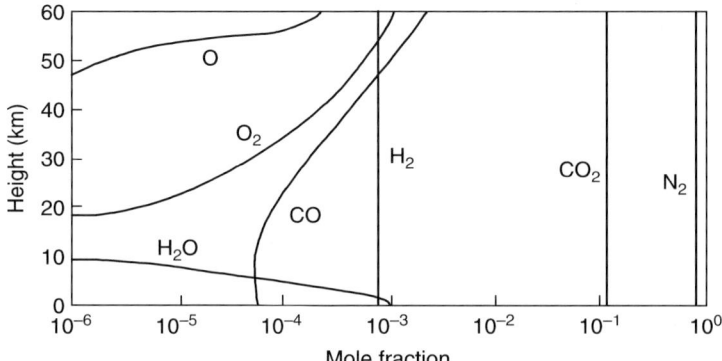

Fig. 2.6 The main components of a typical weakly reducing primeval atmosphere as a function of the altitude above the Earth's surface. The "mole fraction" refers to the mixing ratio of the atmospheric mixture at an assumed surface pressure of one atmosphere. After Kasting (1993)

Our knowledge of the processes which led to the formation of the primeval Earth's atmosphere has increased considerably. However, estimates of its percentage composition are still extremely tentative. The uncertainty is underscored by recent work, according to which the young Earth's atmosphere may have been (weakly?) reducing after all. Since a redox-neutral composition of the primeval atmosphere does not favour prebiotic chemistry, a reducing atmosphere would have had a much more positive influence on the synthesis of biomolecules and their predecessors (Kasting and Eggler, 2002; Schwartz, 2002).

A model of the primeval Earth atmosphere presented by Tian and co-workers supports these ideas. It has caused lively discussion, as the hydrogen content is suggested to be two orders of magnitude higher than that previously assumed. According to this model, the atmosphere was rich in carbon dioxide and thus not a methane-rich Miller–Urey atmosphere, but it contained around 30% hydrogen. The model, which is a hydrodynamic escape model, is based on the hydrogen volcanic outgassing levels observed today, taking into account the (relatively low) additional amount due to the higher geological activity of the young Earth. For a hydrogen-rich atmosphere, hydrogen escape into space is limited by the availability of external UV irradiation (EUV) from the sun; a lower hydrogen escape naturally leads to a higher atmospheric hydrogen content (Tian et al., 2005; Chyba, 2005).

Such a thought-provoking model was naturally subject to criticism; Catling (Department of Earth Science, University of Bristol) considered the calculations to be unrealistic, since (for example) the authors had underestimated the temperatures of the upper layers of the atmosphere. The prompt answer of the authors to these criticisms was quite clear: "Hence, the ancient atmosphere was hydrogen rich" (Catling, 2006; Tian et al., 2006). J. F. Kasting and M. Tazewell (2006) have given a detailed account of the climate of the primeval Earth and the composition of its atmosphere.

2.6 The Primeval Ocean (the Hydrosphere)

It is clear that liquid water is the main prerequisite for all phases of biogenesis. Water is characterized by a series of unusual properties. Its molecular weight alone suggests that, like H_2S, CO_2 and SO_2, it will exist as a gas under normal conditions at the Earth's surface. That it is a liquid is due to the formation of hydrogen bonds between individual H_2O molecules, and its excellent solvent properties are due to their polar nature (Brack, 1993). The interactions between biochemically important species and water are extremely complex in nature. However, water also seems to play an important role in the formation of stars. According to H. Nisini (2000), water in the warm, star-forming regions of the galaxy acts as a coolant in the interstellar gas and removes the excess energy set free in processes involving protostellar collapse. The water may exist as a gas or as ice on interstellar dust particles. The discovery of this phenomenon was made by the Infrared Space Observatory (ISO), which recorded IR spectra at 100–200 μm. The synthesis of water in the warm regions of the galaxy probably occurs according to the following reaction mechanism:

$$O + H_2 \rightarrow OH + H \tag{2.4}$$

$$OH + H_2 \rightarrow H_2O + H \tag{2.5}$$

The special position of the Earth among the terrestrial planets is also shown by the availability of free water. On Venus and Mars, it has not until now been possible to detect any free water; there is, however, geological and atmospheric evidence that both planets were either partially or completely covered with water during their formation phase. This can be deduced from certain characteristics of their surfaces and from the composition of their atmospheres. The ratio of deuterium to hydrogen (D/H) is particularly important here; both Mars and Venus have a higher D/H ratio than that of the Earth. For Mars, the enrichment factor is around 5, and in the case of Venus, 100 (de Bergh, 1993).

Water can be found, in all three aggregate states, almost everywhere in the universe: as ice; in the liquid phase on the satellites of the outer solar system, including Saturn's rings and in the gaseous state in the atmospheres of Venus, Mars and Jupiter and in comets (as can be shown, for example, from the IR spectra of Halley's comet). The OH radical has been known for many years as the photodissociation product of water.

But how did water get to the surface of the emerging primeval Earth? There are no clear answers to this important question. Two sources are considered likely:

An internal one: by gas emission after accretion of the Earth, and
An external one, via collisions with comets and asteroids which contained water.

If the starting materials for the primitive nebula from which the planets were formed were not completely homogeneous, it is possible that thermodynamically more stable, hydrated silicates could have been localized closer to the Earth during its formation than to the orbit of Venus. This would have meant that our sister planet would have had much less water available, even during its formation. In the case of water set free by gas emission, the exhalation rate determined the amount of water made available.

The second important source for the hydrosphere and the oceans are asteroids and comets. Estimating the amount of water which was brought to Earth from outer space is not easy. Until 20 years ago, it was believed that the *only* source of water for the hydrosphere was gas emission from volcanoes. The amount of water involved was, however, unknown (Rubey, 1964). First estimates of the enormous magnitude of the bombardment to which the Earth and the other planets were subjected caused researchers to look more closely at the comets and asteroids. New hypotheses on the possible sources of water in the hydrosphere now exist: the astronomer A. H. Delsemme from the University of Toledo, Ohio, considers it likely that the primeval Earth was formed from material in a dust cloud containing anhydrous silicate. If this is correct, *all* the water in today's oceans must be of exogenic origin (Delsemme, 1992).

Comets probably consist of at least 40% water. The hypothesis that the waters of the ocean have their origin in cometary mass is supported by the following result: the D/H ratio in Halley's Comet is 0.6–4.8×10^{-4}, and thus of a similar magnitude to the value of 1.6×10^{-4} found in terrestrial ocean water. Both values agree with those found for meteorites (Chyba and Sagan, 1997). François Robert from the Museum de Minéralogie in Paris has also come to a similar conclusion; he reported a good agreement between the D/H ratios of the ocean and carbonaceous chondrites (Robert, 2001).

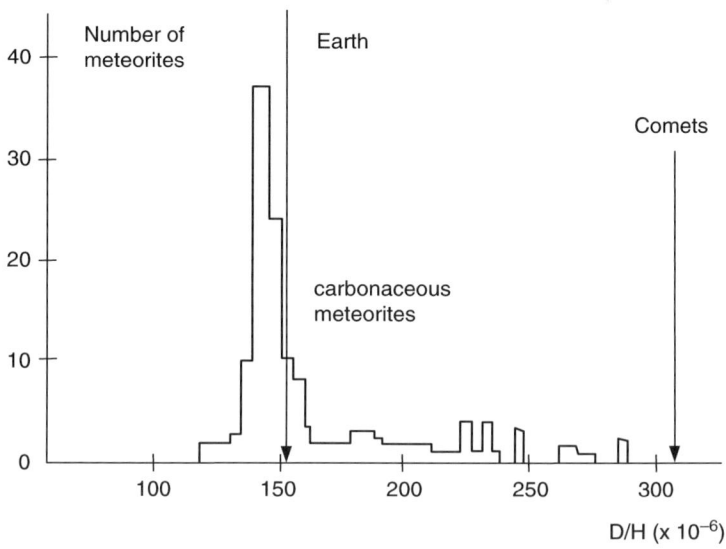

Fig. 2.7 The distribution of the ratio of the two hydrogen isotopes (D/H) in carbonaceous meteorites compared with that on Earth and in the comets. According to this distribution, most of the water on Earth must have had its origin in meteorites. From Robert (2001)

New computer simulations of the accretion process of the protoearth indicate that only a few large bodies with a high water concentration collided with the Earth during the later bombardment. They apparently came from the same region of the asteroid belt as the carbonaceous chondrites.

One of the greatest difficulties in estimating the amount of material which came from asteroids and comets lies in determining the amount of material which would have remained on the Earth's surface after the collisions, in comparison with that which escaped from its gravitational field and disappeared into space. The amount of energy set free in such collisions in turn depends on various parameters, the values of which can only be estimated. Some of these correlate with results on the number and size of the moon's craters (the Lunar Cratering Record), which are themselves subject to a number of uncertainties (Chyba, 1990). Differing rates of impact by extraterrestrial objects have been estimated for the other three terrestrial planets. In the case of Mars, the factor compared to Earth is 0.7 for long-period comets and 2.0–3.6 for short-period comets; the lower mass of Mars has led to greater atmospheric erosion.

Estimates of the mass of the primeval oceans diverge greatly: they lie between 0.2 and 0.7 of the mass of the present oceans. The range of variation of the figures given by the different models show that there are many uncertainties involved in the calculations. One of these lies in water loss due to solar UV irradiation, which would have led to a decomposition of the water in the upper levels of the atmosphere; the hydrogen thus set free would have escaped into space. This process probably occurred mainly during the accretion phase. Its involvement in the fractionation of the elements and the noble gases is indisputable. According to estimates made by some authors, the amounts of water involved might have been as high as several ocean masses, as the intensity of the "extreme solar UV" (EUV) flux in the early periods of the Earth's history would have been 1.3 times greater.

What chemical composition can we assume for the ocean? Unfortunately we have no clear results. Apart from the chemical components, it would be desirable to have information on temperature and pH values. We also do not know whether there was one single primeval ocean, or whether there were several. It is also possible that there were lakes and ponds with differing compositions. We must not forget that huge changes must have taken place on the primeval Earth's surface during the space of a few hundred million years.

If the primeval atmosphere did not contain enough CO_2 to maintain a greenhouse climate, the much lower solar irradiation at that time would have led to frozen oceans. But that would make almost all the assumed synthetic mechanisms for the formation of biomolecules impossible! Bada et al. (1994) consider "external help" as a way out of this dilemma. They assume that the energy from meteor impacts (diameters up to around 100 km), converted into heat, would have sufficed to melt the oceanic ice. If such a process were to have occurred periodically, chemical evolution reactions (see Chap. 4) could have taken place in the ice-free periods and have led finally to biogenesis.

We know nothing of the pH value of the primeval ocean. However, the acidic character of volcanic exhalations must have meant that the young ocean was also acidic. In later phases of the early history of the Earth, however, washing out due to intense rain could have led to neutral pH values. The possibility that the primeval ocean was basic in character has also been discussed (Abelson, 1966). In this case, the water from the erosion of basic regions of the Earth's crust must have changed the pH value. In today's oceans, the pH value is close to 8, and it is possible that this value varied only a little across the many million years of the Earth's history. The salt content of the young ocean was probably higher than today, but again we have no exact information (Wills and Bada, 2000). It is likely that the primeval ocean contained not only dissolved salts, but also substances which were in some cases highly toxic. The cooling process of the Earth's surface, i.e., of the still thin, cooling crust, proceeded very slowly, since the generation of heat by radioactive decomposition was about four times as intense as today (Mason, 1992). The atmospheric pressure was also probably higher than today, so that the boiling point of the ocean would also have been higher, i.e., above 373 K.

According to Summers and Chang from NASA's Ames Research Center, Moffett Field (1993), the oxidation of Fe^{2+} to Fe^{3+} provided a possibility for the reduction of nitrites and nitrates to ammonia. This reaction would have been of great importance, as NH_3 is required in many syntheses of biogenesis precursors. The authors assume that nitrogen was converted to NO in a non-reducing atmosphere, and thence to nitrous and nitric acids. These substances entered the primeval oceans in the form of "acid rain", and here underwent reduction to NH_3 with the help of Fe^{2+}, thus raising the pH of the oceans to 7.3. Temperatures above 298 K favoured this reaction, which can be written as:

$$6\,Fe^{2+} + 7\,H^+ + NO_2^- \rightarrow 6\,Fe^{3+} + 2\,H_2O + NH_3 \qquad (2.6)$$

The question of the prebiotic origin or formation of ammonia has recently been discussed by a group in Jena; they devised a method in which NH_3 is formed from N_2 with the help of H_2S. The presence of freshly precipitated FeS (prepared from $FeSO_4$ by precipitation with Na_2S at room temperature under an argon atmosphere) was found to be vital: aged FeS is inactive. In this reaction, FeS is converted to FeS_2 (iron pyrites). The reaction occurred under mild conditions, i.e., at atmospheric nitrogen pressure and at temperatures between 343 and 353 K. The yield of ammonia (with respect to 3 moles of iron sulphide) was 0.1% (3 mM). The experiments were carried out extremely carefully, so that contamination (e.g., by NO, NO_2, N_2O and NH_3) could be excluded (Dörr et al., 2003). These experimental results support the hypothesis of a chemoautotrophic origin of life (see Sect. 7.3).

References

Abelson PH (1966) Proc Natl Acad Sci USA 55:1365
Atkins PW, Beran JA (1992) General Chemistry, Second Edition. W.H. Freeman and Company.
Bada JL, Bigham C, Miller SL (1994) Proc Natl Acad Sci USA 91:1248
Binzel RP, Barucci MA, Fulchignoni M (1991), Scientific American, October, p 66

Brack A (1993) Orig Life Evol Biosphere 23:3
Catling DC (2006) Science 311:38a
Chyba CF (1990) Nature 343:129
Chyba CF, Sagan C (1997) Comets as a Source of Prebiotic Organic Molecules for the Early Earth.
    In: Thomas PJ, Chyba CF, Mc Kay CP (Eds.) Comets and the Origin and Evolution of Life.
    Springer, Berlin Heidelberg New York, p 147
Chyba CF (2005) Science 308:962
de Bergh C (1993) Orig Life Evol Biosphere 23:12
Decker R and B (1997) Volcanoes. WH Freeman and Company, New York and Oxford
Delsemme AH (1992) Orig Life Evol Biosphere 21:279
Dörr M, Käßbohrer J, Gronert R, Kreisel G, Brand WA, Werner RA, Geilmann H, Apfel C, Robl
    C, Weigand W (2003) Angew Chem 115:1581, Int Ed 42:1540
Ellis GFR (2003) Nature 425:566
Gaffey MJ (1997) Orig Life Evol Biosphere 27:185
Gehrels T (1996) Scientific American, March
Gough DO (1981) Solar Phys 74:21
Halliday AN, Drake MJ (1999) Science 283:1861
Hillebrand W, Ober W (1982) Naturwissenschaften 69:205
Hofmann AW (1997) Science 275:498
Holland HD (1962) in Engel AEJ, James HL, Leonard BF (Eds.) Petrolic Studies. Princeton Uni-
    versity Press, Princeton, NJ, p 447
Holland HD (1984) The Chemical Evolution of the Atmosphere and Oceans. Princeton University
    Press, Princeton, NJ
Joyce GF (1989) Nature 338:217
Kant I (1755) Universal Natural History and Theory of Heaven (English translation by Ian
    C. Johnston, Malaspina University-College, Nanaimo, BC: http://www.mala.bc.ca/~johnstoi/
    kant/kant2e.htm
Kasting JF (1993) Science 259:920
Kasting JF, Eggler D (2002) 13th International Conference on the Origin of Life, Oaxaca, abstr
    p 45
Kasting JF, Howard MT (2006) Phil Trans R Soc B 361:1733
Koerner DW (1997) Orig Life Evol Biosphere 27:157
Larson RB, Bromm V (2001) Scientific American, December
Luminet J-P, Weeks JR, Riazuelo A, Lehoucq R, Uzan J-P (2003) Nature 425:593
Luu JX, Jewitt DL (1996) Scientific American, May
Macdougall JD (1996) A Short History of Planet Earth. John Wiley and Sons, Inc., New York
Maciá E, Hernández MV, Oró J (1997) Orig Life Evol Biosphere 27:459
Maciá E (2005) Chem Soc Rev 34:691
Mason SF (1992) Chemical Evolution. Clarendon Press, Oxford
Miller SL (1953) Science 117:528
Mojzsis SJ, Harrison TM, Pitgeon RT (2001) Nature 409:178
Münker C, Pfänder JA, Weyer S, Büchl A, Kleine T, Mezger K (2003) Science 301:84
Neukum G (1987) Zur Bildung und frühen Entwicklung der Planeten. In: Wilhelm F (Ed.) Gang
    der Evolution, Beck, Munich, p 28
Ninisi B (2000) Science 290:15
Ostriker JR, Steinhardt PJ (2001) Scientific American, January
Owen T, Cess RD (1979) Nature 277:640
Peebles PJE (2001) Scientific American, January
Press F, Siever R (1994) Understanding Earth, WH Freeman and Company, New York Oxford
Rubey WW (1951) Bull Geol Soc Amer 62:1111
Rubey WW (1964) In: Brancazio PJ, Cameron AGW (Eds.) The Origin and Evolution of Atmo-
    spheres and Oceans. Wiley, New York
Robert F (2001) Science 293: 1056
Sagan C, Chyba C (1997) Science 276:1217

Sagan C, Mullen G (1972) Science 177:52

Schidlowski M (1980) In: Hutzinger O (Ed.) The Handbook of Environmental Chemistry, Vol 1/ Part A, S 5 Springer, Berlin Heidelberg New York

Schwartz AW (2002) Orig Life Evol Biosphere 32:399

Sleep NH, Zahnle K, Neuhoff PS (2001) Proc Natl Acad Sci USA 98:3666

Struve O, Zebergs V (1962) Astronomy of the 20th Century. The Macmillan Co., NY

Summers DP, Chang S (1993) Nature 365:630

Taylor GJ (1994) Scientific American, July

Tian F, Toon OB, Pavlov AA, de Sterck H (2005) Science 308:1014

Tian F, Tovn OB, Pavlov AA, de Sterck H (2006) Science 311:38b

Unsöld A, Baschek B (2001) The New Cosmos. Springer Berlin Heidelberg New York

Wänke H (1986) In: Jahrbuch Max-Planck-Gesellschaft. S 450

Walker JCG, Klein C, Schidlowski M, Schopf JW, Stevenson DJ, Walter MR (1983) Environmental Evolution of the Archean-Early Proterozoic Earth. In: Schopf JW (Ed.) Earth's Earliest Biosphere: Its Origin and Evolution. Princeton University Press, Princeton pp 260–290

Weinberg S (1977) The First Three Minutes. A Modern View of the Origin of the Universe. Basic Books, New York

Weissmann PR (1998) Scientific American, September

Wetherhill G (1981) Scientific American, June, p 130

Whittet DCB (1997) Orig Life Evol Biosphere 27:249

Wills C, Bada J (2000) The Spark of Life. Perseus Publishers, Cambridge, Mass

Chapter 3
From the Planets to Interstellar Matter

3.1 Planets and Satellites

The best classification of the development of planetary research is due to Kuiper, the father of modern planetology, who distinguishes three phases:

First, the three centuries of basic, classical discoveries (Galilei, Kepler, Laplace et al.).

The second phase, beginning at the end of the nineteenth century, was linked to the development of astrophysics and astrophotography; this phase was, however, marked by a *decrease* in scientific interest in planetary research.

Phase three, the renaissance of planetology, starting around 1960, caused in particular by the rapid development and successes of space travel.

In this chapter, we will deal particularly with those planets and moons which are relevant to the question of the origin of life.

The planets of the solar system are normally divided into two groups, according to their chemical composition:

The inner, or terrestrial, planets, from Mercury to Mars, including the planetoids. These have masses between 0.06 and 1 Earth masses, densities between 3,000 and 5,500 kg/m^3, and similar structures:

A relatively thin upper layer, the crust
A mantle
A core

The gas giant planets: Jupiter, Saturn, Uranus and Neptune. The planet Pluto has a status of its own, and has recently been renamed a dwarf planet.

3.1.1 Mercury

This planet, the nearest to the sun, has almost no atmosphere; its surface is covered with craters. During the formation of Mercury, planetesimals were able to impact the planet's surface without any resistance. Thus, the lack of erosion processes (due to

H. Rauchfuss, *Chemical Evolution and the Origin of Life*,
© Springer-Verlag Berlin Heidelberg 2008

wind and/or water), which could flatten the surface, left the craters as they originally were. There are great temperature differences between the day and the night side of Mercury's surface, from 600 down to 100 K. Radar mapping (using wavelengths of 3.5 cm) indicates the presence of water ice at the poles, in craters which have probably never been reached by the sun's rays (Slade et al., 1992). According to one hypothesis, Mercury was once a moon of Venus and was shifted to a new orbit around the sun by an unknown event.

Until now, Mercury has only been studied more closely by one spacecraft (Mariner 10, 1974), since its nearness to the sun means that spacecraft approaching it are subject to particularly extreme conditions. NASA's MESSENGER (Mercury Surface, Space, Environment, Geochemistry and Ranging) was launched in 2004 and is planned to reach Mercury in March 2011, and then to orbit the planet. The main tasks of the MESSENGER mission are to map the planet, to make measurements of its magnetic field and to collect data relevant to its geological and tectonic history (Solomon, 2007).

3.1.2 Venus

Apart from the sun and the moon, Venus is the brightest heavenly body. It is the only satellite of the sun which is greeted during an emotional aria in a well-known opera: Wolfram von Eschenbach serenades the brightly shining Venus in Wagner's *Tannhäuser* with the words "O Du mein holder Abendstern wohl grüss' ich Dich so gern" ("O my fair evening star, I always gladly greeted thee").

The surface of Venus is hidden under an unbroken layer of clouds 45–60 km above it. Recently, the planet has been subjected to a complete cartography by radar satellites. Its atmosphere contains 96% CO_2 by volume, the remainder consisting of N_2, SO_2, sulphur particles, H_2SO_4 droplets, various reaction products and a trace of water vapour. The water is probably subject to photolytic decomposition. Noble gases are more abundant than on Earth: ^{36}Ar by a factor of 500, neon by a factor of 2,700, and D (deuterium) by a factor of 400.

Because of the CO_2 greenhouse effect, the annual average temperature at the surface of Venus is around 733 K, and there is intense atmospheric activity. Results from the Cassini spacecraft did not confirm the earlier assumption that lightning is very frequent. According to Gurnett et al. (2001), flashes of lightning either occur very seldom, or are completely different from terrestrial electrical discharges. The turbulences at the surface of Venus are extremely vehement: wind speeds of up to 360 km/h have been measured, which means that the cloud layer moves 60 times faster than the planetary surface. The surface pressure is 90 times greater than that at sea level on Earth. New model calculations show that the climate of Venus has changed in a significant manner across only a few hundred million years (Prinn, 2001). According to the Bullock-Greenspoon model, Venus was colder between 600 and 1,100 million years ago. Two main processes now control the climate of Venus:

Global warming, mainly determined by the CO_2 greenhouse effect

Cooling, caused by reflection of solar irradiation due to the presence of thick clouds of sulphuric acid

There are now doubts as to whether Venus is in fact extremely hostile to life. An audacious theory suggests that the cloud cover in the Venusian atmosphere could have provided a refuge for microbial life forms. As the hot planet lost its oceans, these primitive life forms could have adapted to the dry, acid atmosphere. However, the intensity of the UV radiation is very puzzling. The authors suggest that sulphur allotropes such as S_8 act on the one hand as a UV umbrella and on the other as an energy-converting pigment (Schulze-Makuch et al., 2004).

The Venus Express spacecraft launched by the European Space Agency (ESA) in November 2005 reached its goal in April 2006. Its main purpose was to find out more about the (still not understood) super-rotation of the Venusian atmosphere, which causes clouds to circulate the planet in about four earth days. Venus takes 243 earth days to rotate about its own axis.

The VIRTIS apparatus (Visible Infrared Thermal Imaging Spectrometer) on board can observe the atmosphere and the cloud layers at various depths (on both the day and the night side of the planet). VIRTIS has also provided data for the first temperature map of the hot Venusian surface. These data have led to the identification of "hot spots" and thus provided evidence for possible volcanic activity (www.esa.int/specials/venusexpress).

3.1.3 Mars

Western man has had a special relationship with the planet Mars for many centuries. The Romans venerated the red planet as the god of war, and in Italy it later became the god of fruitfulness and the god of the peasants. The astronomer Tycho Brahe (1546–1601), born in the then Danish and now Swedish province of Skåne, determined the exact position of Mars by means of precise observations of the heavens. The discovery of the telescope by the Dutch physicist and mathematician Christiaan Huygens (1629–1695) made it possible to determine the rate of rotation of the planet. Huygens determined the period of rotation as 24.5 hours; the value accepted today is 24.623 hours, making clear how great an achievement his measurements, carried out 250 years ago, in fact were! The Italian astronomer Giovanni Schiaparelli (1835–1910) discovered *canale* on the surface of Mars; in Italian, "canale" means not only canals, but other water-bearing systems. The "canal hypothesis" was the subject of great interest, not only among scientists but also among authors of fiction.

The planet Mars is smaller than the Earth: its diameter is 6,762 km, compared with the Earth's 12,760 km. Our neighbour planet has only a very thin atmosphere (surface pressure 0.005–0.010 atm), so its surface can easily be observed. The atmosphere consists of the following (volume percentages given):

$\sim 95\% CO_2$

$\sim 2.5\% N_2$

$\sim 1.5\% Ar$

$\sim 1\%$ other noble gases

$\sim 0.1\% O_2$ and CO, formed from CO_2 by photodissociation

Recently, sulphur has also been found on the surface of Mars; it was probably deposited from the atmosphere and originated in volcanic activity. Sulphur was also found in meteorites which probably originated on Mars (Farquhar et al., 2000). The mean surface temperature is ~ 210 K (at night 150 K and during the day 270 K).

Ice has been found at the poles: new measurements of Mars' southern polar region indicate the presence of extensive frozen water. The polar region contains enough frozen water to cover the whole planet with a layer of liquid approximately 36 ft deep. A joint NASA–Italian Space Agency instrument on the European Space Agency's Mars Express spacecraft provided these data (NASA press release, 15 March 2007). It must be assumed that volcanic exhalations contained large amounts of water.

The planet Mars provides another sensation: it has the highest and largest volcano in the solar system, Mons Olympus (25 km high). Volcanologists disagree about the formation of this huge volcano, but there are several models which attempt to explain its formation. Not only the extreme size and height of the volcano, but also the almost circular high escarpment which surrounds it make the volcano unique (Helgason, 1999).

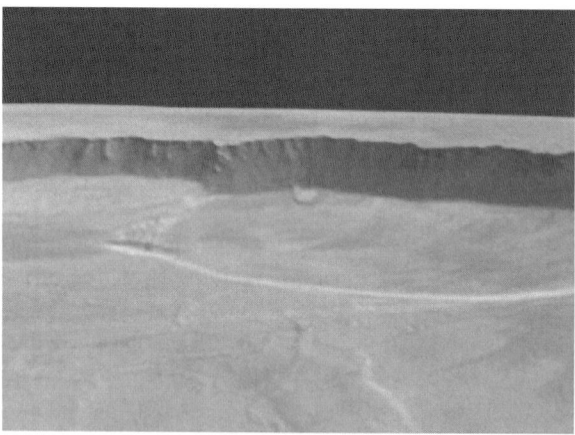

Fig. 3.1 Perspective view of part of the caldera of Olympus Mons on Mars. This view was obtained from the digital altitude model derived from the stereo channels, from the nadir channel (vertical perspective) and the colour channels on the Mars Express Orbiter. The photograph was taken on 21 January 2004 from a height of 273 km. The vertical face is about 2.5 km high, i.e., about 700 m higher than the north face of the Eiger mountain (Switzerland). With permission of the DLR

The surface of Mars is covered by meteorite craters, some up to 200 km in diameter. The question as to whether water exists on Mars has been the subject of scientific controversy for many years (see Chap. 11). Costly Mars missions with the goal of mapping the surface have afforded important results on now dry river valleys. The weather on Mars is characterized by ground-level fog, thin ice clouds and (often very violent) dust storms, which vary not only seasonally but also daily. The question as to whether our neighbour planet harbours life (of any kind), or if it ever did so, gave rise to a media sensation at a NASA press conference on August 7, 1996. The researchers, who had been studying the 1.9 kg Mars meteorite ALH 84001, came to the conclusion that it bore clear evidence of previous life forms:

A certain carbonate species with magnetite and iron deposits: microorganisms could have been involved in its formation.

Organic compounds: polycyclic aromatic hydrocarbons, in particular phenanthrene ($C_{14}H_{10}$), pyrene ($C_{16}H_{10}$) and chrysene ($C_{18}H_{12}$), which were detected using high resolution mass spectrometry.

Structures which showed similarities with microorganisms (McKay, 1996).

There was, however, much criticism of these optimistic results, in particular from the palaeontologist William Schopf from UCLA.

Not only would proof of the existence of life on Mars be a great sensation, but even the discovery of precursors of life, such as biomolecules or building blocks for their formation, would change our perspective greatly (see also Chap. 11).

3.1.4 Jupiter

The planet Jupiter occupies a special position in the solar system. It is the largest and heaviest planet, with a mass of 1/1,047 that of the sun. Jupiter consists almost solely of hydrogen and helium with a ratio similar to that found in the sun itself: He:H ≈ 1:10. Small amounts of some heavier elements are present, such as B, N, P, S, C and Ge. The density of Jupiter has been calculated as 1,300 kg/m^3. Its atmosphere can be divided into three zones (starting from the outermost):

– The zone of the ammonia clouds (temperature ∼140 K),
– The zone of the NH_4SH clouds, also containing NH_3, H_2 and He (∼200 K) and
– The zone of the ice clouds, consisting of water ice crystals and H_2/He gas (∼270 K).

The planet does not have a real surface; instead, there is a gradual transition from the H_2/He mixture to the central body, which consists of molecular hydrogen. Since there is no actual surface, temperatures can only be expressed in terms of their corresponding pressures.

Around 85% of the total amount of hydrogen is present as a metallic phase. It is assumed that there is a silicate rock core with a temperature estimated to be

24,000 K. This mass of rocky material probably formed the nucleus for condensation and attracted large amounts of the H_2/He-rich solar material around 4.5 billion years ago, acting as a galactic vacuum cleaner.

Research on Jupiter has progressed greatly in the last decades. The Galileo mission, which started in 1989, provided important data on Jupiter and its moons. The Galileo spacecraft had a special probe on board, which left the mother craft and entered Jupiter's atmosphere on December 7, 1995. A great deal of heat was generated in the process: temperatures of around 16,000 K were measured at around the 5 mbar level (Seiff et al., 1997). The probe flew for about an hour before disintegrating in the depths of the Jovian atmosphere. The data which it transmitted to the Galileo spacecraft provided information on temperature, pressure, chemical composition (and also isotope ratios) of the atmosphere, its water content, and on electrical discharges (Young, 1996). Surprisingly, the water concentration measured by the probe was only about 10% of that expected; it is unclear whether this was merely a local phenomenon, or whether it is characteristic for the whole atmosphere of Jupiter. The giant planet rotates around its axis in about ten hours, and thus compensates for its lack of mass by its enormous rotary motion. The latter causes the outer visible atmosphere to be very dynamic: it exhibits conspicuous, complex zones parallel to the degrees of latitude. Rapidly rotating planets are characterized by many regular dynamic phenomena, which are due to the Coriolis force.

In 1994, a unique incident occurred: the impact of the Shoemaker-Levy comet on the Jovian atmosphere. The strong gravitational field of Jupiter caused the comet to break up before it could enter the atmosphere, and the parts of the comet crashed separately into the atmosphere one after the other. This unique spectacle was observed by many observatories and also by the Galileo spacecraft and the Hubble telescope. It led to the discovery of yet another phenomenon: the most intensive aurora effects in the solar system, observed at Jupiter's poles. Astronomers assume that the energy for these comes from the planet's rotation, possibly with a contribution from the solar wind. This process differs from that of the origin of the aurora on Earth, where the phenomenon is caused by interactions between the solar wind and the Earth's magnetic field.

One more important property of Jupiter must be mentioned: the Earth owes its relatively "quiet periods" (in geological terms) to the huge gravitational force of the giant planet. Jupiter attracts most of the comets and asteroids orbiting in its vicinity, thus protecting the Earth from impact catastrophes!

3.1.5 Jupiter's Moons

The four brightest and largest Jovian moons are also called the "Galilean moons", as Galileo Galilei discovered them in 1610 and gave them their names: Io, Europa, Ganymede and Callisto, in the order of their orbits around Jupiter. The system of Jupiter and its four moons has many similarities to the solar system as a whole, in particular the extreme regularity and the planar orbits (Stevenson, 2001).

3.1.5.1 Io

As already mentioned, the moon Io is the innermost of Jupiter's satellites. Its density is $3,550 \, kg/m^3$, similar to that of the terrestrial planets, and the colour of its surface ranges from yellow gray to orange red. The latter colour may be due to the presence of S_2O, as sulphur and sulphur compounds play a vital role in the chemistry of Io, the surface rocks of which contain large amounts of potassium and sodium compounds. The greatest sensation was the discovery of an active volcano by one of the Voyager missions in 1979. The eruption of the volcano, known as "Prometheus", spews out matter to a height of 100 km above the moon's surface. Nine other volcanoes have since been discovered on Io, which during their active phases emit sulphur- and oxygen-containing gases as well as molten sulphur and sulphur dioxide; these exhalations reach heights of up to 250 km. Because of its volcanic activity, the surface of Io is relatively flat; impact crater structures are hardly visible. An answer to the question of the energy source for the volcanic activity of Io was soon found: like the other Galileic moons, but in particular because it is so near to Jupiter, the orbit of Io is elliptic. This leads to tidal forces which generate frictional heat in its interior. It seems likely that the moon has remained in its present state for about the last two billion years. The atmospheric pressure of Io is around 10^{-10} bar of SO_2 at temperatures between 60 and 120 K.

The SSI (solid-state imaging) camera on board the Galileo spacecraft transmitted impressive high-resolution pictures of Io's volcanic activity. Active lava lakes, lava "curtains", calderas, mountains and plateaus can be seen (McEwen et al., 2000). The Hubble telescope detected both S_2 gas and SO_2 in a SO_2 to S_2 ratio of 1:4 in the smoke trail of the volcano Pele. This value suggests an equilibrium between silicate magmas in the neighbourhood of the quartz–fayalite–magnetite buffer (see Sect. 7.2.2).

Io is one of the most interesting objects in planetary research. However, it is completely irrelevant to the biogenesis problem, in complete contrast to the Jovian moon Europa.

3.1.5.2 Europa

Jupiter's moon Europa has only been the subject of intense scientific investigation in recent years; it is considered to be a member of that small group of heavenly bodies which could perhaps accommodate life (or a precursor of life). About 20 years ago, the Voyager passes afforded sensational pictures of Europa. These showed a network of linear bands, of differing breadths, on a very bright surface. The mean density was calculated as $3,018 \pm 35 \, kg/m^3$, and the surface temperature measured was 90–95 K. Circumstantial evidence points to either a surface consisting of water ice, or the presence of liquid water or "warm ice" under the surface. Three models were proposed (Oró et al., 1992):

– *The thin ice model*: the silicates are mainly hydrated, so there is a thin layer (a few kilometres) of water ice above the silicates.

- *The ice-ocean model*: Europa's core consists of dehydrated silicates, since heat
  production made dehydration possible. Around the core, there is a thick layer of
  liquid water (about 100 km), and above that a thin layer (about 10 km) of water
  ice.
- *The thick ice model*: enough heat was generated in the interior of the moon to
  dehydrate the silicates. The water set free froze to give an ice layer about 100 km
  thick.

The first model appears unlikely, since it would entail the presence of more ancient
collision craters on the moon's surface. The decision as to whether the second or
third model is favoured depends on the question as to whether the amount of heat
generated by tidal friction was low enough to allow the water mass to freeze completely. Theoretical considerations and calculations suggest that the second model is
probably most likely to be correct. More recent results, in particular from the Galileo
mission in October 1996, provided pictures of the moon's surface with much higher
resolution than before. They showed episodic separation of the surface plates, with
the crevasses being filled up by material from lower layers (either ice or water)
(Sullivan et al., 1998). Convection of the solid ice layer may be involved, i.e., the
formation of ice domes and ridges caused by the motion engendered by the upthrust-
ing, warm ice masses (Pappalardo et al., 1998).

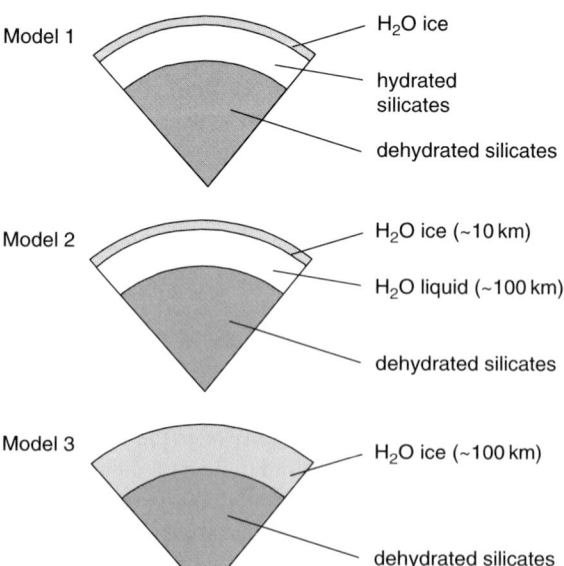

Fig. 3.2 The three possible
models for the inner structure
of the Jovian moon Europa:
model 1 has a thin layer of ice
at the surface, model 2 is the
ice-water model and model 3
involves a thick ice layer

A further piece of evidence for the presence of an ocean below the ice surface was found by Carr et al. (1998) during their analysis of pictures with reso-
lutions of 1.2 km, 180 m and 54 m per pixel: local icebergs are visible. A more

exact morphology indicates that liquid water is present under the ice surface. Using data from the Galileo NIMS (Near-Infrared Mapping Spectrometer), McCord et al. (1998) from the University of Hawaii found evidence for the presence of salts on the moon's surface. The water absorption bands recorded at 1–2.5 μm showed the presence of hydrated minerals (magnesium sulphates, sodium carbonates and mixtures of the two). These can be detected in the surface lines (ridges) and the optically denser regions of the surface. New IR and UV spectral data from the Galileo probe (Carlson et al., 1999a) and model measurements under simulated "Europa conditions" in the laboratory can only be interpreted in terms of the presence of hydrogen peroxide (H_2O_2). This is probably formed by radiolysis at the water surface, since the Jovian moon is subjected to violent bombardment, originating from Jupiter's magnetosphere, by energy-rich electrons, protons and S and O ions.

Carlson et al. (1999b) compared laboratory spectra and Galileo data and suggested that hydrated sulphuric acid is the main component in the dark surface material, which probably also contains sulphur polymers modified by radiation chemistry. A sulphur cycle involving three sulphur species is suggested: sulphuric acid (H_2SO_4), sulphur dioxide (SO_2) and sulphur polymers (S_x).

Before data from the Galileo mission became available, the interior structure of the moon was still basically unknown. The data obtained during two encounters of the probe with Europa (E4 and E6) on December 19, 1996, and February 20, 1997 (Anderson et al., 1997), indicated the presence of an inner core with a density of 4,000 kg/m^3. This could be a metal core with a radius of about 40% of that of the moon, surrounded by a rock mantle with a density of 3,000–3,500 kg/m^3. Two further approaches of the probe to Europa made refinement of the model possible (Anderson et al., 1998), and they concluded that the moon's interior may consist of a mixture of silicates and metals. If the moon does in fact have a metallic core, estimation of its diameter is not possible because of its unknown composition.

Moore (1998) suggested that the data available could be interpreted in terms of an ice crust 10–15 km thick. Christopher Chyba from the SETI Institute (Mountain View, California) has published articles in *Nature* (2000), the *Proceedings of the National Academy of Sciences* (2001a) and in *Science* (2001b) in which he suggests that a detailed study of this Jovian moon is necessary: he discusses the possibility of a complex ecosystem, nourished by the radiation coming from outer space, on or in the ice layers of the moon. The planned Europa orbiter mission may provide certainty on this, but at least another five years of uncertainty lie ahead. The use of a submersible robot to study the (possible) ocean layer and its floor has been discussed.

Such a mission would require successful drilling through the moon's surface ice layer (Rummel, 2000; de Morais, 2000): testing of a new apparatus required for the study of Europa's ice could be done in the subglacial Antarctic Lake Vostok, under the Antarctic ice. It does not, however, seem appropriate to test such technologies in this extremely sensitive environmental situation. However, Russian scientists are carrying out drilling studies on Lake Vostok (Inman, 2006).

Fig. 3.3 An artist's impression of the originally planned "hydrobot" mission to Europa. The robot has bored through the ice layer in the moon's intermediate aqueous layer and is investigating the ocean floor. From NASA

Recent work suggests that there may have been a period in Europa's history when an extreme greenhouse effect led to temperatures which would have sufficed for reactions necessary for chemical evolution. According to this (unproven) hypothesis, building blocks for biomolecules or even primitive life forms could have existed. The authors assume that there is a high probability that bioelements could have been "delivered" by comets (Chyba and Phillips, 2002).

3.1.5.3 Ganymede and Callisto

These two Jovian moons are in some respects quite similar. They probably consist of rocky material and frozen water (in a ratio close to 1:1) and, in contrast to Europa, are covered by a large number of craters caused by collisions with other heavenly bodies.

Ganymede has a diameter of 5,268 km and is thus the largest moon in the solar system; it is in fact larger than the planet Mercury. Reflection spectra provided by the NIMS apparatus on the Galileo spacecraft suggest that the surface of Ganymede contains aqueous material (McCord et al., 2001). As on Europa, it is likely that this material is in fact frozen $MgSO_4$ sols formed in liquid layers under the surface. A careful evaluation of the pictures of the moon's surface led to a great deal of speculation; thus, some authors discuss a secondary encrustation of the moon's surface due to tectonic or volcanic processes (Schenk et al., 2001). Volcanic eruptions could have brought liquid water or solid water ice to the surface. The tectonic activity on Ganymede may have been much greater than has previously been assumed (Kerr, 2001).

Callisto orbits Jupiter at a distance of 1.9 million kilometres; its surface probably consists of silicate materials and water ice. There are only a few small craters (diameter less than a kilometre), but large so-called multi-ring basins are also present. In contrast to previous models, new determinations of the moon's magnetic field suggest the presence of an ocean under the moon's surface. It is unclear where the necessary energy comes from: neither the sun's radiation nor tidal friction could explain this phenomenon. Ruiz (2001) suggests that the ice layers are much more closely packed and resistant to heat release than has previously been assumed. He considers it possible that the ice viscosities present can minimize heat radiation to outer space. This example shows the complex physical properties of water: up to now, twelve different crystallographic structures and two non-crystalline amorphous forms are known! Under the extreme conditions present in outer space, frozen water may well exist in modifications with as yet completely unknown properties.

3.1.6 Saturn and Its Moon Titan

The giant planet Saturn is in many ways similar in its chemical and physical properties to Jupiter. However, it has the lowest density of all the bodies in the solar system. The cloud structure and the chemistry of Saturn's atmosphere resemble those of Jupiter, but the structures on the ring planet appear more diffuse and less clear, because of the presence of a layer of haze. The best-known feature of Saturn is the ring discovered by Christiaan Huygens in 1659, which is 278,000 kilometres in diameter and whose fine structure was determined only in 1978, 1980 and 1981 by the Pioneer 11, Voyager 1 and Voyager 2 missions. The material in the ring probably has its origin in a former Saturnian moon which came too close to the planet and was torn apart. It appears that the ring system is only about three kilometres thick and that its total mass is only about one millionth of that of Saturn itself.

Titan is certainly the most interesting and most important moon in terms of the subject of this book. It was discovered by Christiaan Huygens in 1655 and is a highly unusual planetary satellite: it is the only moon in the solar system which has a real atmosphere. The only two bodies which are surrounded by a thick layer of nitrogen are Titan and the Earth. Titan is the second largest moon in the solar system, and with a diameter of 5,150 km, it is larger than the planet Mercury. Its mass is sufficient to bind the nitrogen atmosphere, but not to retain hydrogen. The Voyager mission had provided data on Titan's atmosphere, and these were complemented on July 3, 1989, when Titan eclipsed the giant star 28 Sagitarii (Sicardy et al., 1990; Hubbard, 1990). The pressure at the surface of Titan is around 1.5 atm, and the atmosphere contains, by volume, 90% nitrogen; in 1944 Kuiper found that methane was also present. Titan's atmosphere has regions of haze which are between 200 and 300 kilometres thick. The IR spectrometer aboard the Voyager spacecraft detected the following carbon compounds: HCN, C_3H_8, methylacetylene, diacetylene (C_4H_2), cyanoacetylene (HC_3N), cyanogen (C_2N_2), CO and CO_2.

Why does only Titan have such a massive atmosphere, in contrast to the other similarly sized Jovian moons (which are closer to the sun, but have an escape velocity of the same magnitude)? One explanation is that the orbit of the Jovian moons lies within the sphere of influence of Jupiter's strong magnetosphere, whereas Titan is only slightly affected by the magnetosphere of Saturn. Its greater distance from the sun could also be important, since lower temperatures favour the incorporation of volatile gases into clathrates (cage compounds) and thus bind them to the moon. Titan's temperature is between 70 and 180 K, the minimum occurring at a height of about 70 km; the surface temperature is about 94 K. The planet's density suggests the presence of approximately equal amounts of water ice and rocky material. The information presently available indicates that Titan consists of a core made of rocky material, which is surrounded by layers of water/ammonia and water/methane clathrates. Its distance from the sun is 9.5 AU, i.e., it is subject to only about 1% of the amount of solar radiation which reaches the Earth.

Studies carried out on Earth, for example, by the NASA infrared telescope on Mauna Kea (Hawaii), showed albedo variations which indicated the presence of "holes" in the Titanian cloud formations (Griffith, 1993). It is, however, still unclear as to whether these inhomogeneities result from differences in the surface composition. Lorenz et al. (1997) reported large variations in Titan's atmosphere due to photochemical processes. The methane contained in the dense nitrogen atmosphere is decomposed by solar and thermal radiation, and its content may be replenished from methane lakes or from clathrates.

The common properties of Titan and our Earth have led to great scientific interest in this Saturnian satellite, which can be considered as a type of "extraterrestrial laboratory" in which a series of chemical and physical processes occur which are similar to those involved in chemical evolution on the primeval Earth.

Table 3.1 Chemical composition of the Titan stratosphere at a height of 80–140 km (Raulin, 1998)

Compound or element	Percentage composition in the stratosphere
Nitrogen (N_2)	0.90–0.99
Methane (CH_4)	0.017–0.045
Hydrogen (H_2)	0.0006–0.00014
Ethane (C_2H_6)	1.3×10^{-5} (equator)
Ethyne (C_2H_2)	2.2×10^{-6} (equator)
Propane (C_3H_8)	7.0×10^{-7} (equator)
Hydrocyanic acid (HCN)	6.0×10^{-7} (north pole)
Carbon monoxide (CO)	2.0×10^{-5}

Several laboratories, including that of F. Raulin in Paris (Coll et al., 1998) and of J. Ferris in the USA (Clarke and Ferris, 1997) have carried out experiments on simulated Titan atmospheres; these indicate that methane and nitrogen can exist side by side (Table 3.1).

While the presence of methane indicates a reducing atmosphere, that of nitrogen fits better into a (weakly) oxidising environment. It is believed that the present composition of Titan's atmosphere is the result of chemical or radiation-induced reactions.

Laboratory simulation experiments involve several problems. The mixing ratios of the reacting gases depend strongly on the height of the assumed reaction space above Titan's surface and thus on the gas pressures and the corresponding temperatures. An additional problem is provided by "reactor wall effects" and the incomplete exclusion of impurities such as oxygen. Both factors are absent in outer space but can lead to huge errors in laboratory simulations.

As can be seen from Table 3.1, the Titanian atmosphere contains a relatively large amount of ethane. Laboratory results show that methyl radicals ($H_3C\cdot$), which are primary products of methane photolysis, may be present in the upper reaches of the atmosphere:

$$2CH_4 \overset{hv}{\rightarrow} 2CH_3 + 2H\cdot \qquad (3.1)$$

$$2CH_3 \rightarrow C_2H_6 \qquad (3.2)$$

The short-wavelength radiation necessary for this decomposition is absent in the lower layers of the atmosphere; it is likely that the photolysis of ethyne occurs via C–H cleavage to give radicals, which react with methane to give methyl radicals, the recombination of which affords ethane:

$$C_2H_2 \overset{hv}{\rightarrow} H\cdot + \cdot C_2H \qquad (3.3)$$

$$\cdot C_2H + CH_4 \rightarrow C_2H_2 + \cdot CH_3 \qquad (3.4)$$

$$2 \cdot CH_3 + M \rightarrow C_2H_6 + M \qquad (3.5)$$

$$(M = \text{catalyst})$$

Calculations showed that this indirect photolysis occurs between 2.5 and 4 times faster than the processes occurring in the upper atmosphere. Analogous reactions were described by Clarke and Ferris (1997):

$$\text{For } C_4H_2 : C_4H_2 \overset{hv}{\rightarrow} H\cdot + \cdot C_4H \qquad (3.6)$$

$$\cdot C_4H + CH_4 \rightarrow C_4H_2 + \cdot CH_3 \qquad (3.7)$$

$$\text{For } HCN : HCN \overset{hv}{\rightarrow} H\cdot + \cdot CN \qquad (3.8)$$

$$\cdot CN + CH_4 \rightarrow HCN + \cdot CH_3 \qquad (3.9)$$

$$\text{For } HC_3N : HC_3N \overset{hv}{\rightarrow} H\cdot + \cdot C_3N \qquad (3.10)$$

$$\cdot C_3N + CH_4 \rightarrow HC_3N + \cdot CH_3 \qquad (3.11)$$

$$\text{For } C_2N_2 : C_2N_2 \overset{hv}{\rightarrow} 2\cdot CN \qquad (3.12)$$

$$\cdot CN + CH_4 \rightarrow HCN + \cdot CH_3 \qquad (3.13)$$

The hydrogen set free can add to unsaturated compounds; these reactions occur in the lower reaches of the Titanian atmosphere. Hydrogen cannot escape from the upper atmosphere before it reacts. The authors suggest a catalytic scheme in which reactive hydrogen atoms are converted into molecular hydrogen (H_2) without a net loss of unsaturated compound (here C_4H_2):

$$C_4H_2 + H\cdot + M \rightarrow \cdot C_4H_3 + M \qquad (3.14)$$

$$\cdot C_4H_3 + H\cdot \rightarrow C_4H_2 + H_2 \qquad (3.15)$$

The photochemistry of Titan's atmosphere can be summarized as follows: the unsaturated compounds are formed from HCN and C_2H_2, which is derived from CH_4. Methane decomposition leads to further ethane formation.

Two important substances have so far *not* been found on Titan: the noble gas argon and water. The analysis of the results of the successful Cassini mission may soon shed light on this mystery.

Joseph et al. (2000), from the laboratory of J. Ferris, used a new type of flow reactor for simulation experiments designed to investigate the reason for the haze formation in the Titanian atmosphere; this apparatus made it possible to use very small amounts of gases, so that concentration ratios close to those actually present on Titan could be reached. Thus, extrapolation was no longer necessary, and the undesirable reactor wall effects were negligible. Mixtures containing N_2, CH_4, H_2, C_2H_2, C_2H_4 and HC_3N were used. The analysis of the volatile reaction products formed on irradiation were carried out using IR and nuclear magnetic resonance spectroscopy, and IR was also used to study the solid products (haze and dust particles). Size distribution and morphology were determined using scanning electron microscopy.

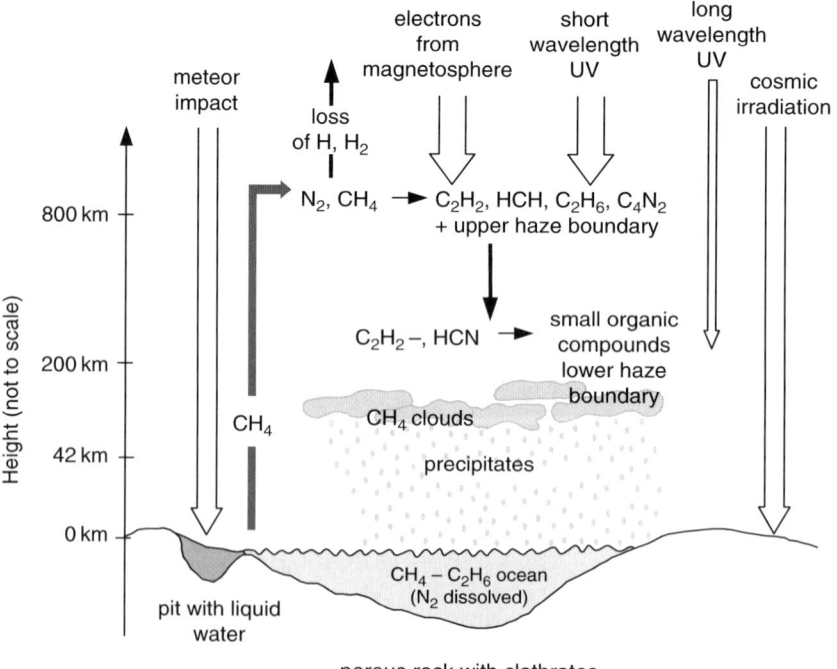

Fig. 3.4 Summary of the processes which may occur on Saturn's moon Titan (Clarke and Ferris, 1997)

Some of the data thus obtained are necessary for the interpretation and analysis of information which will be provided by the Huygens spacecraft. The Cassini–Huygens project, which is carried out jointly by NASA and ESA, started in 1997 with the launch of a Titan IV/Centaur rocket. After passing close to Venus, Earth and Jupiter, the spacecraft was brought into orbit around Saturn on July 1, 2004. Apart from an extensive research program, which includes studies of Saturn's magnetic field as well as a close look at its rings, the Titan project was one of the most spectacular. After orbiting several times around Titan, the Huygens probe landed on its surface at 12:34 GMT on January 14, 2005.

The problem of the seasonal changes in Titan's atmosphere was studied by T. Tokano from the Institute of Geophysics and Meteorology at the University of Cologne using a general circulation model (Tokano et al., 1999; Tokano, 2000). As expected, methane plays an important role, although its origin is unknown, since this hydrocarbon is rapidly decomposed by photochemical processes, as discussed above. Loveday et al. (2001) reported on the thermodynamic behaviour of methane hydrates, which may well be present in large amounts on the surface of this Saturnian moon, perhaps as a methane clathrate layer 100 kilometres thick. Such methane/ice clathrates also exist on Earth, particularly in the ocean depths and in permafrost regions. One cubic metre of such a clathrate can theoretically set free 164 cubic metres of methane and 0.8 cubic metres of water. However, if methane hydrates are present on Titan, they will be subject to much more complex conditions than they are on Earth.

The structures of the thick layers of haze which surround Titan, and which are in some ways comparable to the smog we know so well on Earth, are a mystery to scientists. It is possible that a numeric simulation model has solved the problem (Rannou et al., 2002): their results suggest that winds are responsible for the seasonal variations of the haze structures. The tiny particles which form the haze move from one pole to the other during a Titanian year (which corresponds to 4 years on Earth). This new model also explains the formation of a second separate haze layer above the main layer: this is formed from small particles which are blown to the poles and separate from the main haze layer before later returning to it.

The most recent results from the successful Cassini–Huygens mission will be discussed in Sect. 11.1.1.3.

3.1.7 Uranus and Neptune

Although Uranus and Neptune also belong to the group of gas giant planets, they are constructed differently from Jupiter and Saturn:

> They are smaller than the two giant planets and
> They contain, by weight, only about 15–20% hydrogen and helium. The greater part of the planetary mass consists of rocky material and water ice (a mixture of H_2O, NH_3 and CH_4).

Uranus The temperature in the Uranus atmosphere, which consists of molecular hydrogen containing around 12% helium, is close to 60 K. A methane cloud layer has been detected in the lower layers of this atmosphere. The planet is surrounded by a magnetosphere which extends into space for about ten times the diameter of Uranus. The planet has 27 moons of various sizes and is surrounded by a ring system which consists of thin dark rings. The planet is unusual in two respects: its tilted axis and retrograde rotation.

Neptune Small amounts of methane colour the H_2/He mixture of the Neptunian atmosphere blue. Energy sources in its interior are probably responsible for the fact that Neptune radiates 2.6 times as much energy as it receives from the distant sun. Triton is the largest of the eight moons and has a clearly structured surface, a world which compares to no other (Kinoshita, 1989). Voyager 2 has given us remarkable pictures of Triton's surface from a distance of only 38,000 kilometres. An icecap consisting of frozen methane and nitrogen was found in the southern polar region: its temperature is 37 K, which makes it the coldest object ever detected in the solar system. Trails of "smoke", which seem to come from geyser-like eruptions, were also detected. The material ejected from the surface may consist of a mixture of water and liquid methane; nitrogen in liquid or vapour form has also been suggested (Söderblom et al., 1990). In these exhalations, dark-coloured material is flung to a height of 8 kilometres. The moon is surrounded by a layer of nitrogen, which is 700–800 kilometres thick and contains about 0.01% methane. The density of the Triton atmosphere is, however, very low: the atmospheric pressure at the moon's surface is only about 1/70,000 of that at sea level on Earth.

3.1.8 The Dwarf Planet Pluto and Its Moon, Charon

In August 2006, the International Astronomical Union redefined the term "planet" and decided that the former ninth planet in the solar system should be referred to as a "dwarf planet" with the number 134340. The dwarf planet Pluto and its moon, Charon, are the brightest heavenly bodies in the Kuiper belt (Young, 2000). The ratio of the mass of the planet to that of its moon is 11:1, so the two can almost be considered as a double planet system. They are, however, quite disparate in their composition: while Pluto consists of about 75% rocky material and 25% ice, Charon probably contains only water ice with a small amount of rocky material. The ice on Pluto is probably made up mainly of N_2 ice with some CH_4 ice and traces of NH_3 ice. The fact that Pluto and Charon are quite similar in some respects may indicate that they have a common origin. Brown and Calvin (2000), as well as others, were able to obtain separate spectra of the dwarf planet and its moon, although the distance between the two is only about 19,000 kilometres. Crystalline water and ammonia ice were identified on Charon; it seems likely that ammonia hydrates are present.

3.2 Comets

The appearance of a comet in the sky is something which fascinates many people; the comet's long tail of luminescent material moving rapidly in the dark night sky has been the subject of much speculation across the centuries. Aristotle mentions comets, which he considers to consist of substances which evaporate from the Earth's surface and ignite when they reach great heights. In the Middle Ages, comets induced fear and trepidation: their appearance was considered to herald catastrophic events to come. Tycho Brahe is seen as the father of modern cometary research. He realized that the comet which appeared in 1557 had an orbit which took it beyond that of the moon; thus, Aristotle's theory, which was still adhered to at that time, must have been wrong. Brahe assumed that the comet orbited around the sun, so that the old, geocentric model of the universe could not possibly be correct. However, he did not abandon this model, but instead merely modified it.

We have learned a great deal about comets in the intervening centuries, but there still remain some unanswered questions.

3.2.1 The Origin of the Comets

Comets, like planetoids and meteorites, belong to the group of small heavenly bodies. According to the nature of their orbits, we distinguish two groups:

Long-period comets: their extended ellipsoidal orbits reach far outside our solar system (up to half the distance to the next fixed star). This group includes the comet Kohoutek, discovered in the 1970s, which requires about 75,000 years for a single orbit.

Short-period comets: these display a strong tendency for their farthest point from the sun (aphelia) to coincide with a giant planet's orbital radius, so that we can distinguish so-called "comet families". The Jupiter family of comets is the largest and numbers around 70 comets. The shortest orbital period known is that of the short-period comet Encke—about 3.3 years.

According to Delsemme (1998), the two groups of comets originated as follows:

Short-period comets are thought to have originated in the Kuiper belt (Luu and Jewitt, 1996).

The source of long-period comets is thought to be the Oort cloud (Weissmann, 1998).

The latter group was probably responsible for the early bombardment of the protoplanets. Delsemme believes that the cometary nuclei of the members of the Jupiter family never experienced temperatures greater than 225 K. The values suggested for the others are: Saturn family, 150 K; Uranus family, 75 K; Neptune family, 50 K. During many million years, these comets got mixed together in the Oort cloud (which has a diameter of around 50,000 AU).

It has recently been suggested that the comets also went through a number of subtle, but important, evolutionary processes in the Oort cloud and the Kuiper belt. Thus, their present nature is probably not the "original" one, as was previously thought (Stern, 2003). The assumption that the material which comets contain is in the same state as it was when the solar system was formed must be revised or modified. The evolutionary mechanisms to which they were subjected are likely to have changed their chemical composition.

The following mechanisms have been suggested:

The evaporation of volatile components by heat from supernovae or passing stars

Collisions with other heavenly bodies

Radiation chemical processes involving cosmic and UV irradiation

The extremely low density of material in interstellar space (ISM gas and ISM nuclei), which could affect the cometary material in the course of millions of years

According to these research results, comets can no longer be considered as genuine relics (unchanged material witnesses) of the period 4–4.5 billion years ago (Stern, 2003).

3.2.2 The Structure of the Comets

Comets consist of three elements: the nucleus, the coma and the tail.

The cometary nucleus This is not normally visible. Nuclear diameters lie in the range of 1–15 km, with masses of 10^{12}–10^{15} kg. The American astronomer F. L. Wipple (1950) developed the now generally accepted model of the "dirty snowball", according to which the nucleus consists of various types of ice: in particular, water ice, methane ice and carbon dioxide ice. The ice contains dust particles with differing compositions, about a third being organic material. These particles are of great importance for the issue of biogenesis.

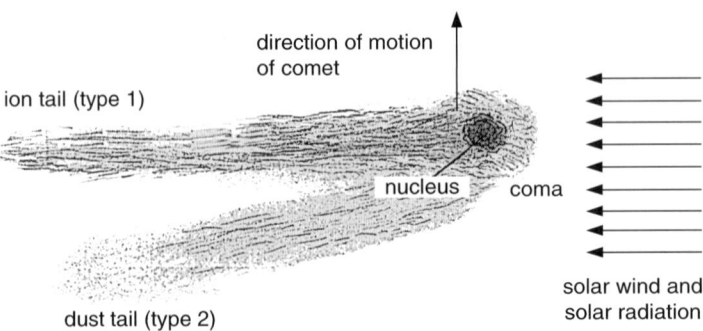

Fig. 3.5 Structure of a comet

The cometary coma The coma and the nucleus form the head of the comet; the streams of dust and gas released by the comet form a very large, extremely tenuous atmosphere called the coma, which can have a spread up to around 10^4–10^5 km. The coma is not developed when the comet is a long way from the sun, but when it comes closer (at around 5 AU), the ice mixture begins to sublime and is ejected as a gas stream. Dust particles are entrained at a velocity of around one kilometre per second.

The comet's tail The tail only develops when the comet is inside the orbit of Mars and can reach a length of between 10^7 km and one AU. It is not always straight but is often curved. This happens when the comet is subject to strong solar winds, i.e., during periods of greater solar activity. Two types of tail can be distinguished:

Type 1: long, thin tails of gas, mainly containing molecular and radical ions such as N_2^+, CO^+, OH^+, CH^+, CN^+, CO_2^+ and H_2O^+.

Type 2: these consist of dust particles around 1 μm in diameter. They are influenced strongly by the radiation pressure of the sun, which is, however, weaker than the pressure of the solar wind.

3.2.3 Halley's Comet

A great deal of information on the structures of comets was obtained during the investigations of Halley's Comet carried out in 1986. A total of six spacecraft were involved: Giotto (Europe), Vega 1 and 2 (USSR), Suisi and Sakigake (Japan) and ICE (USA). The Giotto spacecraft came as close as 600 km to the comet, while Vega 1 and 2 passed it at distances of 9,000 and 18,000 km, respectively. The first three spacecraft contained mass spectrometers for the analysis of the gas and dust, as well as other sensors. The Giotto spacecraft built by ESA had the following instruments on board: camera, gas and ion mass spectrometer, particle impact mass spectrometer, particle impact detector, optical photometer, ion sensor, electron analyser, ion cluster analyser, analyser for high-energy particles and magnetometer. The spacecraft was able to determine the dimensions of the cometary nucleus, which were $16 \times 8 \times 8$ km; it thus has the form of a rotation ellipsoid, similar to that of a peanut. Its brightness varied in 2.2 and 7.4 days respectively, so that rotation around both the long and the short axis must be assumed. The dark nucleus probably contains carbonaceous material, with a very low reflectivity and a surface temperature of around 330 K.

The mass of Halley's Comet is about 10^{14} kg, and thus its mean density is only 200 kg/m^3. The rate of loss of material has been estimated as 5,000 kg/s. The nucleus is loosely packed and exhibits point craters and chasms from which gas and dust escape. These emissions consist mainly of water vapour (∼80% by volume) as well as 6% CO, < 3% CO_2, ∼2.5% CH_4, ∼1.2% NH_3 and < 6% N_2 (Flechtig and Keller, 1987). At the point where Giotto came nearest to the comet, the estimated amount of water being ejected was close to 15,000 kg/s, while that of dust particles was between 6,000 and 10,000 kg/s. Ions derived from water were detected in the vicinity of the comet's head, e.g., $[H(H_2O)_n]$ + (n = 0, 1, 2, 3) as well as H^+, H_2^+, C^+ and CH^+ (Mason, 1992). The cometary dust consisted of particles between 10^{-4} and 2 mm in diameter, but mostly below 10^{-2} mm. The analysis carried out by the dust particle mass spectrometer showed the presence of much more of the light elements H, C and N than found in the primitive meteorite class C1 (see Sect. 3.3.1), and indeed the dust spectra of the comet indicated that the light elements are present in amounts almost as great as in the sun, i.e., the cometary material has a similar composition to that of the primeval solar nebula.

The analytical data obtained, particularly by the PUMA mass spectrometer on board Vega 1 during the flyby, indicate the presence of a large number of linear and cyclic carbon compounds, such as olefins, alkynes, imines, nitriles, aldehydes and carboxylic acids, but also heterocyclic compounds (pyridines, pyrroles, purines and pyrimidines) and some benzene derivatives; no amino acids, alcohols or saturated hydrocarbons are, however, present (Kissel and Krueger, 1987; Krueger and Kissel, 1987).

3.2.4 Comets and Biogenesis

More than 45 years ago, the chemist John Oró from the University of Houston, Texas, suggested that biomolecules or their precursors could have been formed in space and brought to our Earth by comets (Oró, 1961). Delsemme made similar suggestions at the ISSOL Conference in Mainz in 1983 (Delsemme, 1984).

Which results led to the idea that comets are important in the evolution of life? For more than ten years, some scientists have believed that life has (possibly) existed on Earth for more than 3.5 billion years; recently, however, doubts have arisen as to whether this is really the case. It does seem clear that the heavy bombardment of the primeval Earth slowly started to decrease about 3.8 billion years ago. Many biogenesis researchers believe that a period of about 300 million years after the bombardment ceased would not have been long enough for life to evolve from inanimate systems. Thus the idea that comets (or perhaps even meteorites) played a role in the biogenesis process on Earth is quite appealing. Three possibilities are under discussion:

> *Life itself* was brought to Earth from somewhere in the universe.
> The heavenly bodies which landed on Earth already had *biomolecules* "on board".
> These bodies brought *building blocks* to Earth for the synthesis of biomolecules.

Until a few years ago, it was considered impossible that biomolecules or their precursors could have survived the huge temperatures which would have been generated when comets hit the Earth. Today it seems possible that about 0.1% of such substances could have remained unchanged. A comet with a diameter of around 3 km may contain around 10^{27} dust particles. If 0.1% were to reach Earth unchanged, there would still be 10^{24} intact particles around 1 mm in size.

Similar arguments led Bernstein (1999a) to conclude that organic molecules formed in outer space could have been brought to Earth by comets. Laboratory experiments under simulated outer space conditions showed that polycyclic aromatic hydrocarbons (PAH) stabilized by an ice matrix led, under oxidising conditions and with UV irradiation, to the synthesis of aromatic alcohols, ketones and ethers. As expected, reducing atmospheres caused the formation of hydrogenated aromatic hydrocarbons. The product analysis was carried out by IR spectroscopy and mass spectrometry (Bernstein, 1999b). Such experiments under simulated deep space conditions indicate the fundamental importance of water ice, and thus of its various modifications. As we have already noted, at least 13 different forms of water ice are known. Several ice modifications may be present under deep space conditions, i.e., extremely low pressures, low temperatures (particularly in the range of 10–65 K), and the influence of strong radiation (Blake and Jenniskens, 2001). Cosmic UV irradiation leads to the formation of high-density, amorphous ice, which can flow like water. It is assumed that organic molecules can be formed within this ice modification.

The March 2002 issue of *Nature* contains two articles which report the synthesis of amino acids during UV irradiation of ice under cosmic conditions, one from Europe and one from the USA. Bernstein et al. (2002) report the synthesis of the three amino acids glycine, serine and alanine when a mixture of water, methanol, ammonia and hydrocyanic acid (in a ratio of 20:2:1:1) is irradiated at temperatures below 15 K. The European group (Muñoz Caro et al., 2002) was able to synthesize 16 amino acids as well as some other substances. They used a 2:1:1:1:1 mixture of water, methanol, ammonia, carbon monoxide and carbon dioxide. The reaction conditions used were similar to those of Bernstein's group. It is surprising that the composition of the starting materials has such a great influence on the product mixture. The Muñoz Caro approach uses less water and gives more amino acids, including bis(amino acids). The relative amounts of the mono(amino acids) synthesized are similar to those detected in the analysis of the Murchison meteorite, though the latter did not contain any bis(amino acids).

While accepting the high quality of these results, Everett L. Shock from the Department of Earth and Planetary Research of Washington University, St. Louis, poses the critical question as to whether the many simulation experiments really help us in answering the question of the origin of life on Earth (Shock, 2002).

Missions to investigate comets have taken place or are planned.

Rosetta The Rosetta mission was planned to reach the comet Wirtanen in 2013, to orbit it for eleven months and then to land and study the comet's surface. The start, planned for 2003, had to be postponed until 2004. This ESA mission involves the use of a lander ("Philae") developed by the German DLR, which is to take and analyse samples from the surface of the cometary nucleus. In May 2003, the scientific committee of ESA decided that the mission's goal should be changed to the comet 67P/Churyumov-Gerasimenko. The start, using an Ariane 5G+ rocket, took place at the Kourou space centre (French Guiana) in February 2004. Rosetta is expected to reach its new goal after a flight lasting ten years and ten months. The lander had to be modified to take into account the different gravitational force of the larger comet (www.esa.de).

Although Rosetta left Earth in 2004, it had still not covered even a half of its journey by 2007; the swingby of Mars in February 2007 was successful, and two more swingby manoeuvres will follow.

Stardust February 7, 1999, saw the start of NASA's Stardust mission: the cometary probe, the first mission to collect cosmic dust and return the sample to Earth, has a time-of-flight mass spectrometer (CIDA, Cometary and Interstellar Dust Analyser) on board. This analyses the ions which are formed when cosmic dust particles hit the instrument's surface. In June 2004, the probe reached its goal, the comet 81P/Wild 2, getting as close as 236 km! The CIDA instrument, which was developed at the Max Planck Institute for Extraterrestrial Physics in Garching (near Munich), studied both cometary dust and interstellar star dust.

The spacecraft landed in Utah on January 15, 2006, carrying valuable freight. It had collected several tens of thousands of particles, between 1 and 300 μm in size, from the vicinity of 81P/Wild 2. Initial studies indicated that the material was of both presolar and solar origin. A high proportion of the silicate particles (olivine, anorthite, and diopside) were larger than expected; they consist of high-temperature minerals which were formed in the inner regions of the solar nebula. The cometary dust was collected in two ways: with aluminium foil and with aerogel, an extremely low-density material.

More detailed investigations will take some time; polycyclic aromatic compounds have already been detected, some of which contain nitrogen and oxygen. Methylamine and ethylamine were also found. Contamination from the spacecraft seems unlikely. This first mission to return from the depths of the cosmos was a complete success (Brownlee et al., 2006; Hoerz et al., 2006).

The interstellar dust was shown to contain quinone derivatives as well as oxygen-rich condensed aromatic compounds; the quinones were present in both hydrated and carboxylated form. Very little nitrogen was present in the compounds detected. The cometary material, however, contained condensed nitrogen heterocycles. Hardly any oxygen was detected in the solid phase of the cometary dust: it possibly evaporates from the tail of the comet in the form of water or oxidized carbon compounds. The authors assume that these analytical results could lead to a reconsideration of the current biogenesis models (Kissel et al., 2004; Brownlee, 2004).

Mission Deep Impact In July 2005, NASA steered a projectile, about 370 kg in weight, at the comet 9F/Tempel (dimensions $4 \times 4 \times 14$ km), in order to obtain more exact information on its structure and composition. The impact was visible from Earth; the Rosetta spacecraft discussed above also sent pictures to Earth. The dust/ice ratio determined after the impact is very probably greater than unity, so that comets are probably "icy dustballs" rather than (as had previously been surmised) "dirty snowballs". The density of the cometary nucleus, which seems to consist of porous material, is roughly equal to that of ice. The impact set free around 19 GJ of energy, corresponding to the explosion of 4.5 tons of TNT; between ten and twenty thousand tons of cometary material were split off, of which between three and six thousand were dust. The large amount of dust prevented observation of the impact crater, which was estimated to be about 30 m deep and may have had a diameter of 100 m (Kueppers et al., 2005; Feldman, 2005; A'Hearn, 2006 and Burnett, 2006).

Fig. 3.6 Artist's impression of the planned approach of "Rosetta" to the comet 67P/Churyumov/Gerasimenko in the year 2014. ESA picture

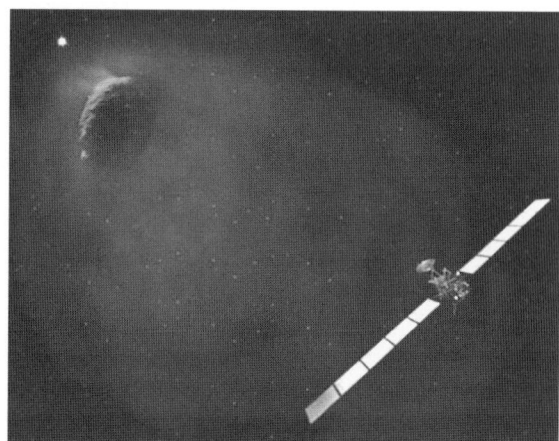

3.3 Meteorites

Although this was contrary to popular, and also scientific, belief at the time, the German physicist Ernst Florens Friedrich Chladni (1756–1827) postulated that rocks could in fact fall from the heavens. His statement was supported by eyewitnesses who had observed the descent of meteorites. In France, Jean Baptiste Biot (1774–1862) was able to convince the Academy of Sciences in Paris that they should revise the memorandum which they had published ten years previously and agree that the meteorite fragments which had been found could in fact have their origin in outer space.

Alexander von Humboldt (1769–1859) recognised meteorites as being a source of extraterrestrial material. Several well-known chemists carried out analyses of material from meteorites, starting at the beginning of the nineteenth century. Thus Louis-Jacques Thenard (1777–1857) found carbon in Alais meteorites; these results were confirmed in 1834 by Jöns Jacob Berzelius, who by dint of very careful work was also able to detect water of crystallisation in meteoritic material.

Today, there is consensus that meteorites are the most important source of material from outer space. Their study is interesting from two points of view:

They contain the oldest material from precursors of the Earth and the solar system.

Their impact on Earth may possibly have delivered important biomolecules (or their precursors).

Fig. 3.7 Jöns Jacob Berzelius (1779–1848), professor of chemistry in Stockholm and discoverer of the elements selenium, silicon, thorium and zirconium. He introduced the modern chemical symbols and also the term "organic chemistry". From the book *Berzelius, Europaresenären* by C. G. Bernhard; with kind permission of the Royal Swedish Academy of Sciences

3.3.1 The Classification of Meteorites

There are two types of meteorites:

Undifferentiated meteorites: these are derived from asteroids which never underwent the heating which leads to fusion. They consist of millimetre-sized spherules (chondrules) embedded in a matrix.

Differentiated meteorites: they come from asteroids which have been through a fusion process which led to a more or less clear separation into nucleus, mantle and crust.

According to the *Catalogue of Meteorites* (1985), there are four main groups of meteorites:

Chondrites
Achondrites
Stony iron meteorites
Iron meteorites

Only the chondrites are undifferentiated.

The chondrules contained in the chondrites contain olivine, pyroxene, plagioklase, troilite and nickel-iron; they can make up 40–90% of the chondrites. Chondrules are silicate spheroids, fused drops from the primeval solar nebula. Because of their differing constitution, chondrites are further subdivided: one group in particular is important for the question of the origin of life, and has thus been intensively studied—that of the carbonaceous chondrites.

3.3.2 Carbonaceous Chondrites

Carbonaceous chondrites (C-chondrites) account for only 2–3% of the meteorites so far found, but the amount of research carried out on them is considerable. C-chondrites contain carbon both in elemental form and as compounds. They are without doubt the oldest relics of primeval solar matter, which has been changed only slightly or not at all by metamorphosis. C-chondrites contain all the components of the primeval solar nebula, apart from those which are volatile; they are often referred to as "primitive meteorites".

The C-chondrites are subdivided further into eight subgroups. The Orgeuil meteorite, which fell in the nineteenth century in France, belongs to the group CI 1, while the Allende meteorite, which fell near the Mexican village Pueblita de Allende, is of type CV 3. Both meteorites were carefully collected only a few weeks after their impact on Earth (avoiding contamination as far as possible) and passed on to scientific institutions. The element carbon occurs not only as carbonates; "exotic" forms such as diamond, graphite and silicon carbide have also been detected (Hoppe, 1996). These latter three species are considered to provide indications of the connections between stellar materials in the presolar nebula (Lugmair, 1999).

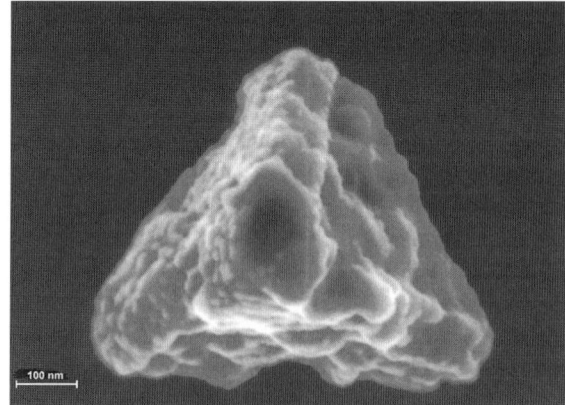

Fig. 3.8 A grain of silicon carbide (smaller than a micrometre) more than 4.57 billion years old, as seen under a scanning electron microscope. The grain was found in the Murchison meteorite and was formed in the presolar nebula (Lugmair, 1999)

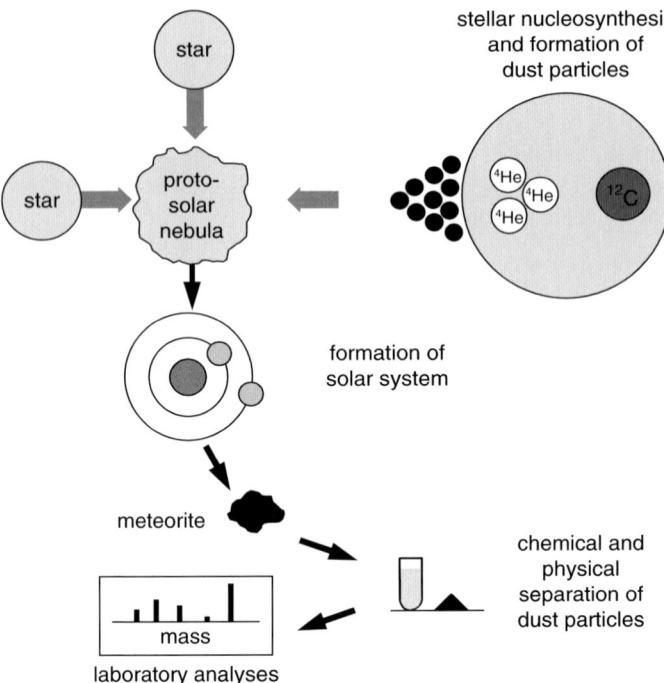

Fig. 3.9 Greatly simplified representation of the path taken by the material under study, beginning with nucleosynthesis and ending with laboratory analysis. Circumstellar dust (a component of the primeval presolar nebula) which was contained in asteroids or comets came to Earth in meteorites and was then available for exact study (Lugmair, 1999)

Table 3.2 The commonness of elements (log n) in the solar system, in the sun and in carbonaceous chondrites of type C1, with respect to hydrogen log $n(H) = 12$, i.e., $n(H) = 10^{12}$ (Unsöld and Baschek, 2001)

Atomic number		Sun	C1-chondrite
1	H	12.0	–
2	He	11.0	–
9	F	4.6	4.5
11	Na	6.3	6.4
12	Mg	7.5	7.6
14	Si	7.6	7.6
16	S	7.2	7.3
19	K	5.1	5.2
26	Fe	7.5	7.5

3.3.2.1 The C-Chondrites

The pulverized meteorite material is extracted using a series of solvents of differing polarity. The extracts contain mixtures of discrete compounds, such as amino acids or hydrocarbons; these make up about 10–20% of the meteorite material, the remainder being referred to as "meteorite polymer" or "kerogen-like material".

All meteorite analyses are made more difficult because of the problem of contamination. Thus one group (Kvenholden, 1970) reported the presence of polycyclic aliphatic compounds, while a second (Studier, 1972) found straight-chain alkanes to be the dominant species. The latter result was often cited and taken as evidence that processes similar to the Fischer-Tropsch synthesis must have occurred in nebula regions of the cosmos.

Aromatic hydrocarbons were found in more recent analyses: pyrene, fluoranthrene, phenanthrene and naphthalene in the ratio of 10:10:5:1 (Cronin, 1998). The majority (around 70%) of the hydrocarbons extracted from the Murchison meteorite are polar compounds such as:

Aryl alkyl ketones,
Aromatic ketones and diketones,
N,S-heterocycles and
Substituted pyridines, quinolines, purines and pyrimidines.

Light hydrocarbons probably only remain in the CM chondrites because they are immobilized (e.g., trapped in crystals) or strongly adsorbed on grain surfaces. In the series from methane to the butanes, the hydrocarbons are isotopically heavier than the corresponding analogs on Earth; these compounds are thus suggested to have an extraterrestrial origin. The $^{13}C/^{12}C$ ratio falls as the chain length increases, an observation which has also been made in the monocarboxylic acids found in meteorites. In total, 17 saturated aliphatic carboxylic acids with chain lengths from 1 to 8 have been detected. Similar results were presented by a Japanese research group, which identified 21 different carboxylic acids (with up to 12 carbon atoms) in meteorite material from the Antarctic (CM-chondrites; Shimoyama, 1989).

In the early days of meteorite analysis, it was difficult to detect N-heterocycles; later, the Murchison meteorite was shown to contain xanthine, hypoxanthine, guanine, adenine and uracil (about 1.3 ppm in total). This meteorite seems to contain various classes of basic and neutral N-heterocycles, as well as isomeric alkyl derivatives.

3.3.2.2 Amino Acids in C-Chondrites

The analysis of extraterrestrial matter is concentrated on the detection of nucleic acid and protein building blocks, i.e., N-heterocycles and amino acids. The search for such compounds began immediately after the fall of the Murchison meteorite. Twenty-two amino acids were detected in it as early as 1974: eight of them proteinogenic, ten which hardly ever occurred in biological material, and four which were unknown in the biosphere. Up to now, about 70 amino acids have been identified (Cronin, 1998), the most common being glycine and α-aminoisobutyric acid. The latter is a branched-chain amino acid with the smallest possible number of carbon atoms. The most frequently found amino acids occur in concentrations of

100 nmol/g. The total concentration is about 700 nmol/g; there is a large variation in the local concentration in any one meteorite.

An important, but as yet unsolved, problem is provided by the chirality of the amino acids (see Sect. 9.4). The main question is whether the enantiomeric selection of the amino acids took place before or after the evolution of life. The initial gas–liquid partition chromatography (GLPC) analyses of meteorite material showed the amino acids to be present as racemic mixtures (within the experimental limits of error) (Kvenholden, 1970; Oró, 1971; Pollock, 1975). These results were taken as convincing proof of the extraterrestrial origin of the amino acids, since an excess of L-amino acids would have indicated contamination. A fierce scientific controversy came about when one research group found an excess of the L-enantiomers of five proteinogenic amino acids in samples from the Murchison meteorite (Engel and Nagy, 1982).

The question was whether impurities were present in the samples analysed (Bada et al., 1983). In a more recent publication, Cronin and Pizzarello (1997) reported amino acid analyses using Murchison material in which an excess of L-enantiomers was present. Contamination with terrestrial biological material can be ruled out, as the amino acids in question are not proteinogenic: α-methylamino acids, which occur either extremely seldom or not at all in terrestrial life forms, were detected. GLPC/mass spectrometry (MS) analysis gave the following enantiomeric excess (ee) values:

- 7.0% α-methylisoleucine
- 9.1% α-methylalloisoleucine
- 8.4% isovaline
- 2.8% α-methylnorvaline

Such enantiomeric excesses have not been observed in analyses of the corresponding α-H-α-aminoalkanoic acids. According to the authors, the excess of the L-forms could be due to a partial photocleavage of the racemic amino acid mixture as a result of the influence of circularly polarized UV light in a presolar cloud (Cronin and Pizzarello, 2000).

New analyses of material from the interior of the Orgeuil and Ivona meteorites show the presence of β-alanine, glycine and γ-amino-n-butyric acid as the main components (0.6–2.0 ppm); traces of other amino acids were also detected. The amino acids were present as racemic mixtures, i.e., D/L = 1, so that an extraterrestrial origin can be assumed (Ehrenfreund et al., 2001).

Now and then, projectiles from outer space cause excitement and surprises, as in January 2000, when a meteorite impacted the frozen surface of Lake Targish in Canada. It was a new type of C-chondrite with a carbon concentration of 4–5%, and probably came from a D-type asteroid (Hiroi et al., 2001). More exact analysis of the Targish meteorite showed the presence of a series of mono- and dicarboxylic acids as well as aliphatic and aromatic hydrocarbons (Pizzarello et al., 2001). Aromatic compounds and fullerenes were detected in the insoluble fraction from the extraction; this contained "planetary" helium and argon, i.e., the ^{3}He/^{36}Ar ratio was about 0.01, equal to that assumed for the planetary system. The sudden appearance of the "super-C-containing" Targish Lake meteorite shows how quickly new facts or events can change or extend our knowledge.

We might think that the Murchison meteorite would have been studied thoroughly enough in the years since its arrival on Earth. But the results obtained always depend on the performance of the technical resources available: in this case, the analytical methods and the apparatus. Thus, it is not really surprising that a new class of amino acid has been discovered in Murchison material: diamino acids, such as DL-2,3-diaminopropionic acid, DL-2,4-diaminobutanoic acid etc. These were identified using a new enantioselective GLPC/MS method, which is also being used in analysis of material from the Rosetta mission.

It remains an open question as to whether this newly identified species of amino acid was important in the construction of peptide nucleic acids (see Sect. 6.7) in the prebiotic chemistry of the RNA world phase of biogenesis (Meierheinrich et al., 2004).

3.3.3 Micrometeorites

The number of scientific articles published on meteorites has increased dramatically in the last few years; few of these, however, concern themselves with small meteorites, the size of which lies between that of the normal meteorites (from centimetres to metres in size) and that of interplanetary dust particles. In the course of an Antarctic expedition, scientists (mainly from French institutions) collected micrometeorites from 100 tons of Antarctic blue ice (Maurette et al., 1991). These micrometeorites were only 100–400 μm in size; five samples, each consisting of 30–35 particles, were studied to determine the amount of the extraterrestrial amino acids α-aminoisobutyric acid (AIBS) and isovaline—both of which are extremely rare on Earth—which they contained. The analysis was carried out using a well-tested and extremely sensitive HPLC system at the Scripps Institute, La Jolla. Although the micrometeorites came from an extremely clean environment, the samples must have been contaminated, as they all showed traces of L-amino acids. Only one sample showed a significantly higher concentration of AIBS (about 280 ppm). The AIBS/isovaline ratio in the samples also lay considerably above that previously found in CM-chondrites.

The authors are of the opinion that these concentrations of organic compounds could have sufficed to support chemical evolution, considering the number of meteorites which would have collided with Earth over billions of years (Brington et al., 1998). The C-containing chondrites in the size range 50–500 μm, which generally survive entry into the atmosphere without suffering damage, deliver the largest amounts of extraterrestrial organic compounds. The French astrophysicist Michel Maurette considers micrometeorites as being able to function as a type of microscopic "chondritic" chemical reactor, as soon as the particles come into contact with reactive gases and water. The traces of metals, or of metal oxides or sulphides, present in the particles could act as catalysts. This process probably took place either in or on the asteroids or in the final planetary environment, e.g., Earth (Maurette, 1998).

Direct measurement of the amounts of cosmic micrometeorites which impacted the Earth was carried out with the help of a satellite. The Long Duration Exposure Facility (LDEF) satellite orbited the Earth for around $5^{1}/_{2}$ years at a height of 480–331 km and was brought back to Earth by a space shuttle. The analysis of the impaction traces left on an external plate (made from a special aluminium alloy) showed that about $40 \pm 20 \times 10^{6}$ small particles per year fall to Earth. The large uncertainty is due to the fact that we do not know how many particles actually *reach* the Earth's surface. The above value agrees with a previous estimate derived from the osmium isotope ratio (^{187}Os/^{186}Os) in deep sea sediments. The particles in the mass range 10^{-9}–10^{-4} g are 0.1–1 mm in size, the peak of the distribution lying at about 1.5×10^{-5} g, with a diameter of about 200 μm (Love and Brownlee, 1993).

3.4 Interstellar Matter

About a hundred years ago, it was still thought that interstellar space was completely empty, except for the cosmic nebulae, which were already known at that time. The presence of matter in interstellar space was shown by the fact that in certain regions of the sky, light from distant stars was either scattered or absorbed: in other words, dark, star-free regions are present.

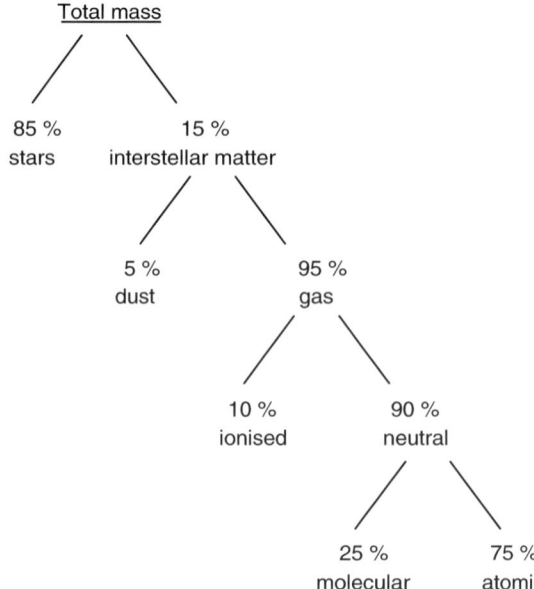

Fig. 3.10 Distribution of matter in the Milky Way. From Feitzinger, J. V., Die Milchstraße – Innenansichten unserer Galaxie, © Spektrum Akademischer Verlag, Heidelberg Berlin 2002

Interstellar matter (ISM) consists mainly of gaseous components, with only a minor fraction existing in particulate form. These few particles, however, cause the light from the stars to be blotted out, as they interact much more strongly with the visible light than do molecules or atoms in the gaseous state. Spectroscopic observations also point to the existence of ISM: Fig. 3.10 gives a survey of the distribution of cosmic matter.

The mean particle density of ISM is 10^6 particles per cubic meter; there are, however, great variations from this mean value. Between the spiral arms of the Milky Way, there are between 10^4 and 10^5 hydrogen atoms per cubic metre; in the dark clouds and the H II regions, there are 10^8–10^{10}. Up to 10^{12}–10^{14} hydrogen atoms per cubic metre are present in regions with OH sources and in certain infrared objects.

3.4.1 Interstellar Dust

The dark clouds were responsible for the discovery of ISM, as they absorb the light from stars which lies behind these clouds of interstellar matter. It is difficult to obtain reliable information on the dust particles. They are probably about 0.1 μm in diameter, consisting of a silicate nucleus and an envelope of compounds containing the elements C, O and N, which, with H and He, are the main elements present in interstellar space. There are only two sources of information for more exact characterisation of the dust particles:

Electromagnetic stellar radiation which has passed through regions containing interstellar dust particles. The values of the absorbed or scattered wavelengths make it possible to draw conclusions on the chemical composition or the physical properties of the particles.

Laboratory simulation experiments with a helium cryostat at temperatures of around 10 K and pressures of 10^{-8} mmHg, which have been carried out in particular by Majo Greenberg at the University of Leiden in Holland (Greenberg, 1983).

Dust and gas are always found together in interstellar space, although in differing ratios. How do we manage to obtain information on the dust particles in the cosmos? The interactions between the particles and light depend both on particle size and on the wavelength of the light interacting with the particles.

If the particles are larger than the wavelength of visible light, all wavelengths are equally scattered, i.e., the light intensity is decreased, and there is no colour change. If, however, the particle size corresponds to a certain light wavelength, these particular particles scatter the incident light. The stars then appear somewhat redder than they actually are. This phenomenon can be seen in the morning or the evening (at sunrise and sunset), when the sun is just above the horizon. The sun's light appears redder, since its rays must pass through the haze and dust layers of the atmosphere and are thus scattered (Greenberg, 2000).

The extinction curve of light which has passed through dust clouds tells us which particles are present in the cosmic dust:

   Around 10% of the dust consists of carbon particles which are about 0.005 μm in diameter. These are amorphous, carbon-containing solids, probably also containing hydrogen.

   The main fraction of the dust (around 80%) consists of coated dust particles (diameter around 0.34 μm).

   The remaining dust constituents (around 10%) are polycyclic aromatic hydrocarbons (PAHs) or PAH-like compounds, 0.002 μm in size (Greenberg, 2001).

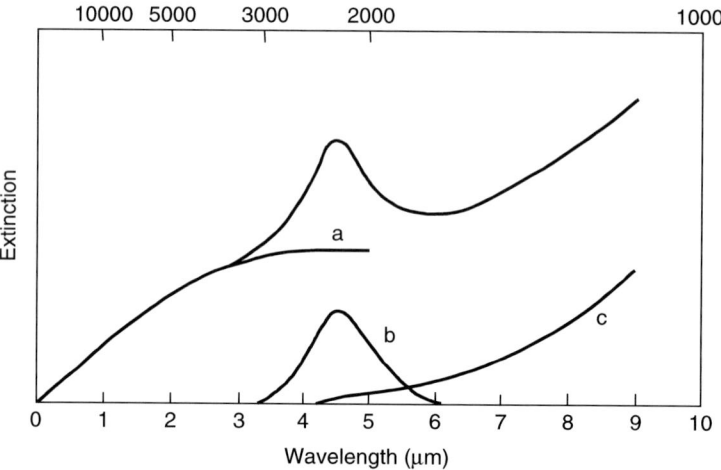

Fig. 3.11 Interstellar dust particles cause the extinction of starlight by the selective scattering of certain light wavelengths. Far IR is on the left, far UV on the right. Satellite data suggest that the extinction curve consists of three components:

   a The extinction is due to particles with a radius of around 10^{-5} cm.
   b Particles about ten times smaller give the curve **b**. They probably consist of pure carbon.
   c Particles similar in size to **b**, but consisting of silicates.

The scale at the top of the figure gives the wavelength in ångstroms (Å) (Greenberg, 1983)

The interpretation of the curves in Fig. 3.11 is not indisputable, but they provide useful information. The Swedish astronomer Bertil Lindblad (1935) was the first to suggest that the interstellar dust particles represent condensates in the universe. Only in the early 1970s was it realized that silicate particles surround the huge, cold M giants. Radiation pressure drives these particles into space, where they cool and act as condensation nuclei, to which the molecules in the gas and interstellar dust adhere. The latter then freeze out and form an ice coating around the silicate nucleus. Experiments to simulate such processes were carried out successfully by Greenberg over many years and permit a better understanding of the processes which could have taken place in outer space. Instead of the silicate condensation nuclei, Greenberg used a cold finger. A second compromise was necessary: instead of the ultrahigh vacuum present in interstellar space, the best laboratory vacuum available was several orders of magnitude lower. The reaction vessel contained CH_4, CO, CO_2, NH_3, N_2, O_2 and H_2O. The mixture was subjected to UV irradiation at temperatures around $10\,K$. Simultaneous measurements of the IR absorption of the sample at 2.5 and $25\,\mu m$, of pressure, of luminescence, of the molecular masses of the gas molecules and of the absorption of visible light were carried out during the reaction.

A further change in the experimental conditions with respect to the natural processes was in the timescale: one hour in the laboratory corresponded to around 1,000 years in space! Such comparisons make clear the limits of simulation experiments: the interpretation of their results must be carried out with great care! In Greenberg's laboratory in Leiden, an amorphous, blue, glassy layer was slowly formed, probably containing formyl radicals. When the temperature was slowly raised, the substance emitted blue green light, probably due to recombination of the formyl radicals to give glyoxal. However, the light emission does not represent the total recombination energy, as the sample can in fact explode when heated up rapidly! The final product is a small amount of a mixture which is stable at room temperature, and which Greenberg calls the "yellow stuff" (Greenberg, 1983). It is a mixture of glycerine, certain amino acids (glycine, alanine and serine among others) and some other types of molecules (Greenberg, 2000).

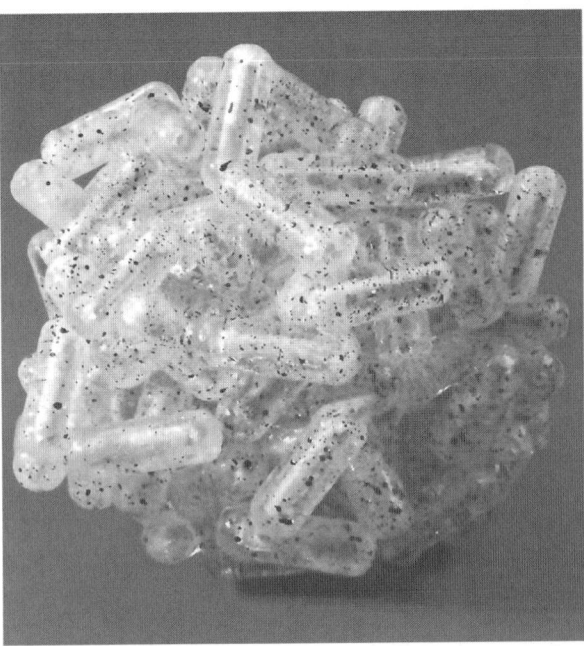

Fig. 3.12 Model of an agglomerate consisting of many small interstellar dust particles. Each of the rod-shaped particles consists of a silicate nucleus surrounded by yellowish organic material. A further coating consists of ice formed from condensed gases, such as water, ammonia, methanol, carbon dioxide and carbon monoxide. Photograph: Gisela Krüger, University of Bremen

The yellow substance was studied under deep space conditions. After being subjected to UV-irradiation for about four months, it changed colour to brown, possibly due to enrichment in carbon or carbon-containing compounds. IR analysis showed that the new "brown substance" showed the same absorption lines as those observed in interstellar dust. A special IR study carried out at Stanford University indicated that the brown substance contained many PAHs.

M. Greenberg (2000) suggested that there is a cycle through which the cosmic dust particles can pass up to 50 times before disintegrating under the influence of supernova shock fronts. It is also assumed that agglomerates of coated dust particles are also present in cometary dust: up to 100 dust particles can build up such agglomerates, around 3 μm across. A confirmation of Greenberg's model apparently comes from the analysis of the data obtained from dust in Halley's Comet, where conglomerates of particles with silicate nuclei and organic coatings were discovered.

Hill and Nuth (2003) from the Goddard Space Center in Greenbelt, Maryland, have investigated whether dust particles in cosmic nebulae could have been involved in prebiotic outer space chemistry. They carried out simulation experiments using a catalyst analogue consisting of iron silicate condensates, which are assumed to exist in such dust particles. In an atmosphere consisting of CO and H_2, these particles catalysed the formation of methane and water (Fischer-Tropsch synthesis), while ammonia was formed from nitrogen and hydrogen (Haber-Bosch synthesis). If CO, N_2 and H_2 are allowed to react simultaneously, nitrogen-containing organic substances such as methylamine (CH_3NH_2), acetonitrile (CH_3CN) and methylmethyleneimine (H_3CNCH_2) are formed. These experiments, carried out at temperatures between 500 and 900 K, indicated that the reactions observed could also take place under outer space conditions (Hill and Nuth, 2003). The planned NASA and ESA missions should be able to help clear up some open questions in the area of ISM.

Interstellar dust is also important for the formation and development of stars. Although the dust particle component is only a minor one in ISM, it acts as a "cooling agent" for collapsing clouds, thus preventing the buildup of an effective thermodynamic counterpressure.

3.4.2 Interstellar Gas

The gas component, which is the most important form in which matter is found in outer space (up to 98–99%), is dominated by the element hydrogen, which makes up 70% of the mass and 90% of the particles. In ionized form (the H-II regions) the gas can be recognized by its recombination and fluorescence light emission. The hydrogen is mainly present in neutral form (H-I regions), at a mean density of 2×10^7 particles per cubic metre and a mean temperature of about 80 K.

It was not possible to detect neutral hydrogen until the 21-cm spectral line (in the radio region), predicted by van de Hulst, was discovered in 1951. Since this radiation is unaffected by dust, regions become available which cannot be studied using optical methods. The capacity of the 21-cm line of interstellar hydrogen for emission depends almost solely on the density of the hydrogen, but not on its temperature. The radiation intensity coming from a certain direction is thus a measure of the total amount of hydrogen present.

The chemical reactions which take place under the extreme conditions of outer space are complex and not always comparable to those which can be simulated under laboratory conditions: Eric Herbst (1990) provides a survey.

We will take one single species and discuss it as an example of the interesting atoms, molecules and ions in the ISM: the hydrogen molecular ion H_3^+, the simplest stable multi-atom species, which was also detected in the Jovian atmosphere (Dalgarno, 1991). This species was discovered by J. J. Thomson 96 years ago. It is the most common interstellar molecule after hydrogen (H_2), but because of its high chemical reactivity, it occurs in outer space in only relatively low steady state concentrations. H_3^+ was first detected in outer space many years later. It is a universal proton donor and thus leads to ion–molecule reactions, which are the precondition for the formation of many complex molecular species in ISM. The importance of this hydrogen species was soon realized, and "Physics, Chemistry, and Astronomy of H_3^+" was the subject of a conference held at the Royal Society in January 2006 (Geballe and Oka, 2006; Oka, 2006).

The high-energy cosmic radiation, which consists of energy-rich protons, ionizes neutral atoms and molecules as it passes through the cosmos, for example, as follows:

$$H_2 + \text{cosmic radiation} \rightarrow H_2^+ \tag{3.16}$$

The ion reacts further:

$$H_2^+ + H_2 \rightarrow H_3^+ + H \tag{3.17}$$

The molecular ion can react further, for example:

$$H_3^+ + O \rightarrow OH^+ + H_2 \tag{3.18}$$

$$OH^+ + H_2 \rightarrow H_2O^+ + H \tag{3.19}$$

$$H_2O^+ + H_2 \rightarrow H_3O^+ + H \tag{3.20}$$

These reactions take place very slowly under the extreme ultra-high vacuum conditions present in outer space, so that the chemistry is different to that observed in the laboratory. Thus there are still some open questions regarding the "mysterious" interstellar ion H_3^+, particularly with respect to its occurrence in diffuse clouds and its rate of decomposition (Suzor-Weiner and Schneider, 2001; Kokoruline et al., 2001).

3.4.3 Interstellar Molecules

In the early days of biogenesis research, only the primeval Earth atmosphere was regarded as being the place where biomolecules or their precursors could have been formed (the Miller era). Later on, the Earth's surface and, after the discovery of the hydrothermal vents, the deep regions of the oceans, were also considered possibilities.

The detection of biomolecules in meteorite material, and of larger molecules in interstellar space, led to the assumption that the molecules required for biogenesis (or simple precursors for them) could have arrived on the young Earth from space.

The observation of the cosmos using radio, millimetre, submillimetre and IR wavelengths led to the observation of more than 100 molecular species in the interstellar clouds and circumstellar gas and dust envelopes. In many cases, the formation of interstellar molecules is not completely understood, as the synthetic processes leading to their formation are complex. There are great differences in the physical parameters of the regions of the universe: they reach from extremely hot zones (around 10^6 K) to extremely "dilute" regions with only a few atoms or molecules per cubic metre. Kinetic considerations show that the formation of molecules from single atoms in the gas phase (e.g., of H_2 or CO) is unfavourable, so it must be assumed that synthetic reactions occur at the surface of interstellar dust particles.

In spite of considerable accomplishments in finding new molecules in outer space, the complex interstellar chemistry still leaves many open questions. The elements hydrogen and helium make up more than 99% of the matter in the cosmos. The others which have been shown to occur are (in decreasing order of importance) oxygen, carbon, nitrogen, magnesium, silicon, sulphur and iron. However, the molecules detected do not by any means follow this order: the molecules containing six and more atoms all contain carbon. Of those with four or five, only H_3O^+, NH_3 and SiH_4 can be classed as inorganic (Klemperer, 2006).

Several of the compounds identified in ISM have not so far been synthesized in the laboratory; however, two of them have now been obtained. The cyclic compound cyclopropylidene (C_3H_2), first detected in ISM in 1985 and later more frequently, was considered to be too unstable to exist on Earth under laboratory conditions. A derivative of this carbocycle, stabilized by amino groups which serve as π-donors, has now been reported. X-ray crystallography shows that the presence of the amino groups has little effect on the molecular geometry as calculated for the unsubstituted cyclopropylidene (Lavallo et al., 2006).

The synthesis and characterisation of a further compound which had been detected in outer space, cyanobutadiyne (2,4-pentadiynenitrile), has been carried out at the University of Rennes in France. This molecule belongs to a class in which one or more C–C triple bonds and only one carbonitrile substituent are present $(H–(C\equiv C)_n–C\equiv N)$. These are often found in ISM: the simplest, cyanoacetylene $(H–C\equiv C–C\equiv N)$, has been detected several times in experiments intended to simulate the Titanian atmosphere. The largest molecule ever found in outer space, $HC_{11}N$, is also a member of this family (Trolez and Guillemin, 2005).

Lewis E. Snyder (2006) has reported on the importance and efficiency of interferometry for the study of complex (and in particular biochemically interesting) molecules in certain areas of ISM, such as interstellar molecule clouds.

The paramount importance of carbon in the cosmos is shown by the fact that more than 75% of the approximately 120 interstellar and circumstellar molecules so far identified are carbon containing (Henning and Salama, 1998). Molecules apparently travel from the ISM via protoplanetary discs to the planetesimals and from there, via accretion, to the heavenly bodies formed. The molecules so far identified in ISM come from quite different types of compounds:

Simple organic compounds such as CH_4, CH_3OH, CO_2
Simple inorganic compounds such as SO, SiO, HF, KCl
Unsaturated hydrocarbon chains such as HC_3N, HC_5N
Polycyclic aromatic hydrocarbons

These different classes of molecules reflect the variety of reaction conditions which are present in ISM. Thus ion–molecule reactions dominate in the dark clouds; these reactions are driven by ionisation processes and lead to unsaturated molecules, ions and radicals.

Most of the molecules presently known were detected by radioastronomers using wavelengths between about 1 mm to around 6 cm. Lines due to CO, H_2 and HD were observed in the UV region, while the optical region showed the presence of CH and CN.

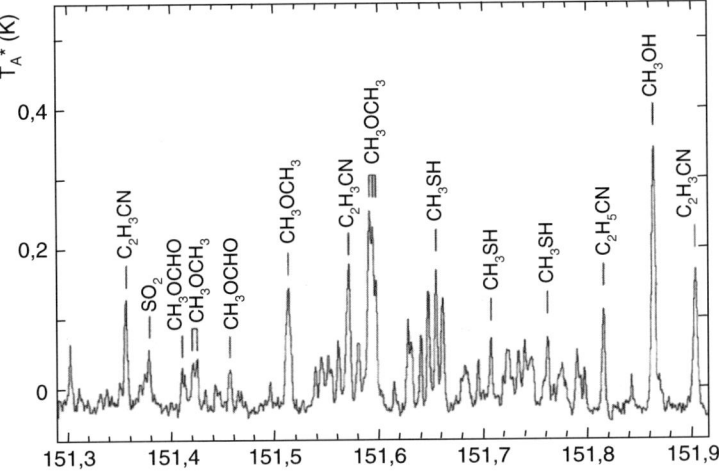

Fig. 3.13 151 GHz Spectrum (excerpt) from the nucleus of the molecular cloud G 327.3-0.6 in the southern sky. The molecular species identified are shown. Peaks which are not labelled are not yet clearly identified. (The numbers used to label the molecular cloud, G 327.3-0.6, are galactic coordinates). With personal permission of Prof. Hjalmarsson, Chalmers University of Technology, Göteborg, Sweden, from the Proceedings of the First European Workshop on Exo/Astrobiology, Frascati, 21-23 May 2001, ESA SP-496

The molecules detected in ISM can also be classified according to the number of atoms they contain (Table 3.3). Compounds containing different isotopes (isotopomers) are not listed.

Some types of molecule, such as CO, CH, OH and H_2CO, are very common. Carbon monoxide is almost always detected if the particle density is greater than 10^9 per cubic metre, and its presence points to that of molecular hydrogen.

As an example, we shall discuss the interstellar synthesis of a compound which is produced on Earth in millions of tons per year: methanol. This simplest alcohol was obtained by Robert Boyle in 1661 from the dry distillation of wood. In the molecular clouds of the universe, it is likely that hydrogenation of CO on the surface of dust particles occurs according to the following scheme (Tielens and Charnley, 1997):

$$CO + H \rightarrow HCO + H \rightarrow H_2CO + H \rightarrow H_3CO + H \rightarrow CH_3OH \quad (3.21)$$

The important reactions of atomic hydrogen with CO and H_2CO are subject to a considerable activation barrier in the gas phase ($\sim 1,000$ K). It has been proposed

Table 3.3 Interstellar and circumstellar molecules (Irvine et al., 2003; Thaddeus, 2006)

2 Atoms	3 Atoms	4 Atoms	5 Atoms	7 Atoms	9 Atoms
H_2	C_2H	C_2H_2	C_4H	C_6H	CH_3C_4H
C_2	CH_2	C_3H	C_3H_2	HC_5N	CH_3OCH_3
CH	HCN	c-C_3H^b	H_2CCC	CH_2CHCN	CH_3CH_2CN
CH^+	HNC	NH_3	HCOOH	CH_3C_2H	CH_3CH_2OH
CN	HCO	HNCO	CH_2CO	CH_3CHO	HC_7N
CO	HCO^+	$HOCO^+$	HC_3N	CH_3NH_2	C_8H
CS	HOC^+	$HCNH^+$	CH_2CN	c-$CH_2OCH_2^b$	
OH	N_2H^+	HNCS	NH_2CN	CH_2CHOH^a	
NH	NH_2	C_3N	CH_2NH		**10 Atoms**
NO	H_2O	HC_2N	CH_4		CH_3COCH_3
NS	HCS^+	H_2CS	HCCNC	**8 Atoms**	$CH_3C_4CN^a$
SiO	H_2S	C_3S	HNCCC	CH_3COOH	$HOCH_2CH_2OH^a$
SiS	OCS	H_3O^+	HCO_2H^+	$HCOOCH_3$	$H_2C(NH_2)COOH$
SO	N_2O	H_2CN	SiH_4	CH_3C_3N	
HCl	SO_2	H_2CO		H_2C_6	**11 Atoms**
SO^+	C_2S	CH_3	**6 Atoms**	CH_2OHCHO^a	$H(C\equiv C)CN$
PN	C_2O	c-SiC_3	H_2CCCC	c-$C_2H_4NH^{ab}$	CH_3C_6H
CO^+	C_3	CH_2D^+	CH_3OH	C_7H	
NaCl	HNO		CH_3CN	HC_6H	**12 Atoms**
HF^a	H_3^+		CH_3NC		c-C_6H_6
AlCl	SiH_2		CH_3SH		
AlF	c-SiC_2		NH_2CHO		**13 Atoms**
SiN	MgNC		HC_3HO		$H(C\equiv C)_5CN$
SiC	NaCN		C_5H		
SH	MgCN		HC_3NH^+		
CF	CO_2		HC_2CHO		
N_2	SiCN		$CH_2=CH_2$		
	AlCN		HC_4H		
	SiNC		C_5S		
	KCN				

[a] Preliminary assignment
[b] Carbocyclic molecule

that atomic hydrogen can "tunnel" through this barrier at the surface of the dust particles. It is not clear whether all four hydrogenation steps take place as shown above. The initial reaction has been detected in Greenberg's laboratory on ice matrices at 10–30 K. The hydrogenation steps of HCO and H_3CO appear feasible on dust particles, whereas the addition of H to H_2CO is still subject to discussion.

The interesting list of molecules in Table 3.3 includes one amino acid in the group of molecules with more than 10 atoms: it is the simplest proteinogenic amino acid, glycine.

The search for glycine in ISM started about 30 years ago when the first laboratory spectra became available. Glycine can exist in three configurations. Radioastronomers have concentrated their search on the lowest-energy configuration I and the next highest II, which has a larger dipole moment (Snyder, 1997).

Four years of study led to the discovery of glycine in the millimetre wavelength range in the hot molecular clouds of Sagittarius (around 81,500 light years away), Orion KL and W51. We can only conjecture as to the mechanism of its formation. Ion–molecule reactions in the gas phase, as well as UV photolytic processes in molecular ice, have been discussed.

The discovery of this amino acid in outer space supports the theory that organic molecules from space must have played a central role in prebiotic chemistry on Earth (Kuan et al., 2003).

A new type of "chemical laboratory" was recently discovered in outer space: the "circumstellar envelopes" of evolved stars. Such envelopes are found around carbon- and oxygen-rich giant stars and red supergiants. More than 50 different chemical compounds have been identified, including "exotic" species such as C_8H, C_3S, SiC_3 and AlNC. The molecules found here are quite distinct from those found in molecular clouds, the most prevalent containing long carbon chains, silicon and metals such as magnesium, sodium, and aluminium.

The highly complex physical conditions in these envelopes indicate that further surprises are still to come (Ziurys, 2006).

References

A'Hearn MF (2006) Science 314:1708
Anderson JD, Schubert G, Jacobson RA, Lau EL, Moore WB, Sjögren WL (1998) Science 281:2019
Anderson JD, Lau EL, Sjögren WL, Schubert G, Moore WB (1997) Science 276:1263
Bada JL, Cronin JR, Ho M-S, Kvenvolden KA, Lawless JG, Miller SL, Oró J, Steinberg S (1983) Nature 301:494
Bernstein MP, Sandford SA, Allamandola LJ (1999a) Scientific American, July
Bernstein MP, Sandford SA, Allamandola LJ, Gillette JS, Clemett SJ, Zare RN (1999b) Science 283:1135
Bernstein MP, Dworkin JP, Sandford SA, Cooper GW, Allamandola LJ (2002) Nature 416:401
Blake DF, Jenniskens P (2001) Scientific American, August
Brington KLF, Engrand C, Galvin DP, Bada JL, Maurette M (1998) Orig Life Evol Biosphere 28:413

Brown ME, Calvin WM (2000) Science 287:107

Brownlee DE, Horz F, Newburn RL, Zolensky M, Duxbury TC, Sandford S, Sekanina Z, Tsou P, Hanner MS, Clark BC, Green SF, Kissel J (2004) Science 304:1764

Brownlee D, Tsou P, Aléon J, Alexander CMO'D, Araki T, Bajt S, Baratta GA, Bastien R, Bland P, Bleuet P, Borg J, Bradley JP, Brearley A, Brenker F, Brennan S, Bridges JC, Browning ND, Brucato JR, Bullock E, Burchell MJ, Busemann H, Butterworth A, Chaussidon M, Cheuvront A, Chi M, Cintala MJ, Clark BC, Clemett SJ, Cody G, Colangeli L, Cooper G, Cordier P, Daghlian C, Dai Z, D'Hendecourt L, Djouadi Z, Dominguez G, Duxbury T, Dwarkin JP, Ebel DS, Economou E, Fakra S, Fairley SAJ, Fallon S, Ferrini G, Ferroir T, Fleckenstein H, Floss C, Flynn G, Franchi IA, Fries M, Gainsforth Z, Gallien J-P, Genge M, Gilles MK, Gillet P, Gilmour J, Glavin DP, Gounelle M, Grady MM, Graham GA, Grant PG, Green SF, Grossemy F, Grossman L, Grossman JN, Guan Y, Hagiya K, Harvey R, Heck P, Herzog GF, Hoppe P, Hörz F, Huth J, Hutcheon ID, Ignatyev K, Ishii H, Ito M, Jacob D, Jacobsen C, Jacobsen S, Jones S, Joswiak D, Jurewicz A, Kearsley AT, Keller LP, Khodja H, Kilcoyne ALD, Kissel J, Kroth A, Langenhorst F, Lanzirotti A, Le L, Leshin LA, Leitner J, Lemelle L, Leroux H, Liu M-C, Luening K, Lyon I, MacPherson G, Marcus MA, Marhas K, Marty B, Matrajt G, McKeegan K, Meibom A, Mennella V, Messenger K, Messenger S, Mikouchi T, Mostefaoui S, Nakamura T, Nakano T, Newville M, Nittler LR, Ohnishi I, Ohsumi K, Okudaira K, Papanastassiou DA, Palma R, Palumbo ME, Pepin RO, Perkins D, Perronnet M, Pianetta P, Rao W, Rietmeijer FJM, Robert F, Rost D, Rotundi A, Ryan R, Sandford SA, Schwandt CS, See TH, Schlutter D, Sheffield-Parker J, Simionovici A, Simon S, Sitnitsky Y, Snead CJ, Spencer MK, Stadermann FJ, Steele A, Stephan T, Stroud R, Susini J, Sutton SR, Suzuki Y, Taheri M, Taylor S, Teslich N, Tomeoka K, Tomioka N, Toppani A, Trigo-Rodriguez JM, Troadec D, Tsuchijama A, Tuzzolino AJ, Tyliszczak T, Uesugi K, Velbel M, Vellenga J, Vicenzi E, Vince L, Warren J, Weber I, Weisberg M, Westphal AJ, Wirick S, Wooden D, Wopenka B, Wozniakiewicz P, Wright I, Yabuta H, Yano H, Young ED, Zare RN, Zega T, Ziegler K, Zimmerman L, Zinner E, Zolensky M (2006) Science 314:1711

Burnett DS (2006) Science 314:1709

Calvin M (1969) Chemical Evolution. Oxford at the Clarendon Press

Carlson RW, Anderson MS, Johnson RE, Smythe WD, Hendrix AR, Barth CA, Soderblom LA, Hansen GB, McCord TB, Dalton JB, Clark RN, Shirley JH, Ocampo AC, Matson DL (1999a) Science 283:2062

Carlson RW, Johnson RE, Anderson MS (1999b) Science 286:97

Carr MH, Belton MJS, Chapman CR, Davies ME, Geissler P, Greenberg G, McEwen AS, Tufts BR, Greeley R, Sullivan R, Head JW, Pappalardo RT, Klaasen KP, Johnson TV, Kaufman J, Senske D, Moore J, Neukum G, Schubert G, Burns JA, Thomas P, Veverka J (1998) Nature 391:363

Chyba C (2000) Nature 403:381

Chyba C (2001a) Proc Natl Acad Sci USA 98:801

Chyba C, Hand KP (2001b) Science 292:2026

Chyba C, Phillips CB (2002) Orig Life Evol Biosphere 32:47

Clarke DW, Ferris JP (1997) Orig Life Evol Biosphere 27:225

Cockell CS (2006) Phil Trans R Soc B 361:1845

Coll P, Coscia D, Gazeao M-C, Guez L, Raulin F (1998) Orig Life Evol Biosphere 28:195

Cronin J, Pizzarello S (1997) Science 275:951

Cronin J (1998) Clues from the origin of the Solar System: meteorites. In: Brack A (Ed.) The Molecular Origins of Life. Cambridge University Press, p 119-146

Cronin J, Pizzarello S (2000) Orig Life Evol Biosphere 30:209

Dalgarno A (1991) Nature 353:502

Delsemme (1984) Orig Life Evol Biosphere 14:51

Delsemme AH (1998) Cosmic origin of the biosphere. In: Brack A (Ed.) The Molecular Origins of Life. Cambridge University Press, p 100

de Morais A (2000) Orig Life Evol Biosphere 30:317

Ehrenfreund P, Glavin DP, Botta O, Cooper G, Bada JL (2001) Proc Natl Acad Sci USA 98:2138

Engel MH, Nagy B (1982) Nature 296:837
Farquhar J (2000) Nature 404:50
Feitzinger JV (2002) Die Milchstraße. Spektrum, Heidelberg, p 75
Feldman PD (2005) Nature 437:958
Greenberg M (1983) Scientific American, June p 96
Greenberg M (2000) Scientific American, December
Griffith CA (1993) Nature 364:511
Gurnett DA, Zarka P, Manning R, Kurth WS, Hospodarsky GB, Averkamp TF, Kaiser ML, Farrell
    WM (2001) Nature 409:313
Helgason J (1999) Geology 27:231
Henning T, Salama F (1998) Science 282:2204
Herbst E (1990) Angew Chem 102:627
Hill HGM, Nuth JA (2003) Astrobiol 3:291
Hiroi T, Zolensky ME, Pieters CM (2001) Science 293:2234
Hoppe P (1996) Science 272:1314
Hoerz F, Bastien R, Borg J, Bradley JP, Bridges JC, Brownlee DE, Burchell MJ, Chi M, Cintala
    J, Dai ZR, Djouadi Z, Dominguez G, Economou TE, Fairey SAJ Floss C, Franchi IA, Graham
    GA, Green SF, Heck P, Hoppe P, Huth J, Ishii H, Kearsley AT, Kissel J, Leitner J, Leroux H,
    Marhas K, Messenger K, Schwandt CS, See TH, Snead C, Stadermann FJ, Stephan IT, Stroud
    R, Teslich N, Trigo-Rodriguez JM, Tuzzolino AJ, Troadec D, Tsou P, Warren J, Westphal A,
    Wozniakiewicz P, Wright I, Zinner E (2006) Science 314:1716
Hubbard WB (1990) Nature 343:353
Inman M (2006) Science 310:611
Irvine WM, Carramiñana A, Carrasco L, Schloerb FP (2003) Orig Life Evol Biosphere 33:597
Joseph JC, Clarke DW, Ferris JP (2000) Orig Life Evol Biosphere 30:234
Kerr RA (2001) Science 291:22
Kinoshita I (1990) Scientific American, November
Kissel J, Krueger FR (1987) Nature 326:755
Kissel J, Krueger FR, Silén J, Clark BC (2004) Science 304:1774
Klemperer W (2006) Proc Natl Acad Sci USA 103:12232
Kokoouline V, Greene CH, Esry BD (2001) Nature 412:891
Krueger FR, Kissel J (1987) Naturwissenschaften 74:312
Kuan Y-J, Charnley SB, Huang H-C, Tseng W-L, Kisiel Z (2003) Astrophys J 593:848
Kueppers M, Bertini I, Fornasier S, Gutierez PJ; Hviid SF, Jorda L, Keller HU, Knollenberg J,
    Koschny D, Kramm R, Lara L-M, Sierks AH, Thomas N, Barbieri C, Lamy P, Rickman H,
    Rodrigo R & the OSIRIS team (2005) Nature 437:987
Kvenholden K, Lawless J,Pering K, Peterson E, Flores J, Ponnamperuma C, Kaplan IR, Moore C
    (1970) Nature 228:923
Lavallo V, Canac Y, Donnadieu B, Schoeller WW, Bertrand G (2006) Science 312:722
Lindblad B (1935) Nature 135:133
Lorenz RD, McKay CP, Lunine JI (1997) Science 275:642
Love SG, Brownlee DE (1993) Science 262:550
Loveday JS, Nelmes RJ, Guthrie M, Belmonte SA, Allan DR, Klug DD, Tse JS, Handa YP (2001)
    Nature 410:661
Lugmair GW (1999) Präsolares Siliciumcarbid in Meteoriten. In: Jahrbuch Max-Planck-
    Gesellschaft, S 470
Luu JX, Jewitt DC (1996) Scientific American, May
Mason SF (1992) Chemical Evolution. Clarendon Press Oxford
Maurette M, Olinger C, Michel-Levy MC, Kurat G, Purchet M, Brandstätter F, Bourot-Denise AM
    (1991) Nature 351:44
Maurette M (1998) Micrometeorites on the early earth. In: Brack A (Ed.) The Molecular Origins
    of Life. Cambridge University Press p 147
McCord TB, Hansen GB, Fanale FP, Carlson RW, Matson DL, Johnson TV, Smythe WD, Crow-
    ley JK, Martin PD, Ocampo A, Hibbitts CA, Granahan JC, the NIMS Team (1998) Science
    280:1242

McCord TB, Hansen GB, Hibbitts CA (2001) Science 292:1523
McEwen AS, Belton MJS, Breneman HH, Fagents SA, Geissler P, Greenley R, Head JW, Hoppa
    G, Jaeger WL, Johnson TV, Keszthelyi L, Klaasen KP, Lopes-Gautier R, Magee KP, Milazzo
    MP, Moore JM, Pappalardo RT, Phillips CB, Radebaugh J, Schubert G, Schuster P, Simonelli
    DP, Sullivan R, Thomas CP, Turtle EP, Williams DA (2000) Science 288:1193
McKay DS, Gibson Jr EK, Thomas-Keprta KL, Vali H, Romanek SC, Clemet SJ, Chillier XDF,
    Maechling CR, Zare RN (1996) Science 273:924
Meierhenrich UJ, Muñoz Caro GM, Bredehöft JH, Jesberger EK, Thiemann W H-P (2004) Proc
    Natl Acad Sci USA 101:9182
Moore JM (1998) Icarus 135:127
Muñoz Caro GM, Meierhenrich UJ, Schutte WA, Barbier B, Arcones Segovia A, Rosenbauer H,
    Thiemann WH-P, Brack A, Greenberg JM (2002) Nature 416:403
Oka T (2006) Proc Natl Acad Sci USA 103:12235
Oró J (1961) Nature 190:389
Oró J (1971) Nature 230:105
Oró J, Squyres SW, Reynolds RT, Mills TM (1992) Europa: Prospects for an Ocean and Exobi-
    ological Implications. In: Carle GC, Schwartz DE, Huntington JL (Eds.) Exobiology in Solar
    System Exploration. NASA SP 12, p 103
Pappalardo RT, Head JW, Greeley R, Sullivan RJ, Pilcher C, Schubert G, Moore WB, Carr MH,
    Moore JM, Belton MJS, Goldsby DL (1998) Nature 391:365
Pizzarello S, Huang Y, Becker L, Poreda RJ, Neeman RA, Cooper G, Williams M (2001) Science
    293:2236
Pollock (1975) Geochem Cosmochem Acta 39:1571
Prinn RG (2001) Nature 412:36
Rannou P, Hourdin F, McKay CP (2002) Nature 418:853
Raulin F (1998) Titan. In: Brack A (Ed.) The Molecular Origins of Life, Cambridge University
    Press, p 368
Ruiz J (2001) Nature 412:409
Rummel JD (2000) Orig Life Evol Biosphere 30:345
Schenk PM, McKinnon, Gwynn D, Moore JM (2001) Nature 410:57
Schulze-Makuch D, Grinspoon DH, Abbas O, Irwin NL, Bullock MA (2004) Astrobiogy
    4:11
Seiff A, Kirk DB, Knight TCD, Young LA, Milos FS, Venkatapathy E, Mihalov JD, Blanchard
    RC, Young RE, Schubert G (1997) Science 276:102
Shimoyama (1989) Geochem J 23:181
Shock E (2002) Nature 416:380
Sicardy B, Brahic A, Ferrari C, Gautier D, Lecacheux J, Lellouch E, Roques F, Ariot JE, Colas
    F, Thuillot W, Sèvre F, Vidal JL, Blanco C, Cristaldi S, Bull C, Klotz A, Thouvenot E (1990)
    Nature 243:350
Slade MA, Butler BJ, Muhleman DO (1992) Science 258:635
Snyder LE (1997) Orig Live Evol Biosphere 27:115
Snyder LE (2006) Proc Natl Acad Sci USA 103:12243
Söderblom LA, Kieffer SV, Becker TL, Brown RH, Cook AF, Hansen CJ, Johnson TV, Kirk RL,
    Shoemaker EM (1990) Science 250:401
Solomon SC (2007) Science 316:702
Spencer JR, Yessop KL, McGrath MA, Ballester GE, Yelle R (2000) Science 288:1208
Stern SA (2003) Nature 424:639
Stevenson DJ (2001) Science 294:71
Studier MH, Hayatsu R, Anders E (1972) Geochim Cosmochim Acta 36:189
Sullivan R, Moore J, Thomas P, Greeley R, Homan K, Klemaszewski J, Chapman CR, Tufts R,
    Head JW, Pappalardo R, Galileo Imaging Team (1998) Nature 391:371
Suzor-Weiner A, Schneider IF (2001) Nature 412:871
Tielens AGGM, Charnley SB (1997) Orig Life Evol Biosphere 27:23
Tokano T, Neubauer FM, Laube M, Mc Kay CP (1999) Planetary and Space Science 47:493

Tokano T (2000) Mitteilungen aus dem Institut für Geophysik und Meteorologie der Universität zu Köln, Heft 139, Cologne
Trolez Y, Guillemin J-C (2005) Angew Chem 117:7390, Int Ed 44:7224
Unsöld A, Baschek B (2001) The New Cosmos, Springer Berlin Heidelberg New York
Walte Jr JH (2001) Nature 410:787
Weissman PR (1998) Scientific American, September
Whipple FL (1950) Astrophys J. 111:375
Young E (1996) Science 272:837
Young E (2000) Science 287:53
Ziurys LM (2006) Proc Natl Acad Sci, USA 103:12274

Chapter 4
"Chemical Evolution"

The term "chemical evolution" was introduced by the Nobel Prize winner Melvin
Calvin and refers to the process of the synthesis of biochemically important
molecules from small molecules and certain chemical elements under the (hypothetical) conditions present on prebiotic Earth. It is assumed that the smaller "building
block" molecules such as amino acids, fatty acids or nucleobases were formed initially, and that these underwent polycondensation to give macromolecules in later
stages of development.

The biomolecules from which the first living systems on the primeval Earth developed could have had various origins:

> The building block molecules may have been synthesized in the atmosphere,
> the hydrosphere or on the lithosphere of the young Earth from species such
> as CO, CO_2, CH_4, H_2O, N_2 and NH_3 (endogenous synthesis).
> The molecules may not have been formed on Earth but brought there from outer
> space (from within or from outside the solar system) by meteorites or comets
> (exogenous synthesis).
> Or perhaps there was a combination of the two: substances from outer space
> undergoing further reactions on Earth.

Three compatible methods for the scientific investigation of synthetic possibilities
are available (Raulin, 2000):

> Simulation experiments in the laboratory
> Theoretical (computer) models
> The observation of planets, moons and comets with analytical instruments, ei-
> ther from Earth or in situ.

Critical discussion of the results so far available on biogenesis indicates that we still
do not know which of these three will lead to the most convincing answers to the
many questions which still remain unanswered.

4.1 The Miller–Urey Model Experiments

Very few chemical experiments resulted in as much publicity as the first synthesis of biomolecules under prebiotic conditions carried out by the doctoral candidate

H. Rauchfuss, *Chemical Evolution and the Origin of Life*,
© Springer-Verlag Berlin Heidelberg 2008

Stanley Miller at the University of Chicago more than 50 years ago. This experiment (in fact, of course, many were carried out prior to the successful one) is probably as well known as the Wöhler synthesis of urea! Miller's doctoral supervisor, Harold Urey (winner of the Nobel Prize in 1934), had suggested to Miller that he simulate a reducing primeval Earth atmosphere (as required by the Oparin–Haldane hypothesis) to electrical discharges "and see what happens". Urey apparently expected that such an experiment would lead to a huge variety of organic compounds.

Miller used a mixture of methane, ammonia, water and hydrogen and provided the system with enough energy to make synthetic reactions possible by subjecting it to electrical discharges. His experiments were successful after he had made various modifications to the apparatus. Apart from a few simple organic molecules, Miller was able to detect the presence of amino acids, in particular glycine and alanine. Thus Urey had been too pessimistic: a relatively small number of organic molecular species were formed, some of which were of biochemical importance! The results were published in *Science* under the title "A Production of Amino Acids Under Possible Primitive Earth Conditions" (Miller, 1953). J.L. Bada and A. Lazcano (2003) provide some very interesting background information to Miller's article.

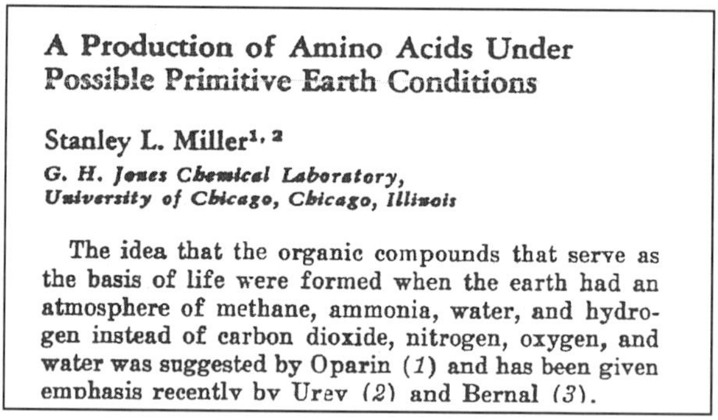

Fig. 4.1 Stanley Miller's first publication in *Science* (1953)

> ### A Production of Amino Acids Under Possible Primitive Earth Conditions
>
> **Stanley L. Miller[1, 2]**
>
> *G. H. Jones Chemical Laboratory,*
> *University of Chicago, Chicago, Illinois*
>
> The idea that the organic compounds that serve as the basis of life were formed when the earth had an atmosphere of methane, ammonia, water, and hydrogen instead of carbon dioxide, nitrogen, oxygen, and water was suggested by Oparin (*1*) and has been given emphasis recently by Urey (*2*) and Bernal (*3*).

These experiments provided the first proof that the question of the origin of life is a scientific problem which can be approached (and possibly solved) by using scientific methodology.

Several institutes throughout the world immediately began to carry out experiments on prebiotic chemistry. At this point, we need to realize that the prebiotic synthesis of protein building blocks is only a first step towards solving the biogenesis problem. Put simply, it is a method for making bricks which will later be used in building a multi-storey office block!

A few facts about the Miller–Urey experiments: the now famous original apparatus was modified and improved by Miller himself, and by other groups, in order to improve product yields. In the reaction vessel, temperatures near the reaction zone were between 350 and 370 K, but as high as 870–920 K at the centre of the reaction. Experiments took between several hours and a whole week. The main products (starting with the highest yields) were formic acid, glycine, lactic acid and alanine. A little of the C_5 amino acid glutamic acid was also formed. It is amazing that this short list includes two protein-forming amino acids, even if these are the simplest two. Soon after Miller's publication appeared, questions were asked about the mechanism which led to these products.

Kinetic studies on reducing gas mixtures showed that the concentration of ammonia falls during the reaction, while that of HCN first rises and then stays almost constant. The amino acid concentration increases steadily as the reaction time increases, while the aldehyde concentration remains constant.

Glycine and the other amino acids are probably formed via the Strecker-cyanohydrin synthesis (which has been known for more than 150 years) from aldehyde, HCN and ammonia, with subsequent hydrolysis (Strecker, 1850, 1854; Miller, 1953).

It is also possible that reactions involving free radicals take place; the formation of such radicals has been clearly demonstrated during electrical discharges.

Abelson (1956) carried out experiments using atmospheres containing CO_2, (CO), N_2, (NH_3), H_2 and H_2O and detected small amounts of simple amino acids, the formation of which was prevented by adding oxygen. Many of the Miller–Urey experiments involved various energy sources and were carried out in the liquid and solid states as well as on gas mixtures.

Although the Miller–Urey experiments of 1953 are of only historic interest today, they do mark the beginning of prebiotic chemistry and modern biogenesis research.

4.2 Other Amino Acid Syntheses

Since hydrogen is the most common element in the universe, many researchers working in the first half of the last century assumed that many compounds were present in their hydrogen-rich form during the formation of the Earth, i.e., in a reduced state (Miller and Urey, 1959). However, since the atmospheres of Earth's two direct neighbours, Venus and Mars, contain between 95 and 98% CO_2, doubts soon arose as to the reality of the assumed reducing properties of the primitive Earth atmosphere. Studies of the oldest sedimentary rocks also indicate that the atmosphere of the very young Earth was rich in CO_2 and N_2 (see Sect. 3.4). Experiments also showed that NH_3 and CH_4, photochemically labile molecules, are readily decomposed by the influence of sunlight and cosmic radiation.

New experiments using weakly reducing or neutral gas atmospheres were conceived and carried out more than 20 years after Miller's first successes (Schlesinger and Miller, 1983). Comparisons of a series of simulated prebiotic atmospheres containing CH_4, CO and CO_2 as carbon sources, using electrical discharges at 298 K, led to the following results:

Mixtures of (a) $CH_4 + H_2 + N_2 + NH_3$, (b) $CH_4 + H_2 + N_2$, with the molar ratio of H_2 to CH_4 varying between 0 and 4: the amino acid yields (with respect to the amount of carbon) varied between 1.2 and 4.7%, almost independent of the H_2/CH_4 ratio and the amount of NH_3 in the reaction mixture.

Mixtures of (a) $CO + H_2 + N_2 + NH_3$, (b) $CO + H_2 + N_2$: The amino acid yields with an H_2/CO ratio of zero are about 0.44% if ammonia is present, but only 0.05% if it is not. The yields increase at higher H_2/CO ratios, up to around 2.7%, but almost only glycine is formed. The curves cross (Fig. 4.2).

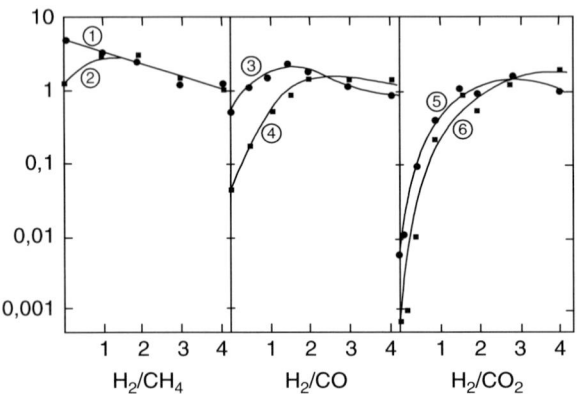

Fig. 4.2 Comparison of amino acid yields using CH_4, CO and CO_2 as carbon sources with the addition of varying amounts of H_2. The yields were calculated on the basis of the amount of carbon present in the reaction mixture. In all cases, the partial pressures of nitrogen, methane, carbon monoxide and carbon dioxide were 100 mmHg. For the reactions using nitrogen, the reaction vessel contained 100 mL of water, but no ammonia. Reactions involving nitrogen and ammonia were carried out using 100 mL of ammonium chloride (0.05 M). The electrical discharge experiments took 48 hours at room temperature (Schlesinger and Miller, 1983)

Mixtures of (a) $CO_2 + H_2 + N_2 + NH_3$, (b) $CO_2 + H_2 + N_2$, with the molar ratio of H_2 to CO_2 varying between 0 and 4: The amino acid yield is 10^{-3}% for $H_2/CO_2 = 0$ (in the presence of ammonia) and 6×10^{-4}% if no ammonia is present (these concentrations are close to the limits of detection). An increase in the H_2/CO_2 ratio leads to increased yields, up to 2% (but again almost solely glycine).

To sum up: all three carbon sources provide amino acids, if the H_2/C ratio is high enough (dependent on the oxidation and reduction state of carbon). Figure 4.3 shows the dependence of the amino acid yield on the reaction time.

After 2 days, the yield approached 60% of the maximum when CH_4 was used and 80% using CO.

The results so far available, and the models derived from them, indicate the following: a reducing atmosphere is more favourable for amino acid synthesis. If, however, the partial pressure of methane on the primeval Earth was either zero or very low, a relatively high H_2/CO or H_2/CO_2 ratio still allowed good rates of amino acid synthesis. It is, however, still an open question as to whether these concepts are realistic, because of the possibility that hydrogen could have escaped into space. It is arguable that in certain areas on the young Earth (and under unknown conditions),

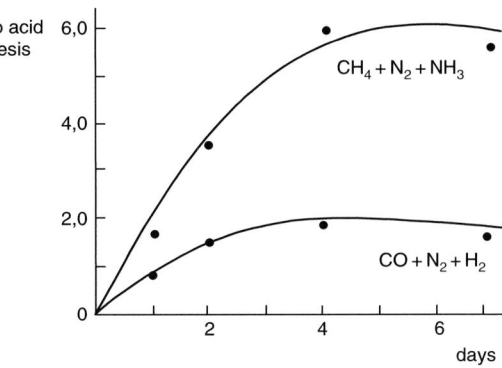

Fig. 4.3 The amino acid yield (in percent) as a function of time. The gas mixture in the experiment using methane did not contain hydrogen. That using CO had an H_2/CO ratio of 1 at a nitrogen partial pressure of 100 mmHg (Schlesinger and Miller, 1983)

relatively high methane concentrations could have been present, so that amino acid syntheses such as those observed in the Miller–Urey experiment may have occurred.

More recent work shows that the amino acid yield (in particular of glycine) in a $CO–N_2–H_2O$ atmosphere subject to irradiation with high energy particles is much higher than if spark discharges are used (Kobayaski et al., 1998). These authors thus assume that cosmic radiation could have been an important source of energy for prebiotic syntheses on Earth.

The variety of prebiotic organic reactions seems to be almost unlimited. Strasdeit et al. (2002) from the University of Hohenheim (Germany) reported the synthesis of zinc and calcium complexes of the amino acids valine and isovaline. They assume that these could have had a certain significance on the mineral-rich primeval Earth: on heating to 593 K under nitrogen, valine was converted to the corresponding cyclic dipeptide.

Plankensteiner et al. (2004) from the B. M. Rode group in Innsbruck (Austria) reported recently on a simulation of the formation of glycine and alanine in a neutral (or weakly oxidized) mixture of carbon dioxide, nitrogen and water vapour. Energy was provided by electrical discharges, and the reaction vessel (at 353 K) was half full of water. The tungsten cathode was above the surface of the water, the copper anode below it, and CO_2 and N_2 were added continuously. No reducing agent was added, not even a weak one. After a reaction time of 2 weeks, the yield of amino acids (determined by HPLC using two different gradients) was not particularly high, but detectable; it seems possible that it could be improved by modification of the procedure.

However, the amino acids were not necessarily formed in the gas phase. John Oró, one of the pioneers in prebiotic chemistry, carried out syntheses in the liquid phase by reacting HCN, NH_3 and H_2O at 353 K. The results were confirmed by Löwe et al. (1963) and developed further; ten years later, Jim Ferris took them up and did considerable further work (Ferris et al., 1973, 1974). In all these simulation experiments, the simplest amino acids (glycine, alanine and in small amounts aspartic acid and α-aminobutyric acid) are, as expected, the main products; the yields of glycine are around 1%, those of the other amino acids much lower.

Morowitz et al. (1995) chose a completely different route. They were able to synthesize glutamic acid from α-ketoglutarate, ammonia and formic acid in an aqueous medium without using enzymes. It is noteworthy that the reducing agent, formic acid, is one of the main products in the Miller–Urey experiment.

The existence of the huge aerosol layer on Titan has caused biogenesis researchers to become interested in a possible aerosol layer on Earth. Experiments with electrical discharges in a reducing atmosphere (nitrogen, methane, hydrogen) in the presence of an aqueous aerosol led to the synthesis of several amino acids (in particular alanine, glycine and β-alanine), but also to that of adenine and other purines; in addition, some hydroxyl acids (e.g., glycolic acid) were formed. Control experiments in the absence of aerosol showed either much lower yields or even values below the limits of measurement (Ruiz-Bermejo et al., 2007).

The discovery of hydrothermal vents on the ocean floor has led some biogenesis researchers to turn their attention to the hydrosphere (see Sect. 7.2) and to the processes occurring there at a depth of 2–3 km.

4.3 Prebiotic Syntheses of Nucleobases

Only a few years after the Miller–Urey experiment was published, J. Oró was able to synthesize one of the most important biomolecules, adenine. This purine derivative is not only a component of the nucleic acids, but as ATP, adenosine triphosphate (in combination with ribose and three phosphate residues), it plays a key role in the metabolism of all living creatures. The chemical formula of adenine is $C_5H_5N_5$, or expressed in another way, $(HCN)_5$.

It has been known for many years that hydrocyanic acid molecules are found in interstellar space and in the tails of comets. The question thus arises as to whether there is a connection between the immense importance of adenine in living beings and the occurrence in the cosmos of a building block for its formation.

As A. and B. Pullman showed more than 40 years ago, the purine base adenine occupies a unique situation in the purine family; in comparison to the other purines, it has the greatest resonance energy per π-electron, i.e., it is more stable, and thus likely to have been incorporated preferentially into biomolecules (Pullman, 1972).

Oró's experiment (Oró, 1960) was very simple: he heated ammonium cyanide to 343 K and was able to detect adenine after a few days. Shortly afterwards, he showed that five molecules of HCN combined to give one of adenine (Oró, 1961). An intermediate, aminomalonitrile, can be converted to 4-aminoimidazole-5-carboxamide in two different ways (Sanchez et al., 1966a, b) (Fig. 4.4):

(a) Via a direct reaction with formamidine or
(b) Via a photochemically induced rearrangement of the HCN tetramer (diaminomaleindinitrile).

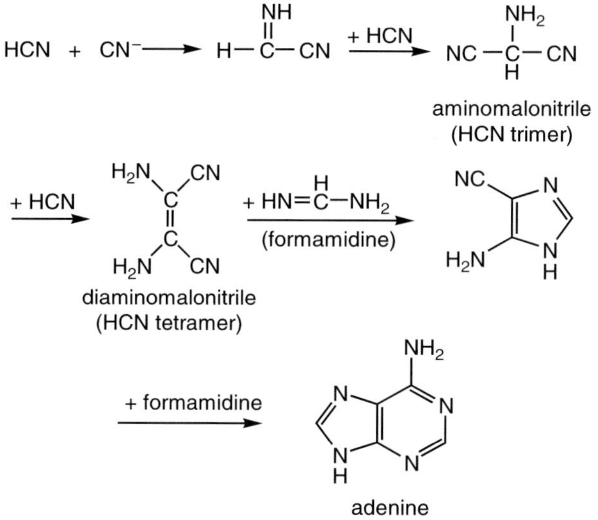

$$HCN + CN^- \longrightarrow H-\underset{\underset{}{\overset{\overset{NH}{\|}}{C}}}{}-CN \xrightarrow{+ HCN} NC-\underset{\underset{H}{|}}{\overset{\overset{NH_2}{|}}{C}}-CN$$

aminomalonitrile
(HCN trimer)

$$\xrightarrow{+ HCN} \quad \underset{\underset{H_2N}{}\overset{H_2N}{}}{C}=\underset{\underset{}{}}{C}\underset{}{} \overset{CN}{\underset{CN}{}} \xrightarrow[\text{(formamidine)}]{+ HN=\overset{\overset{H}{|}}{C}-NH_2}$$

diaminomalonitrile
(HCN tetramer)

4-amino-imidazole-5-carbonitrile structure: NC, N, H_2N, N–H

+ formamidine →

adenine

NH_2 / purine ring (adenine)

Fig. 4.4 Schematic representation of a possible synthetic pathway which leads from HCN (and formamidine) to adenine (Sanchez et al., 1967,1968)

The hydrolysis leads to 4-aminoimidazole-5-carboxamide, which under certain conditions can react with various partners (e.g., HCN, dicyan or formamidine) to give purines (i.e., adenine, guanine, hypoxanthine and diaminopurine).

There may also have been periods of extreme cold on the primeval Earth. Experiments showed that the formation of tetramers of HCN can take place in eutectic aqueous solutions of HCN (Sanchez et al., 1966a). Yields of more than 10% were obtained when such solutions were allowed to stand for several months at temperatures between about 243 and 263 K.

Many years later, Schwartz (Schwartz and Goverde, 1982; Voet and Schwartz, 1983) discovered that the synthesis of adenine via polymerisation of HCN can be accelerated by adding formaldehyde and other aldehydes. Reactions in the gas phase (nitrogen/methane atmosphere) promoted by electrical discharges led to the formation of cyanoacetylene in relatively good yields; the latter reacts with urea to give various products, including cytosine (Sanchez et al., 1968).

Until the 1980s, yields of nucleobases obtained in prebiotic syntheses were very small. Thus, some scientists assumed that in earlier phases of molecular evolution, the nucleic acids used other bases in their information-transmitting substances. Piccirilli et al. (1990) suggested isocytosine and diaminopyridine, while Wächtershäuser (1988) suggested that the first genetic material possibly consisted only of purines. However, pyrimidine (about a fifth of the total amount of purines present) had been detected in the Murchison meteorite, so that an effective pyrimidine synthesis should have been possible.

Robertson et al. (1996) were successful in synthesising cytosine and uracil in amazingly high yields. They started from cyanoacetaldehyde (obtained via hydrolysis of cyanoacetylene) and, again, urea. Cytosine is not formed in detectable concentrations unless concentrated urea solutions are used: then, however, the yields are between 30 and 50%. Uracil was obtained by hydrolysis of cytosine solutions.

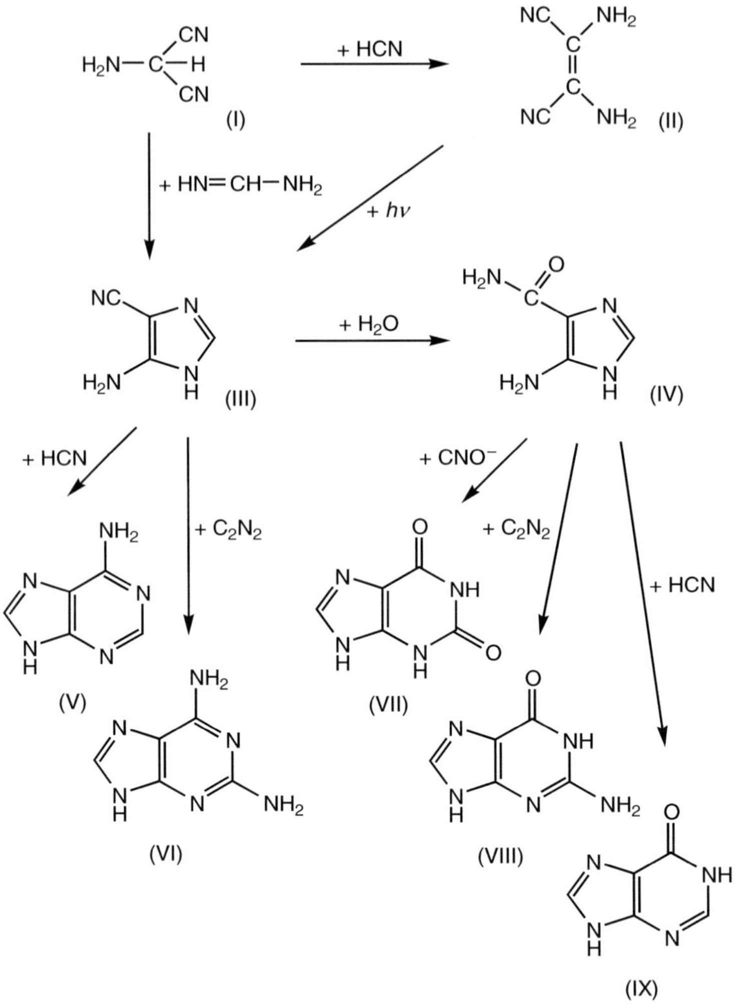

Fig. 4.5 The adenine synthesis can be varied to give other purine derivatives. Structures **I–IX** are those of **I** aminomalonitrile, **II** HCN tetramer, **III** aminoimidazole-carbonitrile, **IV** 4-aminoimidazole-5-carboxamide, **V** adenine, **VI** diaminopurine, **VII** xanthine, **VIII** guanine and **IX** hypoxanthine (Sanchez et al., 1966a)

This success does, however, have a drawback: the urea concentrations required could not possibly have existed in the primeval ocean. Thus, as in the case of other condensation reactions, one has to assume that there were ponds or lagoons, in which the necessary reagent concentrations could build up via evaporation of water.

Roberts et al. (1966) described a similar synthesis: cyanoacetaldehyde and guanidine hydrochloride gave 40–80% yields of 2,4-diaminopyrimidine under the conditions of the "lagoon model" mentioned above. Hydrolysis of diaminopyrimidine leads to cytosine, isocytosine and uracil. Thiourea reacts with cyanoacetylene to give 2-thiocytosine: however, the yield is considerably lower than with urea or

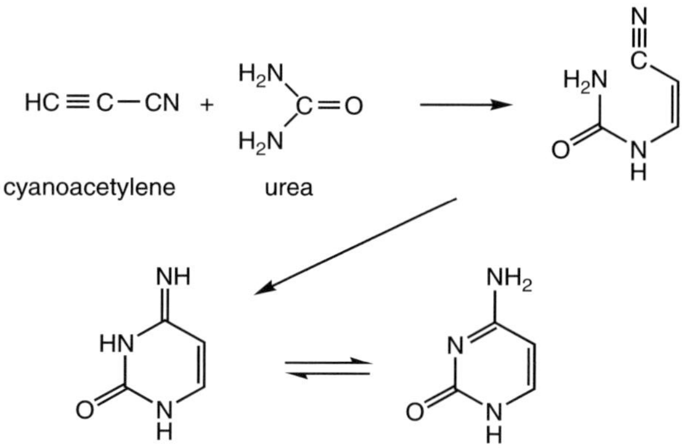

Fig. 4.6 A relatively simple synthetic route, using readily available starting materials, cyanoacetylene and urea, leads to pyrimidines, including cytosine, under various conditions (Sanchez et al., 1966b)

guanidine hydrochloride. Thiocytosine hydrolyzes to give thiouracil, which in turn reacts further to give uracil.

Whether the adsorption of molecules at the surface of minerals is a curse or a blessing for the adsorbed substances depends on many parameters. Experiments showed very different adsorption behaviour of adenine on different minerals. Active minerals are of particular importance for hydrothermal processes (see Sect. 7.2). The surface concentration of adenine on pyrites is fifteen times, that on quartz five times, and on pyrrhotite three and a half times as high as in a starting solution whose concentration is $20\,\mu M$ (Cohn, 2002).

Does adsorption at the surface of minerals provide protection for important biomolecules against hydrolysis or cosmic irradiation?

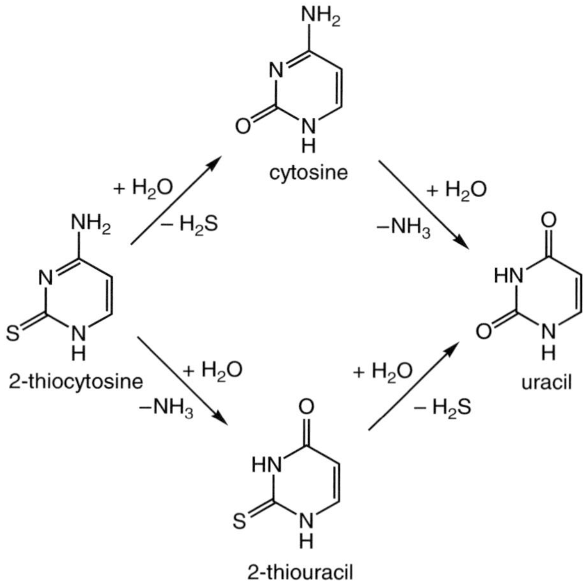

Fig. 4.7 The reaction of cyanoacetaldehyde with thiourea gives 2-thiocytosine, which can react further to give cytosine or thiouracil (and further to uracil) (Robertson et al., 1996)

F. G. Mosqueira et al. (1996) investigated this question; they studied the effect of ionising radiation on nucleobases adsorbed at mineral surfaces. Such studies were important; it had been assumed that the effect of irradiation on biomolecules would be prevented or reduced by adsorption. The Mexican research group showed that there is an important radiolytic effect of long-lived radionuclides, in particular ^{40}K, but also to a lesser extent ^{238}U and ^{232}Th, on nucleobases and their derivatives. The rate of decomposition due to irradiation could have equalled or exceeded the rate of synthesis of these compounds. A relatively homogeneous distribution of ^{40}K in sediments and in the primeval ocean has been assumed, but this isotope could have been present in higher concentration in acidic, magmatic rocks and in sedimentary aluminium silicates. It has been estimated that over a period of a thousand years, around 1.8×10^{-7} mol of substance per kilogram of clay would have been modified by ionising radiation (mainly from ^{40}K). Thus, molecules adsorbed on clay minerals would have not been as well-protected as had previously been assumed! The necessary precondition is of course that the structure of the Earth and its composition has not changed greatly in the past 3.8 billion years.

The prebiotic chemistry of the nucleic acid bases is still the subject of debate among experts. One of the most mindful critics is Robert Shapiro, professor of chemistry at New York University and a DNA expert. His book *Origins–A Sceptic's Guide to Creation of Life on Earth* includes a critical analysis of the results previously obtained in biogenesis research (Shapiro 1986). Shapiro's has been the critical voice in the community of biogenesis researchers for many years. He identifies the weak points in some of the audacious hypotheses, which are often raised to the status of theories even though they involve many open questions.

Shapiro (1995) studied the prebiotic role of adenine, i.e., the question as to whether or not this nucleobase could have been involved in the hypothetical RNA world (see Chap. 6). He lists a series of problems:

> The synthesis of adenine requires HCN concentrations of at least 0.01 M. Were such concentrations possible on the young Earth?
> The sensitivity of adenine to hydrolysis, with a half-life for deamination of 80 years at pH 7 and 310 K (at 273 K, the half-life is 4,000 years). Ring-opening reactions are also possible.
> The interaction between adenine and uracil occurs via two hydrogen bonds, while in the case of cytosine and guanine, there are three.

Shapiro concludes that adenine could not have played the outstanding role in the initial phases of biogenesis which it plays in cells today.

The same problem, the stability of the nucleobases, was taken up by Levi and Miller (1998). They wanted to show that a synthesis of these compounds at high temperatures is unrealistic, and thus they took a critical look at the high temperature biogenesis theories, such as the formation of biomolecules at hydrothermal vents (see Sect. 7.2). The half-life of adenine and guanine at 373 K is about a year, that of uracil about 12 years and of the labile cytosine only 19 days. Such temperatures could have easily been reached when planetoids impacted the primeval ocean.

If the Earth's surface was frozen (temperatures near 273 K), the half-life of organic bases would have been around a million years (except for the more labile cytosine). Thus Miller and Levi suggest that biogenesis models should be developed under the premise that the earth and its atmosphere were cold, rather than hot.

In another study of the problem of the prebiotic adenine synthesis, Leslie Orgel concludes that, in spite of many uncertainties, the formation of these complex heterocyclic bases probably required three preconditions:

The presence of HCN solutions
Conditions on primeval Earth which led to eutectic freezing, and thus to concentration of the HCN solutions
Photochemical reactions of HCN tetramers in such eutectic solutions to give 4-amino-3-cyanoimidazole (Orgel, 2004)

The purine base guanine is also formed in concentrated solutions of ammonium cyanide, i.e., the same substance which became known from Oró's adenine synthesis. Oró, as well as Stanley Miller, was involved in a new series of experiments (Levi et al., 1999). The yield of guanine is, however, 10–40 times lower than that of adenine; surprisingly, the synthesis is just as effective at 253 K as at 353 K. Low temperatures seem conceivable in certain parts of Earth as well as on the Jovian moon Europa (see Sect. 3.1.5) or in the Murchison meteorite.

Shapiro published a critical analysis of the availability of the nucleobase cytosine on the primeval Earth in the highly-regarded *Proceedings of the National Academy of Science* (Shapiro, 1999). Some biogeneticists still believe that all the substances necessary for the synthesis of a nucleic acid were available in the much-cited (but hypothetical) "primeval soup". Shapiro directs these optimists to the following problems:

Cytosine has not been detected in meteorite material.
Reports of the synthesis of cytosine from cyanoacetylene (or its hydrolysis product cyanoacetaldehyde) with cyanate, cyanogens or urea show that these substances react faster with nucleophilic compounds to give side products than to give the required main product. In addition, the formation of cytosine requires concentrations which are unrealistic in prebiotic environments.
The deamination reaction destroys cytosine, the half-life of which is about 340 years at 298 K.

Table 4.1 Percentage yields of adenine and guanine from polymerisation of NH_4CN, based on the initial concentration of HCN (Levy et al., 1999)

	10 M NH_4CN (353 K, 24 h)	0.1 M NH_4CN (253 K, 25 h)	0.1 M NH_4CN (253 K, 2 months)
Adenine	0.028	0.038	5×10^{-4}
Guanine	0.0007	0.0035	1.4×10^{-4}

Thus, one of the four nucleobases would not have been available, and we do not know whether a suitable replacement was present on the primordial Earth.

The reaction to Shapiro's criticisms was quick in coming from the Miller group in La Jolla, California (Nelson et al., 2001). These authors argued that cytosine is easier to prepare from cyanoacetaldehyde and urea at low temperatures than theoretically calculated. They also criticized the incorrect interpretation of the data on the hydrolysis of adenine. As was to be expected, the possibility that solutions could have undergone concentration via evaporation (lagoons or beaches) and freezing out at low temperatures (which also leads to concentration of solutions) are defended.

A more recent, extended study of purine synthesis via polymerisation of ammonium cyanide, described at the beginning of this section, showed that the yield of adenine from the non-hydrolyzed solution was only slightly temperature dependent. Shorter hydrolysis times for the insoluble polymerisation products led to higher adenine yields. When the solution is hydrolyzed at pH 8, the adenine yield is comparable to the value of 0.1% found for acidic hydrolysis (a model for the primeval ocean?). Increasing the hydrolysis time has no effect on the adenine yield because of its greater stability at pH 8. Hydrolysis of the black NH_4CN polymer under acidic or neutral conditions results in an adenine yield of about 0.05% (Borquez et al., 2005).

Among the various theories and hypotheses concerning the prebiotic synthesis of biomolecules, processes occurring at the surfaces of minerals are becoming much more popular. The minerals can serve three different functions:

As a matrix, i.e., for molecular orientation
As a platform for the concentration of molecules
As simple catalysts which reduce the activation energy of reactions

The simple molecule formamide, NH_2CHO, adsorbed on a $TiO_2(001)$ single crystal under high vacuum (10^{-10} mmHg), underwent conversion to a series of nucleoside bases on irradiation with UV light (3.2 eV). The UV light activates the semiconductor surface and has no effect on either the reaction products or the adsorbed formamide molecules, so that accumulation of product molecules is possible. Other "more elegant" prebiotic syntheses do not have this advantage! The hexaatomic molecule formamide is present in interstellar matter (Table 3.3), as are semiconductor particles, so that reactions of this type, leading to purines and pyrimidines, could in principle occur there (Senanayake and Idriss, 2006).

The kinetics of the formation of oligocytidylates [oligo(C)] from the 5'-monophosphorimidazolide of cytidine (ImpC) have been studied at 283–348 K in the presence of Pb(II) as a catalyst. The rate constant for the formation of oligo(C) increases in moving from dimer to tetramer, as in the clay-catalysed formation of oligonucleotides. While the rate constant increases with increasing temperature, the yield of oligo(C) falls with increasing temperature. At even higher temperatures (e.g., in hydrothermal systems), oligonucleotide synthesis may be possible if reaction boosters such as protein-like molecules and/or minerals are present (Kawamura and Maeda, 2007).

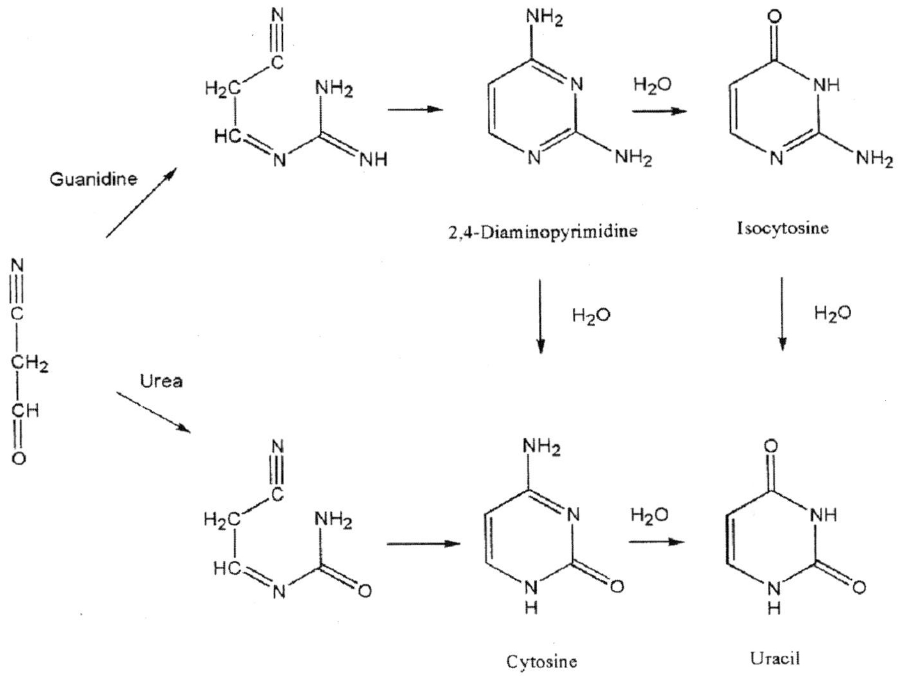

Fig. 4.8 Suggested mechanisms for the synthesis of pyrimidines from cyanoacetaldehyde (CAA) and cyanate, urea or guanidine (Cleaves et al., 2007)

The synthesis of pyrimidines under specific prebiotic conditions, at low temperature, has apparently been carried out successfully (Fig. 4.8). The formation of the DNA building blocks was carried out by freezing out a dilute solution of cyanoacetaldehyde (CAA) and urea or guanidine. The concentration of CAA was only 10^{-3} M, that of guanidine 1 M. The reaction took 2 months at 273 K and a pH of 8.1. Yields were as follows:

Cytosine	0.5%
Uracil	10.8%
Isocytosine	0.75%
2,4-diaminopyrimidine	0.42%

Successes of this type indicate that important prebiotic syntheses may have occurred either in regions on the primordial Earth where temperatures were low, or in a period where temperatures on the whole Earth were low (Cleaves et al., 2006).

All this makes it clear that the discussions on pyrimidine synthesis are by no means over. The question is whether there was an RNA world or not. Recently, attempts have been made to bypass this problem by postulating a phase of development *before* the RNA world: a pre-RNA world (see Sect. 6.7). But this is also hypothetical!

The formation of purines in interstellar space has been considered feasible for some considerable time. A theoretical study (using ab initio methods) on the mechanism of adenine formation from monocyclic HCN pentamers has been reported and has afforded a deeper insight into the gas-phase chemistry of possible purine syntheses. The authors drew the following conclusions from their results:

(a) 5-(N'-formamidinyl)-1H-imidazole-4-carbonitrile can serve as a substrate for proton-catalysed purine formation under photochemical conditions.
(b) N-(4-(iminomethylene)-1H-imidazole-5 (4H)-ylidene) formamidine could be important for non-catalysed purine formation under corresponding conditions.

This study again shows the prime importance of the purine base adenine, whether in the vastness of interstellar space or in the biochemical processes taking place in a single cell (Glaser et al., 2007).

4.4 Carbohydrates and their Derivatives

Carbohydrates are still difficult to deal with in prebiotic chemistry: although the first laboratory syntheses of carbohydrates were described more than 150 years ago, their prebiotic synthesis is still unclear.

Butlerov (1828–1886), professor of chemistry in Kazan and St. Petersburg, carried out the first carbohydrate synthesis. He obtained a sugar mixture from aqueous formaldehyde solutions under alkaline conditions. This "formose reaction", a complicated autocatalytic process, requires inorganic catalysts such as $Ca(OH)_2$ or $CaCO_3$. A number of intermediates are involved: glycolaldehyde, glycerine aldehyde, dihydroxyacetone, the C_4 sugars, the C_5 sugars and the hexoses. Prebiotic syntheses used catalytically active clays such as kaolin in order to obtain sugars from 0.01 M formaldehyde solutions. Ribose is also formed, although in only small amounts (Gabel and Ponnamperuma, 1967; Reid and Orgel, 1967). Although the formose reaction is readily carried out, it has various problems:

> About 40 different sugars are formed. Those required for nucleic acid synthesis, ribose and deoxyribose, are obtained in yields of less than 1%. It is completely unclear whether these could have been separated from the others under primeval Earth conditions (Shapiro, 1988).
> Ribose is relatively unstable, which leads all the critics of an RNA world to look for other models. The half-life of ribose is only 73 minutes at 373 K and pH 7, but increases at 273 K to 44 years.

The above-mentioned facts require that ribose must have undergone further reactions immediately after its formation under prebiotic conditions. More than 20 years ago R. Shapiro (1984) pointed out the immense problems which would have needed to be solved in prebiotic nucleic acid formation.

Larralde et al. (1995) published work on the "ribose problem" in the *Proceedings of the National Academy of Science*. They suggested that the results referred to above on the stability of ribose rule out its use, and that of other sugars, as prebiotic reagents, except under very particular conditions. In 1996, they repeated their arguments at the ISSOL conference in Orleans (Larralde et al., 1996).

Two points thus argue against the participation of ribose in nucleic acid formation: the lability of the molecule and the problems with its synthesis (the concentrations of the starting materials are too high). Other, newer and more effective syntheses seem necessary, whereby prebiotic conditions (although these are not known precisely) strongly limit the possibilities.

A ray of hope appeared when a synthetic route was developed in the laboratory of Albert Eschenmoser in Zürich, leading in good yields to ribose-2,4-diphosphate (in racemic form). The starting material was glycol aldehyde, which was phosphorylated in the 2-position and then incubated with formaldehyde. Unfortunately the synthetic conditions are only those of a modern laboratory, but could the reaction have taken place on the primeval Earth? (Müller et al., 1990).

$$
\begin{array}{ccc}
2\ \ \underset{\substack{\text{glycol aldehyde}\\\text{phosphate}}}{\overset{\displaystyle H\!\!\diagdown_{\!\!C}\!\!\diagup\!O}{\underset{\displaystyle H_2C-O-\overset{\displaystyle O}{\overset{\|}{P}}-O^-}{|}}}\overset{}{\underset{\displaystyle \diagdown O^-}{}}
&
\underset{\substack{\text{(in 2 M NaOH, 7 days}\\\text{at room temperature}\\\text{in the absence of air)}}}{\xrightarrow{\ +\ HCHO\ }}
&
\begin{array}{l}
H\diagdown_{C}\diagup O\\
\ \ |\\
HC-O-PO_3^{2-}\\
\ \ |\\
HC-OH\\
\ \ |\\
HC-O-PO_3^{2-}\\
\ \ |\\
H_2C-OH
\end{array}
\end{array}
$$

rac-ribose-2,4-diphosphate
(main product)

The quest for synthetic routes to ribose continued. The reduction of the high pH in the formose reaction was made possible by adding $Mg(OH)_2$: the aldopentoses obtained were more stable, but the reaction was slower. The search for more effective catalysts led to Pb^{2+} ions. The first mention of this possibility was made by Wolfgang Langenbeck (University of Halle) as early as 1954 (Langenbeck, 1954). He reported that formaldehyde solutions, together with $Mg(OH)_2$ suspensions and Pb^{2+} ions as catalysts, gave aldehyde yields corresponding to the four pentoses (ribose, arabinose, lyxose and xylose) of about 30%. No mention was made of the amount of ribose in the mixture (Zubay, 1998). The question as to whether lead ions were available on the primeval Earth is still an open one. Lead is found in several minerals, e.g., galenite (PbS), cerussite ($PbCO_3$), stolzite ($PbWO_4$) and anglesite $PbSO_4$.

The synthesis of pentose-2,4-diphosphate referred to above gave the best yields of a ribose derivative. Thus, the search for an effective synthesis leading to necessary starting materials such as glycol aldehyde phosphate (GAP) was important: Krishnamurthy et al. (1999, 2000) reported new synthetic routes to GAP: glycol aldehyde is allowed to react with amidotriphosphate (AmTP) in dilute aqueous solution. The triphosphate derivative is formed from trimetaphosphate and NH_4OH.

All attempts to phosphorylate GAP with triphosphate were unsuccessful, so that AmTP is an important phosphorylating agent.

As already mentioned, one of the extremely weak points of the RNA world is the lability of ribose. A short report (only one page) in *Science* showed that it is possible to stabilize ribose with the help of borate minerals. The group of A. Ricardo from the chemistry department of the University of Florida (Gainesville) carried out a ribose synthesis from glycol aldehyde and glycerine aldehyde [with $Ca(OH)_2$, pH around 12]. After an hour at 298 K or only 10 minutes at 318 K, the reaction mixture turned brown, showing the presence of polymeric mixtures. If, however, the borate minerals kernite $\{Na_2[B_4O_6(OH)_2] \cdot 3 H_2O\}$, ulexite $\{NaCa[B_5O_6(OH)_6] \cdot 5 H_2O\}$, or colomanite $\{Ca[B_3O(OH)_3] \cdot H_2O\}$[1] are added under identical reaction conditions, the reaction mixture remains colourless even after two months, and ribose can be detected, apparently stabilized by boric acid. The corresponding ribose–diborate complex (with a molecular weight of 307) has been clearly identified (Ricardo et al., 2004).

The starting aldehyde could have been present on the primeval Earth; the question of the necessary boron derivatives is open, although it seems possible that weathering of the borate minerals referred to above could have set free boron salts, which then stabilized the ribose.

Other studies have supported the reports of considerable stabilisation of carbohydrates by borates: the thermal stability of ribose increases under acidic conditions, while glucose is favoured under basic conditions. Thus, the "pre-RNA world chemistry" of furanosyl borate diesters of ribose could have taken place at high temperatures and low pH. A "pre-metabolism world" at high temperatures and in the basic pH range would have involved glucose borates and anion borates (Scorei and Cimpoiaşu, 2006).

A spectacular news item appeared in the press in December 2001: with the help of the most modern instruments, it had been possible to detect sugar in meteorites. A journalist working for a Swedish daily paper used this for a joky article heading: "Vilket julgodis!" ("What great Christmas candies!"). Sephton (2001) published a comment on the original article in *Nature* under the title "Meteoritics: Life's Sweet Beginnings?" It is surprising that sugar or sugar derivatives had not been detected

---

[1] The structural formulae may differ slightly, depending on the source. Those used here are taken from the Römpp Chemielexikon (19th edition).

earlier in the Murchison meteorite, which had been studied intensively by many research groups. Cooper et al. (2001) (NASA Ames Research Laboratory and University of Pescara, Italy) were able to detect several sugar alcohols and sugar acids in aqueous extracts from the Murchison and Murray meteorites. The smallest was the three-carbon sugar alcohol glycerine; the largest were six-carbon sugar acids such as gluconic acid and some of its isomers. GLPC/MS was used for the analysis; the compounds were derivatized to give the corresponding trimethylsilyl (TMS) and *tert*-butyldimethylsilyl (TBDMS) compounds prior to analysis. As in all meteorite analyses, the question of possible contamination arises. The isotope ratios $^{13}C/^{12}C$ and $^2H/^1H$ indicate an extraterrestrial origin of the polyols. In commenting on these results, Sephton (2001) notes that these measurements represent only an average of a selection of molecules. In order to eliminate the possibility of contamination completely, the isotope ratios would need to be determined separately for each class of compounds detected.

Another example of the ability of proteinogenic amino acids, small peptides, and amines to catalyse the formation of new C–C bonds has been demonstrated by Weber and Pizzarello; they were able to carry out model reactions for the stereospecific synthesis of sugars (tetroses) using homochiral L-dipeptides. The authors achieved a D-enantiomeric excess (ee) of more than 80% using L-Val-L-Val as the peptide catalyst in sugar synthesis (in particular D-erythrose) via self-condensation of glycol aldehyde.

The synthesis of dipeptides under the conditions found on primeval Earth appears highly likely. The discovery that these small molecules can act as catalysts makes it possible to discuss their being involved in basic synthetic reactions occurring in an (as yet hypothetical) RNA world (Weber and Pizzarello, 2006).

Earlier studies showed that reactions of sugars with ammonia lead to small molecules such as amines or organic acids. A. L. Weber has reported important autocatalytic processes occurring when trioses are allowed to react with ammonia: under anaerobic conditions, such reactions provide products which are autocatalytically active. Their autocatalytic activity was determined directly by investigating their effect on an identical triose–ammonia reaction. Both an increase in the triose degradation rate and an increased rate of synthesis of pyruvate, the dehydration product of the triose, were observed. Such processes may have been of importance for prebiotic chemistry occurring on the primeval Earth (Weber, 2007).

4.5 Hydrogen Cyanide and its Derivatives

One of the most important and versatile building blocks for the construction of biomolecules is hydrogen cyanide HCN (also known as prussic acid), which was prepared for the first time by the German-Swedish apothecary Carl-Wilhelm Scheele (1742–1786) in Köping in Sweden. He heated blood with potash and charcoal and obtained what he called "Blutlauge", which he distilled with sulphuric acid (Bauer, 1980; Encycl. Am., 1975).

The compound HCN occupies a position at the border between inorganic and organic chemistry. It is paradoxical that hydrogen cyanide is on the one hand an important starting material for prebiotic syntheses of biomolecules and on the other a deadly poison for living organisms.

As already mentioned, hydrogen cyanide is formed in simulation experiments using reducing primeval atmospheres. CN was discovered in interstellar space as early as 1940 by optical spectroscopy (Breuer, 1974), and later HCN itself (from measurements using millimetre wavelengths). Only a few years after the Miller–Urey experiments, Kotake et al. (1956) obtained HCN in good yields by reacting methane with ammonia over aluminium-silicate contacts:

$$CH_4 + NH_3 \xrightarrow{(Al_2O_3 + SiO_2)} HCN + 3H_2 \tag{4.1}$$

Several research groups have worked extensively with HCN and products arising from it, in particular Clifford N. Matthews (University of Illinois, Chicago) and Jim Ferris, New York. Matthews has been studying this chemistry for more than 45 years. As early as 1962, he and R. M. Kliss, in a more theoretically oriented analysis of previous experiments, described the formal possibilities of HCN dimers in syntheses of various molecules under prebiotic conditions (Kliss and Matthews, 1962). They suggested a mechanism for the formation of polyglycine from HCN, proceeding via the diradical aminocyanocarbene; several years later, this hypothesis was verified experimentally. In general, HCN polymers are synthesized via this overall reaction:

$$x\,HCN \rightarrow (HCN)_x \tag{4.2}$$

This reaction is catalysed by traces of bases. Matthews remained true to HCN chemistry, which is indeed an attractive research area. HCN can polymerize under certain conditions; the process can be induced by heat or radiation, both of which were certainly present the on primeval Earth.

Nearly 50 years ago, T. Völker (1960) reviewed the complex reaction possibilities of polymeric HCN, not from the point of view of prebiotic chemistry, but from that of industrial chemistry.

Moser et al. (1968) (one of the co-authors was Clifford Matthews) reported a "peptide synthesis" using the HCN trimer aminomalonitrile, after pre-treatment in the form of a mild hydrolysis. IR spectra showed the typical nitrile bands $(2,200\,cm^{-1})$ and imino-keto bands $(1,650\,cm^{-1})$. Acid hydrolysis gave only glycine, while alkaline cleavage of the polymer afforded other amino acids, such as arginine, aspartic acid, threonine etc. The formation of the polymer could have occurred according to the scheme shown in Fig. 4.9.

The complexity of the HCN oligomerisation reaction was also studied by Schwartz et al. (1984) in Nijmegen. The reactions worked best at a pH value which lies near the pK value of HCN. A list of the products of oligomerisation includes 38 compounds, including orotic acid, adenine, guanine and glycine, as well as more complex molecules such as isoleucine, glutamic acid, diaminosuccinic acid and guanidinoacetic acid. The synthetic conditions differ considerably: some require conditions which could hardly be considered as prebiotic!

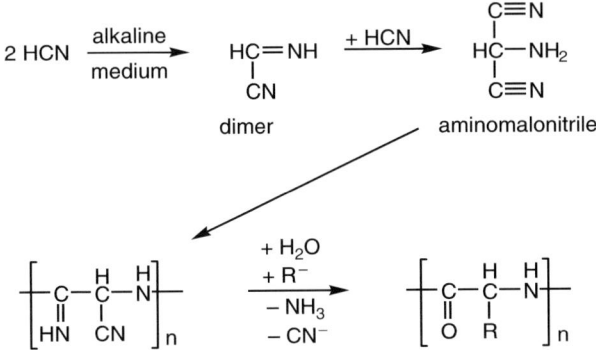

Fig. 4.9 A synthesis resulting in polypeptides (or compounds similar to polypeptides using amino-malonitrile and further adding hydrogen cyanide. In this process, R 15 may represent different side chains of amino acids

Schwartz and Goverde (1982) were able to carry out an adenine synthesis under exceptional conditions, via HCN oligomerisation in the cold: a 0.01 M HCN solution was brought to pH 9.2 by adding ammonia and left to stand at 271 K for 60–100 days. When glycol nitrile was added, the adenine yield increased by a factor of five, reaching 48 μg/L, i.e., 0.02% of the starting concentration of HCN. These experiments were carried out because there may have been longer cold periods in the Earth's history.

The New York research group of Ferris (Ferris et al., 1981) studied the structure of HCN oligomers using NMR spectroscopy and were able to detect carboxamide and urea functionalities. Pyrolysis of the oligomers led to the formation of CO, H_2O, HCN, CH_3CN, $HCONH_2$ and pyridine, in agreement with the spectroscopic results. Hydrazinolysis (similar to cleavage with 6 M HCl) gave about 10% amino acids. Although the biuret test, a standard method for detecting peptide bonds, was positive, it cannot be considered as absolutely definite evidence for the presence of such bonds in the HCN oligomers. Ferris thus suggests that HCN oligomers are not a suitable source for prebiotic peptides.

This interpretation of the experimental results is not accepted by Clifford Matthews, who has for many years defended the following hypothesis: the prebiotic proteins (or peptides) are formed from HCN by polymerisation reactions and not from single α-amino acids (see Chap. 5). The necessary preconditions for poly-condensation of amino acids—high temperatures, acidic conditions and the absence of water—were not present on primeval Earth.

Matthews presented his detailed poster "The HCN World: Formation of Protein-Nucleic Acid Life" at the ISSOL conference in1999 in San Diego, California (Matthews, 2000). He pointed out that mixtures of substances with colours from yellow to black are formed during HCN polymerisation. These could be the main components of the dark matter observed on planetoids, comets and some moons. Laboratory experiments showed the ready formation of polyaminomalonitrile (I), a compound which is constructed solely from HCN molecules. Other reactions of HCN (or other reactive molecules) with the activated nitrile groups of (I) give polyamidine (II, with sidechains R′), which are then converted stepwise to polypeptides (III, with side chains R).

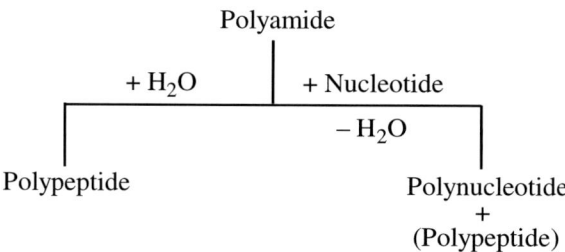

$$(4.3)$$

(I)   (II)   (III)

In other words, a synthetic route to polypeptides which requires only HCN and water is presented; preformed α-amino acids are not necessary. According to Matthews, the pyrolysis of cyanide polymers can give nitrogen-containing heterocycles with purine and pyrimidine structures: in other words, the "HCN world":

Polyamide

+ H_2O + Nucleotide

– H_2O

Polypeptide   Polynucleotide
                                                          +
                                                (Polypeptide)

Polyamides are converted to polypeptides (as just shown), and nucleotides, because of the dehydrating properties of polyamides, to polynucleotides. This hypothetical model can accept the "delivery" of HCN polymers from space (e.g., via meteorites) as well as photochemical reactions in a reducing atmosphere. In spite of some convincing experimental evidence, the "HCN world" favoured by Matthews still awaits further convincing experimental evidence (this is also true for other hypothetical "worlds").

Detailed studies were carried out at the University of Pennsylvania to cast more light on the obscurities of the complex HCN polymers: modern analytical methods were used, including thermochemolysis-GPC/MS. Two suggestions (Fig. 4.10) show possible reaction pathways leading from HCN to HCN polymers and other polymers (Minard et al., 1998).

Thus, there is a great deal of information on the reaction potential of HCN and products derived from it. Gaps in our knowledge may perhaps be closed in the next few years by research results on the chemistry occurring on other planets or in interstellar matter. As early as 1984, Jim Ferris published a review article "HCN and

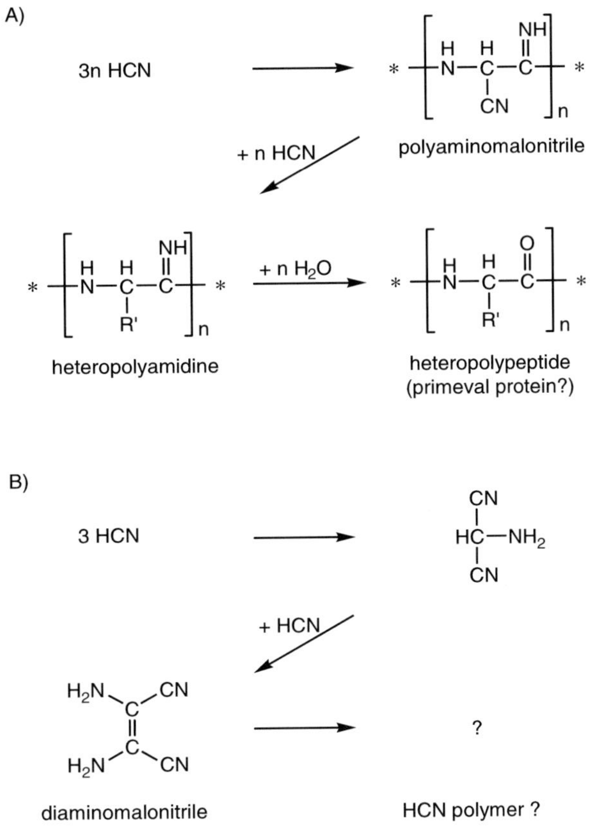

Fig. 4.10 Possible reaction pathways starting from HCN polymers **a** after Matthews and **b** after Ferris (Minard et al., 1998)

A)

3n HCN → polyaminomalonitrile

+ n HCN ↙

heteropolyamidine   + n H₂O →   heteropolypeptide (primeval protein?)

B)

3 HCN →

+ HCN ↙

diaminomalonitrile   →   ?   HCN polymer ?

Chemical Evolution" (Ferris and Hagan, 1984) in which he expressed the hope that the elucidation of the chemistry of HCN would make an extremely important contribution to the solution of the biogenesis problem.

4.6 Energy Sources for Chemical Evolution

The formation of biomolecules from simple molecules requires energy, which must be transferred to the reactants in the form of free or activation energy. This energy can come from various sources, such as the sun's rays or shock waves from planetoid impact; the heat from the Earth's interior, as well as radiation from radioisotopes, can also contribute energy for chemical synthesis. However, the contribution of the various energy sources to the total energy budget of the Earth varies greatly. Exact values for the young Earth are of course not available, although estimates based on realistic assumptions have been made.

Table 4.2 Energy available at the surface of the Earth

Type of energy	Available today (J/cm^2/year)	Assumed values forr primeval Earth (J/cm^2/year)
Total optical solar radiation	1,108,000	711,300
Solar radiation below 200 nm	314	126
Energy-rich radiation (from the Earth's crust to a depth of 35 km)	65	197
Heat from volcanoes	0.65	>0.65
Electrical discharges	17	17

4.6.1 Energy from the Earth's Interior and from Volcanoes

Volcanoes are complex geological systems which perhaps played an active role in biogenesis, although we do not know how.

Sidney Fox and Kaoru Harada, in particular, used simulation experiments to show how volcanism may have been involved in the synthesis of prebiotic molecules. They heated a stream of gas (CH_4, NH_3 and H_2O) to about 1,123 K (using a silicate contact); after cooling, they could detect glycine, alanine, β-alanine and aspartic acid (among others). This experiment was intended to simulate exhalation from the earth's crust, as in volcanoes (Fox and Harada, 1961; Harada and Fox, 1964).

At present there are about 500–600 volcanoes on Earth which are active (i.e., capable of eruption) (Press and Siever, 1994). About 100 of them are considered dangerous, as their eruptions threaten human life. The various undersea thermal sources, which are discussed in detail in Sect. 7.2, are also classed as volcanic.

How important are volcanoes for the processes which led to biogenesis?

Volcanic exhalations provide a great deal of the following starting materials for the synthesis of biomolecules: H_2O, CO_2, CO, CH_4, NH_3 and sulphur compounds.

Volcanoes set energy free in the form of heat and electrical discharges (lightning).

It is clear that the volcanic activity four billion years ago was greater than it is today.

Many catalytically active minerals are present in and around volcanoes.

The huge amounts of energy set free in volcanic eruptions have been estimated by L. M. Mukhin (1976). If we assume that an active volcano emits around $10^9 \, m^3$ of gas during an eruption (90% water vapour and 10% of other substances), then (assuming that the reactions of methane lead to organic compounds in around a 1% yield) around $10^6 \, kg$ of organic compounds should be formed. Even this very crude estimate indicates the importance of volcanic processes on the young Earth. In addition, impacts of planetoids 20–300 km in diameter would have set free huge amounts of matter and energy. It has been estimated that between 1500 and 1914, the volcanoes on Earth emitted around $64 \, km^3$ of lava and $329 \, km^3$ of ash and dust particles. This corresponds to around $1 \, km^3$ of lava and particulate matter per year.

The findings on processes occurring during volcanic eruptions led to simulation experiments, for example with a gas mixture containing water, CO_2, N_2 and NH_3 in a volume ratio of 4:1:1:0.1, under conditions expected for a volcanic ash–gas cloud; this work was reported by G. A. Lavrentier et al. (1984) from the Bach Institute of Biochemistry in Moscow. The gas mixture was subjected to electrical discharges at temperatures between 620 and 800 K, with ash powder from volcanic bombs as a catalyst. Volcanic bombs are globes of molten rock which are ejected from volcanoes during eruptions and take on various forms in flight. The material used by Lavrentier came from the stratovolcano Tolbatchik on the Kamchatka peninsula, which became active again in 1975 after lying dormant for 200 years. Small amounts of nucleobases as well as (in better yields) the synthesis of amino acids ($10^{-4}\%$ with respect to the total amount of gas–vapour mixture passing through the detector) were detected. Yields were clearly dependent on the presence of volcanic ash.

The question of the possible importance of volcanic ash–gas clouds on the primeval Earth for prebiotic chemistry was investigated by V. A. Basiuk (1996) and R. Navarro-Gonzales (1996). These authors assume that volcanic activity on the young Earth was very high, setting free large amounts of energy. They justify this assumption as follows:

The temperature of the Earth's mantle was higher.
The magma contained a higher proportion of volatile substances.
Eruptions were triggered by bombardment of the still young Earth's crust by heavenly bodies.

Large volumes of gas are ejected during volcanic eruptions. Hot, solid fragments expand to give huge ash clouds: the properties of such clouds (pressure, temperature, electric fields) vary greatly. In addition, many lightning flashes occur. Explosive eruptions can cause lightning discharges stretching for up to several hundred metres even today. Various types of lightning flashes are observed:

Flashes between the clouds
Flashes from the clouds to the Earth
Flashes from the Earth to the clouds
Discharges in the air

The maximum rate of lightning flashes observed for the eruptions of Surtsey (1963–1967) off the south coast of Iceland and of Mount Saint Helens (1980) was around 10–16 flashes per minute. The energy of the Surtsey flashes was estimated to be around 10^6 J: this value is lower by a factor of around 10^3 than that involved in typical lightning flashes occurring during thunderstorms (i.e., from a negatively charged cloud to the ground).

Could lightning flashes occurring during volcanic eruptions have been effective energy sources for prebiotic syntheses? An answer to this question would require not only reliable information on the exact chemical composition of the exhalation mixture, but also on the primeval Earth's atmosphere, since dilution effects occur at the boundary between the two gas mixtures.

The chemical composition of the volcanic exhalations was probably extremely variable, depending on the chemical nature and the physical parameters of the part of the Earth's mantle from which the gases emanated. Scientists assume that the gases set free by the Hawaiian volcanoes are similar to those of the primeval Earth, since new isotopic analyses of the noble gases found in volcanic gasses show that these exhalations derive from a primordial, non-degassed reservoir. The following values (in mol%) have been measured for the Hawaiian volcano Kilauea: H_2O = 52.30; CO_2 =30.87; SO_2 = 14.59; CO = 1.00; H_2 = 0.79; H_2S = 0.16, with traces of other compounds. The compositions of such gas mixtures can vary greatly, as Basiuk (1996) found from a survey of several publications; the CH_4 value lay between 0.000000285 and 0.49%, while for H_2, values between 0.00000084 and 4.63 mol% were reported. Here, however, we must take into account the extreme difficulty of obtaining gas samples, which is often only possible because of the reckless dedication of volcanologists.

Laboratory experiments designed to simulate processes occurring in erupting volcanoes were carried out by a Mexican research group (Navarra-Gonzales et al., 2000). They used a mixture containing 50% H_2O, 30% CO, 11% N_2, 4.5% CO and 4.8% H_2, similar to that found in Hawaiian volcanoes. The volcanic lightning flashes were simulated by using a dense, hot plasma, which was generated by passing the gas stream through a microwave discharge tube. The analysis showed that reactive nitrogen was present in the form of NO. It is known that reactive nitrogen is necessary for the synthesis of both amino acids and nucleobases, and estimates suggest that up to 5×10^9 kg NO could have been formed on the primeval Earth.

4.6.2 UV Energy from the Sun

The solar radiation incident on the atmosphere and the Earth's surface represents the largest external energy contribution. The optical irradiation at the upper boundary layers of the atmosphere is $4{,}435 \, kJ/cm^2/year$, of which around $1{,}108 \, kJ/cm^2/year$ reaches the Earth's surface.

The spectral distribution of this radiation is given in Table 4.3, from which we can easily see that radiation with wavelengths below 150 nm represents only a tiny fraction of the total. The energy distribution of the solar radiation corresponds to that from a black body with a temperature of around 5,000 K.

It must, however, be taken into account that the radiation flux has not always been constant during the 4.6 billion years since the formation of the sun. The "solar constant" is in fact not a constant at all, as it depends on the state of the sun's surface. For prebiotic syntheses, it is important to consider the wavelengths which can be absorbed by small molecules such as CO_2, CO, CH_4, N_2, NH_3, H_2O, H_2S etc. The premise here, of course, is that most of the synthetic reactions occurred in the gas phase.

Table 4.3 Spectral distribution of the optical solar radiation in the Earth's atmosphere. Data taken from the Smithsonian Physical Tables (1959)

Wavelength range in nm	% of total	$kJ \cdot cm^{-2} \cdot a^{-1}$
total radiation	100,0	4.435,040
below 150	0,001	0,042
150–200	0,03	1,255
200–250	0,2	8,368
250–300	1,0	41,840
300–350	3,1	138,072
350–400	5,4	238,488
400–700	37,0	1.631,760
700–1000	24,5	1.129,680
über 1.000	29,0	1.255,200

The wavelengths at which most of the components of a primitive Earth atmosphere absorb lie, with few exceptions, under 200 nm. The exceptions include ammonia (< 230 nm), hydrogen sulphide (<260 nm) and ozone (180–300 nm). However, ozone was probably present in the primeval atmosphere only in trace amounts, since free oxygen was only available in extremely low concentrations. The young Earth thus had no protective ozone layer, so short-wavelength UV irradiation could readily penetrate the atmosphere.

4.6.3 High-Energy Radiation

A large fraction of the high-energy radiation which is effectively available on Earth comes from the material forming the Earth's crust and mantle. Since the formation of the primeval Earth, four unstable isotopes have been important energy sources which radiate energy to the cooling planet: ^{40}K, ^{238}U, ^{235}U and ^{232}Th. The heavy elements uranium and thorium decay with emission of α-, β- and γ-radiation, while potassium only emits β-particles. The decomposition products and their half-lives are listed in Table 4.4.

The amount of energy set free in the processes shown in Table 4.4 is, for example, around 3 J/year for 1 g ^{238}U (in equilibrium with its daughter product). ^{235}U emits 18 J/g/year, ^{232}Th 0.8 J/g/year. The average amount of the three elements present in granite and volcanic rocks is shown in Table 4.5.

It is assumed that the average heat generation due to the radioactive processes described above was around 8×10^{-6} J/g of rock (Birch, 1954). If this value is extrapolated to the Earth's crust (to a depth of 35 km), the result is 53 J/cm²/year. It seems clear that the crust of the primeval Earth contained about four times as much ^{40}K as today. The higher half-life of ^{238}U means that the amount of this isotope present about four billion years ago was around twice today's value. The corresponding factor for ^{235}U, with a half-life of 7×10^{8} years, is 64. Thus, the value of

53 J/cm^2/year calculated for the Earth today is much lower than that available four billion years ago (197 J/cm^2/year). Most of this energy was converted to heat, and some simply radiated into space.

Table 4.4 The four unstable radioisotopes and their decomposition products

		half-life in years
^{40} K $\rightarrow$ ^{40}K $\xrightarrow{+1\,\text{Elektron}}$	^{40}Ca + 1 Elektron ^{40}Ar	$1,47 \cdot 10^9$ $11,8 \cdot 10^9$
^{238} U $\rightarrow$	^{206}Pb + 8 α-particles + 6 electrons	$4,468 \cdot 10^9$
^{235} U $\rightarrow$	^{207}Pb + 7 α-particles + 4 electrons	$0,7038 \cdot 10^9$
^{232} Th $\rightarrow$	^{208}Pb + 6 α-particles + 4 electrons	$14,008 \cdot 10^9$

Table 4.5 Amount of uranium, thorium and potassium in granite and basalt rock

Type of rock	U (ppm)	Th (ppm)	K (%)
Granite	4	14	3.5
Basalt	0.6	2	1.0

High-energy radiation from the sun and the cosmos also reaches Earth. It is clear that the young Earth was subjected to such radiation; however, no exact information is available on the radiation intensity four billion years ago, and thus on the possible consequences for chemical evolution; we rely on estimates.

4.6.4 Electrical Discharges

Electrical discharges were and are an effective source of energy for simulation experiments involving gaseous starting materials. In the natural atmosphere, however, electrical discharges are not an original energy source, but represent transformed solar energy. Physically speaking, electrical discharges consist of slow electrons which are active at high temperatures in a strictly limited space and which cover a wide range of the optical spectrum. The electrons formed in discharge reactions are slower than the β-radiation set free in the decomposition of ^{40}K, but they have enough kinetic energy to cause ionisation and activation. If a large amount of the motional energy of the ions and secondary electrons is converted into heat via collisions of the second kind, there can be strong molecular heat motion in limited areas of the reaction volume. Thus "hot" molecules can be converted into free radicals and electronically excited molecules.

The many successful simulation experiments indicate that electrical discharges on prebiotic Earth may have been involved in the synthesis of biomolecules. However, we simply do not know how important they actually were! Studies of energy sources for the formation of prebiotic organic molecules (Chyba and Sagan, 1991) show how greatly estimates for objects under study can differ and how they need revision over time. The authors suggest that the values assumed by Miller and Urey in 1959 for lightning and corona discharges were too high by a factor between 20 and 100!

Table 4.6 Estimated values for the amounts of energy which were available for prebiotic syntheses on the young Earth (Chyba and Sagan, 1991); comparison of Miller, Chyba

Reference	Lightning	Corona discharges
Miller and Urey (1959)	2×10^{19} J/year	6×10^{19} J/year
Chyba and Sagan (1991)	10^{18} J/year	5×10^{17} J/year

Thus, the estimates for the rates of the prebiotic synthesis of biomolecules also need a drastic downward correction; the hypothetical "primeval soup" then becomes much thinner!

4.6.5 Shock Waves

Shock waves can be natural in origin, for example, lightning, volcanic explosions or meteorite impacts. But human activity can also generate shock waves, for example in chemical or nuclear explosions. On the positive side, shock waves can reduce human suffering when used in shock wave lithotripsy (a method for fragmentation of kidney stones and gallstones).

Physically speaking, shock waves are compaction waves with a vertical shock front, which occur in supersonic fluxes or as described above; the pressure reaches a maximum value and then falls rapidly towards zero. Shock waves can also occur in space, which is almost free of matter, via interactions of electrical and magnetic fields (Sagdejev and Kennel, 1991).

More than 40 years ago, Hochstim (1963) showed that organic products are formed in simulation experiments using shock waves. However, no information on yields is available. Using shock wave heating of a reducing gas mixture, Bar-Nun et al. (1970) were able to obtain relatively high yields of amino acids. However, when the work was later repeated, it turned out that the authors had been too optimistic; the yields were lower by a factor of 30 (Bar-Nun and Shaviv, 1975).

According to I. I. Glass from Toronto University (1977), shock waves are more efficient in the synthesis of building block molecules than UV irradiation (by a factor of 10^3–10^9) or lightning discharges (by a factor of 10^3). McKay and Borucki (1997) carried out experiments with shock waves, produced by high energy lasers, to simulate meteoritic impacts. They could show that the effectiveness of this form of energy in the synthesis of organic compounds depended more on the molecular than on the elemental composition of the gas used. Thus the main products using methane-rich gas mixtures were HCN and C_2H_2, with yields of 5×10^{17} molecules per joule. Repeated treatment of gas mixtures with shock waves led to the formation of amino functionalities, so that the optimistic assumption was made that amino acids could in fact also be formed. However, CO_2-rich gas mixtures gave no trace of organic compounds.

The question of the stability of the biomolecules is a vital one. Could they really have survived the tremendous energies which would have been set free (in the form of shock waves and/or heat) on the impact of a meteorite? Blank et al. (2000) developed a special technique to try and answer this question. They used an 80-mm cannon to produce the shock waves; the "shocked" solution contained the two amino acids lysine and norvaline, which had been found in the Murchison meteorite. Small amounts of the amino acids survived the "bombardment", lysine seeming to be a little more robust. In other experiments, the amino acids aminobutyric acid, proline and phenylalanine were subjected to shock waves: the first of the three was most stable, the last the most reactive. The products included amino acid dimers as well as cyclic diketopiperazine. The kinetic behaviour of the amino acids differs: pressure seems to have a greater effect on the reaction pathway than temperature. As had been recognized earlier, the effect of pressure would have slowed down certain decomposition reactions, such as pyrolysis and decarboxylation (Blank et al., 2001).

More recent experiments using even higher shock wave pressures, up to around 40 GPa (produced by a hyper-velocity impact gun) show that the extremely short periods of time (only a few microseconds) for which the pressure is applied have a lower decomposing effect on the amino acids in aqueous solution, and in ice, than had been expected. The exact analysis of the products showed that small amounts of simple peptides were also formed. These results point to the complexity of questions on biogenesis problems. Even cannon can help us in our attempts to reveal the secret of biogenesis!

4.7 The Role of the Phosphates

4.7.1 General Considerations

There is no doubt that the element phosphorus occupies a special position in the family of the elements. The Earth's crust (including the oceans), which is about 16 km thick, contains only about 0.04% phosphorus, compared, for example, with 2.4% potassium; however, phosphorus is present in all the substances necessary for living processes. It does not occur in elemental form because of its high affinity for oxygen and has been known since its discovery by the alchemist Henning Brand in Hamburg in the course of his search for the philosopher's stone. However, it was many years later that Antoine Lavoisier realized that this new, "shining" substance was a chemical element.

The enormous importance of phosphorus for the living cell was not realized until the twentieth century. Phosphorus compounds are active:

In transport processes and information conservation
In energy conversion and transfer
In membrane structures and
In signal transmission.

Phosphorus accounts for 2–4% of the dry weight of living cells. The phosphorus content of the environment can be the life-conserving or life-limiting factor (Karl, 2000).

F. H. Westheimer (1987) has provided a detailed survey of the multifarious ways in which phosphorus derivatives function in living systems (Table 4.7). The particular importance of phosphorus becomes clear when we remember that the daily turnover of adenosine triphosphate (ATP) in the metabolic processes of each human being amounts to several kilograms! Phosphate residues bond two nucleotides or deoxynucleotides in the form of a diester, thus making possible the formation of RNA and DNA; the phosphate always contains an ionic moiety, the negative charge of which stabilizes the diester towards hydrolysis and prevents transfer of these molecules across the lipid membrane.

The central importance of phosphorus and its derivatives in our world leads to the question as to "where" the element came from: phosphorus compounds must have been present on the Earth after its formation, but we do not know their source. Were they already present in planetesimals, or was phosphorus brought to Earth from space (in an elemental form or as compounds)?

Table 4.7 Some important examples of the dominant role played by phosphorus in biochemistry. With permission of F. H. Westheimer (1987)

Phosphates	Acid derivatives
DNA and RNA	Diesters of phosphoric acid
Adenosine triphosphate (ATP)	Phosphoric anhydride
Creatine phosphate	Amide of phosphoric acid
Phosphoenol pyruvate	Phosphoric acid enol ester
Pyridoxal phosphate	Phenol ester of phosphoric acid
Nicotinamide adenine dinucleotide (NAD)	Ester and anhydride of phosphoric acid
Fructose 1,6-diphosphate	Ester of phosphoric acid
Glucose-6-phosphate	Ester of phosphoric acid
Isopentenyl pyrophosphate	Ester of diphosphoric acid
Ribose-6-phosphate-1-pyrophosphate	Ester of phosphoric acid and diphosphoric acid

The phosphorus chemistry occurring in interstellar matter and in the circumstellar regions of the cosmos is not yet understood. We do, however, know that phosphorus compounds are present in meteorites, lunar rocks and Mars meteorites. Oddly enough, the element can be detected nearly everywhere, though only in low concentrations. Phosphate minerals, as well as the anions PO_2^- and PO_3^-, have been found in interstellar dust particles. The presence of alkylphosphonic acids in meteorites shows that outer space contains organic phosphorus derivatives.

According to Maciá et al. (1997), phosphorus-containing minerals are present in the Earth's crust in volcanic rock as well as in metamorphic and sedimentary rocks; their total weight is probably about 8×10^{14} tons, of which perhaps 10% could have been brought to the Earth by meteorites or comets. Accoring to Schwartz (1997), we can assume that an atmosphere which was outgassed at an early stage, or an atmosphere which consisted mainly of volatile materials from impacts, could have contained phosphorus in the form of gaseous compounds, e.g., as phosphine (PH_3), or even initially as elementary phosphorus (P_3 or P_4).

4.7.2 Condensed Phosphates

E. Thilo (1959) describes condensed phosphates as "compounds which contain a certain number of phosphorus atoms, linked via oxygen".

As early as 1816, Berzelius recognized that phosphoric acid was tribasic. Some years later, T. Clark observed that water was eliminated on heating Na_2HPO_4 and $AgNO_3$, then giving a precipitate with the molecular formula $Ag_4P_2O_7$. In 1833 T. Graham heated NaH_2PO_4, obtaining a glassy substance which is now known as "Graham's salt", a high molecular weight polyphosphate.

In Justus von Liebig's laboratory, Fleitmann and Henneberg found that there were several "metaphosphates" with different properties, all of which had the formula $MePO_4$ (Me = monovalent metal ion). The story then becomes complicated, and mistakes in nomenclature led to ambiguities in this class of compounds. As well as "Graham's salt" there is a "Maddrell's" and a "Kurol's salt". All have the same formula, $Na_nH_2P_nO_{3n+1}$, but differ in structure and chain length.

Since no efficient reactions for the synthesis of high-energy phosphates under "prebiotic conditions" had been found, Keefe and Miller (1995) came to a negative, pessimistic conclusion with respect to previous results in this sector of prebiotic chemistry. Some studies on the problem have in fact shown positive results.

However, there are also biogenesis models which do not require phosphate, such as the inorganic hypothesis of the origin of life proposed by Cairns-Smith (see Sect. 7.1), the "thioester world" proposed by de Duve (see Sect. 7.4) or the "sulphur–iron world" suggested by Wächtershäuser (see Sect. 7.3). The "RNA world" (see Chap. 6), however, cannot exist without phosphate.

4.7.3 Experiments on the "Phosphate Problem"

As early as five years after the Miller–Urey experiments, Schramm and Wissmann from the Max Planck Institute for Virus Research in Tübingen reported a successful synthesis of polypeptides using polyphosphate esters. Thus, they were able to prepare the tripeptide alanylglycylglycine (Ala-Gly-Gly), which in turn could undergo polycondensation to give peptides with up to 24 amino acid residues.

During a search for high-energy phosphates, Ponnamperuma was able to synthesize adenosine triphosphate (ATP) from ADP, AMP or adenosine from ethyl metaphosphate in dilute aqueous solution under the influence of UV light. The role of the UV irradiation is unclear, as the phosphate itself is a high-energy species (Ponnamperuma et al., 1963).

Miller and Parris (1964) pointed out that the preparation of ethyl metaphosphate from P_2O_5, diethyl ether and chloroform could hardly be considered "prebiotic"; this criticism is certainly irrefutable!

The search for phosphorylation agents continued in several research groups, on the basis that only those compounds should be studied for which it seemed likely that their formation would have been possible on the young Earth. However, even today it is not clear in which form the element phosphorus was available for the synthesis of biomolecules about four billion years ago. The main source of phosphorus, the mineral apatite $[Ca_5(PO_4)_3F$ or $Ca_5(F, Cl, OH, 1/2 CO_3)(PO_4)_3]$, is only very slightly water soluble. Thus, in the years when the idea of a strongly reducing primeval atmosphere prevailed, it was hypothesized that heating ammonia-containing minerals such as sterkovite ($NH_4NaHPO_4 \cdot 4 H_2O$) or struvite ($MgNH_4PO_4 \cdot 6 H_2O$) could have given condensed phosphates, and thus phosphorylating agents.

Handschuh and Orgel (1973) studied the mineral struvite. It can be precipitated from ocean water in the presence of phosphate if the concentration of NH_4^+ ions in the water is greater than 0.01 M. If struvite is heated with urea, magnesium pyrophosphate is obtained in a yield of about 20% after 10 days at 338 K; if nucleosides are added to the reaction mixture described above, nucleoside diphosphates such as uridine-5'-diphosphate and diuridine-5'-diphosphate are formed in good yields.

Two years earlier, Lohrmann and Orgel (1971) had heated a mixture of $Ca(H_2PO_4)_2$, urea and NH_4Cl at 338–373 K; on addition of nucleosides, they obtained a series of phosphorylated nucleosides in good yields. However, interest in the "phosphate problem" diminished considerably in the following few years.

The "RNA world" hypothesis then rekindled interest in the still open question of the origin of the phosphorus. How could acceptable synthetic routes to the nucleic acids be developed, if it was not even clear which phosphate derivatives were available on the primeval Earth? An RNA world is not possible without reactive phosphates or similarly efficient phosphorus compounds! Thus, the search for effective phosphorus or phosphate sources continued unabated.

The mineral schreibersite $(FeNi)_3P$ is present in meteorites, and in particular in iron meteorites. Polyphosphate minerals have not been found on Earth, except for a few kilograms of a calcium diphosphate-containing mineral in New Jersey (Keefe and Miller, 1995).

The report by Yamagata et al. (1991) that acid basalts, i.e., those with a high SiO_2 content, can set free P_4O_{10} from their apatite component if heated to 1,470 K, appears important. They found concentrations of around 5 μM pyrophosphate and tripolyphosphate in a fumarole in the neighbourhood of the strato-volcano Uzo on the Japanese north island of Hokkaido. The reaction occurring in the basalt rocks could be:

$$4H_3PO_4 \quad \rightarrow P_4O_{10} + 6H_2O \qquad (4.4)$$

Analysis of the Murchison meteorite led to a completely different type of phosphorus compound: the only phosphorus-containing compounds found were alkanephosphonic acids. Spurred on by these results, de Graaf et al. (1995) irradiated mixtures of o-phosphorous acid in the presence of formaldehyde, primary alcohols or acetone with UV light (low pressure Hg lamp, 254 nm with a 185-nm component) and obtained phosphonic acids, including hydroxymethyl and hydroxyethyl phosphonic acids, which had been found in the Murchison meteorite. Alkanephosphonic acids can be derived from phosphorous acid, with a P–H bond being replaced by a P–C bond.

Phosphonic Acid   Alkyl Phosphonic Acid

An interesting reaction was studied by A. Schwartz (1997): UV irradiation of phosphate solutions containing acetylene gave vinylphosphonic acid.

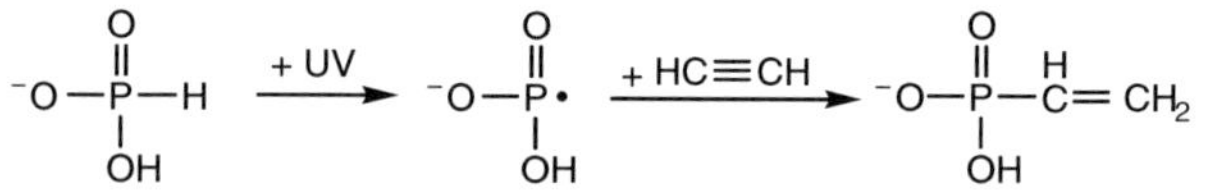

Fig. 4.11 The assumed mechanism of the synthesis of vinylphosphonic acid via addition of the phosphite radical to acetylene (de Graaf et al., 1997)

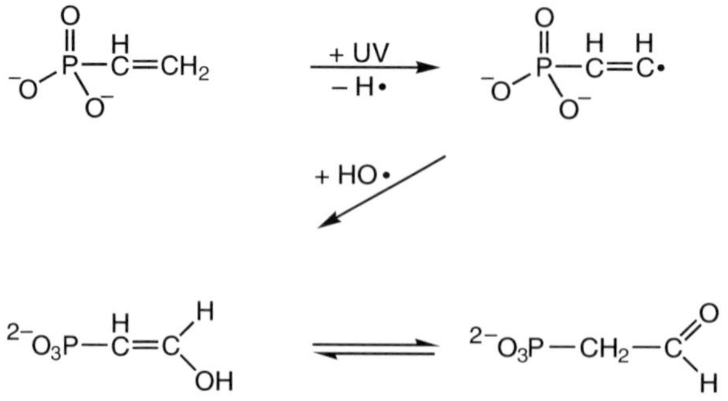

Fig. 4.12 The formation of phosphoaldehyde via recombination of the radical derived from vinylphosphonic acid with a hydroxyl radical (de Graaf et al., 1997)

What importance could vinylphosphonic acid have for the synthesis of important biomolecules? Its photolysis gives many oxidized products, including phosphoacetaldehyde. This analogue of glycol aldehyde phosphate seems to be of interest; its formation involves the recombination of hydroxyl radicals with vinylphosphonic acid radicals.

According to Müller (1990) this aldehyde can give ribose-2,4-diphosphate in the presence of formaldehyde via a two-step, base-catalysed reaction. This reaction provides a route to ribose derivatives, and thus to the nucleic acids.

A different phosphorylation method was chosen by Kolb and Orgel (1996). They reacted glyceric acid with trimetaphosphate in alkaline solution and obtained 2- and 3-phosphoglyceric acid (Fig. 4.13); the total yield of these two compounds was as high as 40%. Briggs et al. (1992) demonstrated the formation of glyceric acid in good yield in photolysis of a $CO:H_2O:NH_3$ mixture (5:5:1) at 10 K; other C_2 and C_3 compounds were also formed (the experiment was planned as a simulation of the conditions present during the formation of comets).

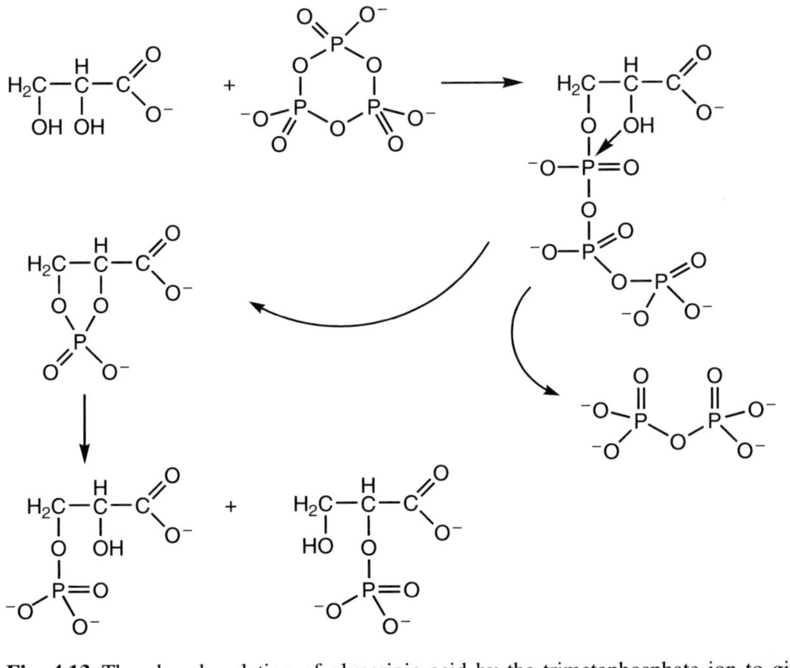

Fig. 4.13 The phosphorylation of glycerinic acid by the trimetaphosphate ion to give 3- or 2-phosphoglycerinic acid. After Kolb and Orgel (1996)

In recent studies, Glindemann et al. (1999, 2000) attempted to obtain definite answers to one of the basic problems of prebiotic phosphorus chemistry: were there realistic enrichment mechanisms for the phosphorus derivatives required in prebiotic syntheses? Only if the phosphorylating agents were present in concentrated form could successful reactions have taken place.

It is probable that the prevalent oxidation states of phosphorus on the young Earth were lower than they are today, so calcium salts with a much better solubility than that of apatite could have been formed. As Glindemann et al. (1999) were able to show in model experiments, up to 11% of the starting material could be converted to phosphite in CH_4/N_2 atmospheres (10% CH_4) using Na_2HPO_4, hydroxyapatite or fluoroapatite sources. Similar processes cannot be excluded for the primeval Earth, for example, under the influence of electrical discharges.

The gas mixtures chosen by Glindemann do not correspond to the present notions of a neutral or weakly reducing atmosphere. Thus de Graaf and Schwartz (2000) used gas mixtures with CO_2 and N_2 as the main components, with a little CO and H_2 (reducing gases) being added. Good phosphite yields were obtained using electrical discharges; the consumption of the apatite depends clearly on the $H_2 + CO$ content of the gas mixture: in the range of 1–5%, the reaction yield is proportional to the CO/H_2 content, i.e., 5% phosphite is obtained with 5% $CO + H_2$.

Fig. 4.14 The conversion of apatite to phosphate under the influence of electrical discharges in a model atmosphere containing 60% CO_2, 22–40% N_2, and various concentrations of CO and H_2. The "reductive ability", i.e., the amount of H_2 and CO (in equal amounts) present, is given on the horizontal axis (de Graaf and Schwartz, 2000)

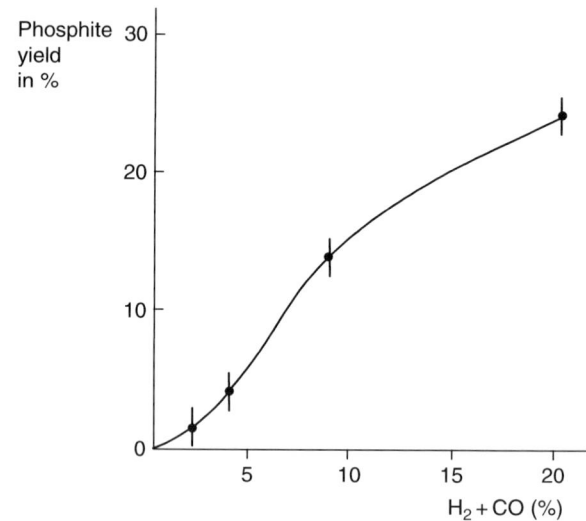

Critics of such experiments may find the concentration of reducing gases too high. It is, however, possible that there were localized areas on Earth where conditions were more strongly reducing for short periods (e.g., after volcanic eruptions). In the search for potential prebiotic syntheses of condensed phosphates, Keefe and Miller (1996) allowed a series of condensation agents to act on o-phosphate or tripolyphosphate, and determined the yields of diphosphate and trimetaphosphate obtained.

The most effective agents for diphosphate formation were $NH_4^+HCOO^-$, H_2O_2 and SCN^-. The formation of the cyclic triphosphate was favoured by maleic anhydride or pantoyl lactone, with Ca^{2+} being added in both cases.

The phosphorylation of adenosine with trimetaphosphate under mild conditions (pH about 7, 413 K) was described by Yamagata (1995). The main product was cyclic $2',3'$-AMP; a little ATP was also formed. The addition of Mg^{2+} to the aqueous reaction mixture is vital. A condensation reaction of diglycine to give Gly_4 and Gly_6 was also reported by Yamagata, again using trimetaphosphate and Mg^{2+}.

Pasek and Lauretta (University of Arizona, Tucson) point to a further important source of phosphate: they studied the corrosion of phosphide minerals from iron meteorites under various conditions in aqueous solution, e.g., using $NaHCO_3$ as a buffer, or in solutions containing $MgCl_2$ and $CaCl_2$, as well as in the presence of ethanol and acetic acid.

In all the experiments, the main decomposition products were phosphonates, which are also stable in concentrated solutions of Mg and Ca chlorides. In some experiments, pyrophosphate, and in smaller amounts triphosphate, could also be detected. The authors thus assume that the primeval ocean contained phosphonates as a source of phosphorus for reactions leading to biochemically relevant molecules. Iron meteorites could have delivered sufficient reduced phosphorus (Fe_3P) to the primeval Earth, so the question of prebiotic phosphorus chemistry should be looked at in more detail in the future (Pasek and Lauretta, 2005).

W. J. Hagan Jr. et al. have reported the cyanate-supported synthesis of pyrophosphate from $CaHPO_4 \cdot 2\,H_2O$. Relatively low concentrations (5–20 mM) of NaOCN were required. The experiments show that the solubility of $CaHPO_4 \cdot 2\,H_2O$ is dependent on the pH value of the solution and the total concentration of Ca^{2+}; NaCl, however, has only a little influence (Hagan et al., 2007).

A. W. Schwartz gave a detailed account of the phosphate problem with the title "Phosphorus in Prebiotic Chemistry" in February 2006 (Schwartz, 2006).

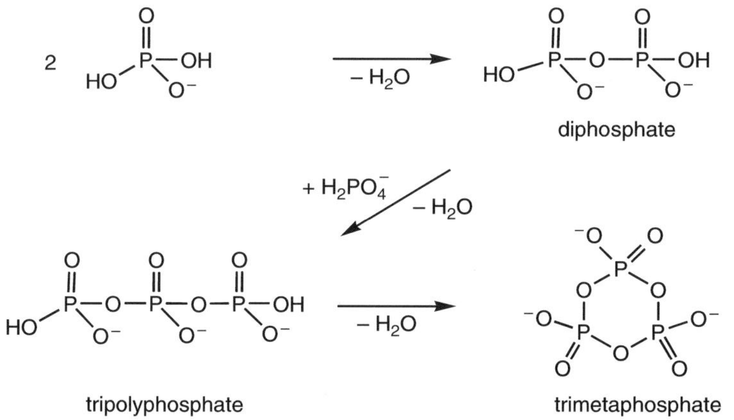

Fig. 4.15 The formation of trimetaphosphate

The often posed question as to whether phosphorus compounds were active in the initial phases of the development of life, or intervened only at a later stage in the evolutionary process, remains unanswered. However, de Duve (1991) was completely right when he stated that "life is basically organized around phosphorus".

References

Abelson PH (1956) Science 124:935
Bada JL, Lazcano A (2003) Science 300:745
Bar-Nun A, Bar-Nun N, Bauer SH, Sagan C (1970) Science 168:470
Bar-Nun A, Shaviv A (1975) Icarus 24:197
Basiuk VA, Navarro-González R (1996) Orig Life Evol Biosphere 26:173
Bauer H (1980) Naturwissenschaften 67:1
Birch F (1954) In: Paul H (Ed.) Nuclear Geology. Wiley, New York
Blank JG, Miller GH, Winans RE (2000) Orig Life Evol Biosphere 30:231
Blank JG, Miller GH, Ahrens MJ, Winans RE (2001) Orig Life Evol Biosphere 31:15
Blank JG (2002) 10. ISSOL-Konferenz, Abstr p 45
Borquez E, Cleaves HJ, Lazcano A, Miller SL (2005) Orig Life Evol Biosphere 35:79
Breuer AD (1974) Angew Chem 86:401, Int Ed 13:370
Briggs R, Ertem G, Ferris JP, Greenberg JM, Mc Cain PJ, Menxosa-Gomez CX, Schutte W (1992)
    Orig Life Evol Biosphere 22:287
Chyba C, Sagan C (1991) Orig Life Evol Biosphere 21:3
Cleaves II HJ, Nelson KE, Miller SL (2006) Naturwissenschaften 93:228
Cohn CA, Hansson TK, Larsson HS, Sowerby SJ, Holm NG, Arrhenius G (2002) Orig Life Evol
    Biosphere 32:412
Cooper G (2001) Nature 414:879
de Duve C (1991) Blueprint for a Cell: The Nature and Origin of Life. Patterson, New York
de Graaf RM, Visscher J, Schwartz AW (1995) Nature 378:474
de Graaf RM, Visscher J, Schwartz AW (1997) J Mol Evol 44:237
de Graaf RM, Schwartz AW (2000) Orig Life Evol Biosphere 30:405
Encyclopedia Americana (1975) 24:343
Ferris JP, Donner DB, Lobo AP (1973) J Mol Biol 74:499
Ferris JP, Wos JD, Nooner DW, Oró J (1974) J Mol Evol 3:225
Ferris JP, Edelson EH, Auyeung JM, Joshi PC (1981) J Mol Evol 17:69
Ferris JP, Hagan WJ Jr (1984) Tetrahedron 40:1093
Fox SW, Harada K (1961) Science 133:1923
Gabel NW, Ponnamperuma C (1967) Nature 216:453
Glaser R, Hodgen B, Farrelly D, McKee E (2007) Astrobiology 7:455
Glass II (1977) Prog Aerospace Sci 17:269
Glindemann D, de Graaf RM, Schwartz AW (1999) Orig Life Evol Biosphere 29:555
Glindemann D, de Graaf RM, Schwartz AW (2000) Orig Life Evol Biosphere 30:121
Handschuh GJ, Orgel LE (1973) Science 179:483
Harada K, Fox SW (1964) Nature 201:335
Hochstim AR (1963) Proc Natl Acad Sci USA 50:200
Karl DM (2000) Nature 406:31
Kawamura K, Maeda J (2007) Orig Life Evol Biosphere 37:153
Keefe AD, Miller SL (1995) J Mol Evol 41:693
Keefe AD, Miller SL (1996) Orig Life Evol Biosphere 26:15
Kliss RM, Matthews, CN (1962) Proc Natl Acad Sci USA 48:1300
Kobayashi K, Saito T, Oshima T (1998) Orig Life Evol Biosphere 28:155
Kolb K, Orgel LE (1996) Orig Life Evol Biosphere 26:7
Kotake M, Nakagaba M, Ohara T, Harada K, Ninomia M (1956) J Chem Soc Japan, Ind Chem Sect
    58:121,151
Krafft F (1969) Angew Chem 81:634
Krishanamurthy R, Arrhenius G, Eschenmoser A (1999) Orig Life Evol Biosphere 29:333
Krishanamurthy R, Arrhenius G, Eschenmoser A (2000) Orig Life Evol Biosphere 30:195
Langenbeck W (1954) Angew Chem 66:151
Larralde R, Robertson MP, Miller SL(1995) Proc Natl Acad Sci USA 92:8158

Larralde R, Robertson MP, Miller SL (1996) Orig Life Evol Biosphere 26:355
Lavrentiev GA, Strigunkova TF, Egorov IA (1984) Orig Life Evol Biosphere 14:205
Levy M, Miller SL (1998) Proc Natl Acad Sci USA 95:7933
Levy M, Miller SL, Oró J (1999) J Mol Evol 49:165
Lohrmann R, Orgel LE (1971) Science 171:490
Löwe CJ, Rees MW, Markham RM (1963) Nature 199:219
Maciá E, Hernández MV, Oró J (1997) Orig Life Evol Biosphere 27:459
Matthews CN (2000) Orig Life Evol Biosphere 30:292
McKay CP, Borucki WJ (1997) Science 276:390
Miller SL (1953) Science 117:528
Miller SL (1955) J Amer Chem Soc 77:2351
Miller SL, Urey HC (1959) Science 130:245
Miller SL, Parris M (1964) Nature 204:1248
Minard RD, Hatcher PG, Gourley RC, Matthews CN (1998) Orig Life Evol Biosphere 28:461
Morowitz H, Petersen E, Chang S (1995) Orig Life Evol Biosphere 25:395
Moser RE, Claggett AR, Matthews CN (1968) Tetrahedron Letters 13:1605
Mosqueira FG, Albarrau G, Negrón-Mendoza A (1996) Orig Life Evol Biosphere 26:75
Müller D, Pitsh S, Kittaka A, Wagner E, Wintner CE, Eschenmoser A (1990) Helv Chim Acta 73:1410
Mukhin LM (1976) Orig Life Evol Biosphere 7:355
Navarro-González R, Basiuk VA (1996) Orig Life Evol Biosphere 26:223
Navarro-González R, Molina MJ, Molina LT (2000) Orig Life Evol Biosphere 30:183
Nelson KE, Robertson MP, Levy M, Miller SL (2001) Orig Life Evol Biosphere 31:221
Orgel LE (2004) Orig Life Evol Biosphere 34:361
Oró J (1960) Biophys Biochem Res Commun 2:407
Oró J (1961) Nature 191:1193
Oró J, Kamat SS (1961) Nature 190:442
Oró J, Kimball AP (1962) Arch Biochem Biophys 96:293
Piccirilli JA, Krauch T, Moroney SE, Benner SA (1990) Nature 343:33
Plankensteiner K, Reiner H, Schranz B, Rode MB (2004) Angew Chem 116:1922, Int Ed 42:1886
Ponamperuma C, Sagan C, Mariner R (1963) Nature 199:222
Press F, Siver R (1994) Understanding Earth, Freeman and Company, New York
Pullman B (1972) Electronic Factors in biochemical evolution. In: Ponnamperuna C (Ed.) Exobiology. North Holland Publishing Company, Amsterdam London, p 140
Raulin F (2000) Orig Life Evol Biosphere 30:116
Reid C, Orgel (1967) Nature 216:216
Ricardo A, Carrigan MA, Olcott AN, Benner SA (2004) Science 303:196
Robertson MP, Miller SL (1995) Nature 375:772
Robertson MP, Levy M, Miller SL (1996) J Mol Evol 43:543
Ruiz-Bermejo M, Menor-Salván C, Osuna-Esteban S, Veintemillas-Verdague S (2007) Orig Life Evol Biosphere 37:123
Sagdejew RZ, Kennel CF (1991) Scientific American, April p 40
Sanchez RA, Ferris JP, Orgel LE (1966a) Science 153:72
Sanchez RA, Ferris JP, Orgel LE (1966b) Science 154:784
Sanchez RA, Ferris JP, Orgel LE (1967) J Mol Biol 30:223
Sanchez RA, Ferris JP, Orgel LE (1968) J Mol Biol 38:121
Schlesinger G, Miller SL (1983) J Mol Biol 19:376
Schramm G, Wissmann H (1958) Chem Ber 91:1073
Schwartz AW, Goverde M (1982) J Mol Evol 18:351
Schwartz AW, Voet AB, Van der Veen M (1984) Orig Life Evol Biosphere 14:91
Schwartz AW (1997) Orig Life Evol Biosphere 27:505
Schwartz AW (2006) Phil Trans Roy Soc B 361:1743
Scorei R, Cimpoiaşu VM (2006) Sephton MA (2001) Orig Life Evol Biosphere Nature 36:1
Senanayake SD, Idriss H (2006) Proc Natl Acad Sci USA 103:1194

Shapiro R (1984) Orig Life Evol Biosphere 14:565
Shapiro R (1986) Origins: A Skeptic's Guide to the Creation of Life on Earth. Summit Books, New York
Shapiro R (1988) Orig Life Evol Biosphere 18:71
Shapiro R (1995) Orig Life Evol Biosphere 25:83
Shapiro R (1999) Planetary Dreams. Wiley & Sons, New York
Shapiro R (1999) Proc Natl Acad Sci USA 96:4396
Strasdeit H, Büsching I, Behrends S, Saak W, Barklage W (2001) Chem Eur J 7:1133
Strecker A (1850) Liebigs Ann Chem 75:27
Strecker A (1854) Liebigs Ann Chem 91:349
Thilo E (1959) Naturwissenschaften 46:367
Völker T (1960) Angew Chem 72:379
Voet AB, Schwartz AW (1983) Bioinorganic Chem 12:8
Wächtershäuser G (1988) Proc Natl Acad Sci USA 85:1134
Weber AL, Pizzarello S (2006) Proc Natl Acad Sci USA 103:12713
Weber AL (2007) Orig Life Evol Biosphere 37:105
Westheimer FH (1987) Science 335:1173
Yamagata Y, Watanabe H, Saltoh M, Namba T (1991) Nature 352:516
Yamagata Y (1995) Orig Life Evol Biosphere 25:47
Yamagata Y (1997) Orig Life Evol Biosphere 27:339
Zubay G (1998) Orig Life Evol Biosphere 28:13

Chapter 5
Peptides and Proteins: the "Protein World"

5.1 Basic Considerations

In today's discussion of the origin of life, the "RNA World" (Chapter 6) is seen as much more important, and is much better publicized, than the "protein world". However, nucleic acids and proteins are of equal importance for the vital metabolic functions in today's life forms. Peptides and proteins are constructed from the same building blocks (monomers), the aminocarboxylic acids (generally known simply as amino acids). The way in which the monomers are linked, the peptide bond, is the same in peptides and proteins. While peptides consist of only a few amino acids (or to be more exact, amino acid residues), proteins can contain many hundreds. The term "protein" (after the Greek *proteuein*, to be the first) was coined by Berzelius in 1838.

5.2 Amino Acids and the Peptide Bond

Hundreds of different natural amino acids have been found and isolated. In some cases, they are quite complex and have a variety of functions. The preparation and isolation of amino acids occurs either from biological material or via chemical synthesis. Several amino acids, such as glutamic acid and methionine, are now prepared on a scale of 100,000 tons per year.

The proteinogenic amino acids are distinguished from all others by two characteristic features. They are:

- α-Amino acids, i.e., the amino group is bonded to the α-C atom of the molecule
- L-amino acids (except for glycine)

$$R-\underset{\underset{NH_2}{|}}{\overset{\overset{H}{|}}{C}}-C\underset{O}{\overset{OH}{\diagup}}$$

The properties of the amino acid are determined by the residue R, which may, for example, confer hydrophilic or hydrophobic character.

H. Rauchfuss, *Chemical Evolution and the Origin of Life*,
© Springer-Verlag Berlin Heidelberg 2008

The linking up of amino acids can be described as a polycondensation reaction involving elimination of water; the result is the formation of peptide bonds.

$$\underset{H}{\overset{O}{\underset{|}{\overset{||}{C}}}}\diagdown N \diagup \quad \longleftrightarrow \quad \underset{H}{\overset{O^-}{\underset{|}{\overset{|}{C}}}}\diagdown \overset{+}{N}\diagup \tag{5.1}$$

The peptide bond is characterized by a fixed planar structure, as was discovered by X-ray crystallography of peptides more than 60 years ago. The arrangement of the atoms in the peptide bond is due to resonance stabilisation: the lowest-energy state of the system is that in which the four atoms forming the peptide linkage lie in a plane, while the C–N bond has partial double bond character.

$$H_3\overset{+}{N}-\underset{R_1}{\overset{H}{\underset{|}{C}}}-COO^- \quad + \quad H_3\overset{+}{N}-\underset{R_2}{\overset{H}{\underset{|}{C}}}-COO^-$$

$$\overset{-H_2O}{\underset{+H_2O}{\rightleftarrows}}$$

$$H_3\overset{+}{N}-\underset{\underset{R_1}{|}}{\overset{H}{\underset{|}{C}}}-\underset{O}{\overset{||}{C}}-\underset{}{\overset{H}{\underset{|}{N}}}-\underset{R_2}{\overset{H}{\underset{|}{C}}}-COO^- \tag{5.2}$$

The formation of a dipeptide from two amino acids via elimination of water (as shown above) can only take place when energy is removed from the system; thus, the starting materials must be converted to a reactive state. The principle is the same for the construction of tri- or tetrapeptides, as well as for the long amino acid chains in proteins. In a 1 M solution of two amino acids at 293 K and a pH value of 7, only about 0.1% exists as the dipeptide, i.e., the equilibrium shown in Eq. 5.2 lies on the side of the free amino acids. The formation of a dipeptide requires more energy than chain lengthening to give higher peptides.

If prebiotic peptides and/or proteins were in fact initially formed in aqueous solution (the hypothesis of biogenesis in the "primeval ocean"), the energy problems referred to above would have needed to be solved in order for peptide synthesis to occur. As discussed in Sect. 5.3, there is some initial experimental evidence indicating that the formation of peptide bonds in aqueous media is possible. An important criterion for the evolutionary development of biomolecules is their stability in the aqueous phase. The half-life of a peptide bond in pure water at room temperature is about seven years. The stability of the peptide bond towards cleavage by aggressive compounds was studied by Synge (1945). The following relative hydrolysis rates were determined experimentally, with the relative rate of hydrolysis for the dipeptide Gly-Gly set equal to unity:

$$\text{Gly-Gly} = 1 \qquad \text{Leu-Gly} = 0.23$$
$$\text{Gly-Ala} = 0.62 \qquad \text{Val-Gly} = 0.015$$

The hydrolysis conditions were as follows: 10 M HCl/glacial acetic acid (ratio 1:1) at 293 K.

It can be seen that the ease with which a peptide bond can be cleaved depends greatly on the character of the partners concerned (i.e., on the amino acid residues R). If dipeptide *cleavage* occurs at different rates, it can be assumed that the *formation* of the peptide bonds also occurs at different rates.

5.3 Activation

The conversion of building block molecules into an energy-rich, reactive form is referred to as activation. This process can be carried out in the laboratory (in vitro) but also occurs in all the cells of all the living organisms on our planet (in vivo).

When derivatives of amino acids are formed, their zwitterionic character is destroyed, a process which requires the supply of energy. We shall first discuss chemical activation, as it is important for the understanding of hypotheses dealing with prebiotic protein formation. In the case of amino acids which are activated at the carbonyl group, the amino group remains unsubstituted. The derivatives are able to react with nucleophilic residues (Y):

$$H_2N-\underset{\underset{R}{|}}{\overset{\overset{H}{|}}{C}}-C\overset{\diagup O}{\diagdown_X} + Y \longrightarrow H_2N-\underset{\underset{R}{|}}{\overset{\overset{H}{|}}{C}}-C\overset{\diagup O}{\diagdown_Y} + X \tag{5.3}$$

If Y is the amino group of a second amino acid, a peptide bond is formed (Wieland and Pfleiderer, 1957).

5.3.1 Chemical Activation

The most important derivatives of the amino acids are halides and esters. The most reactive are the halides, which, as described by Emil Fischer, can be obtained by shaking dry amino acids with phosphorus pentachloride in acetyl chloride.

Apart from these derivatives, the following classes of substance are important for our discussion:

$$H_2N-\underset{\underset{R}{|}}{\overset{\overset{H}{|}}{C}}-C\overset{\diagup O}{\diagdown_{S-R^*}} \qquad\qquad H_2N-\underset{\underset{R}{|}}{\overset{\overset{H}{|}}{C}}-C\overset{\diagup O}{\diagdown_{SH}}$$

These are the amino acid mercaptans (left) and the thioaminoacids.

According to Wieland and Schäfer (1951), aqueous thiophenylglycine undergoes oligomerisation and polymerisation in the presence of bicarbonate to give glycine oligomers and polyglycine. The thioamino acids have been a subject of increased interest in recent years, since the Nobel laureate de Duve (1994, 1995) developed his hypothesis of the "thioester world" (see Sect. 7.4).

A further important group of derivatives is that of amino acids activated by phosphoric acid or its esters. In nature, phosphorylation processes play an important activating role in peptide and protein synthesis.

5.3.2 Biological Activation

The activation of amino acids for the construction of peptides and proteins, which takes place in the cells of all living organisms, is carried out by special enzymes, referred to as "amino acid activating enzymes". Hoogland (1955) did very important work in elucidating the initial steps of protein biosynthesis. The activating enzymes belong to the group of the synthetases; according to the EC (Enzyme Commission) classification, they are ligases with the EC number 6.6.1 (amino acid RNA ligases), and are now known as "aminoacyl-tRNA synthetases".

What is the importance of this enzyme family for the biogenesis problem? These enzymes form the link between the "protein world" and the "nucleic acid world". They catalyse the reaction between amino acids and transfer RNA molecules, which includes an activation step involving ATP. The formation of the peptide bond, i.e., the actual polycondensation reaction, takes place at the ribosome and involves mRNA participation and process control via codon–anticodon interaction.

In 1994, a conference with the title "Aminoacyl-tRNA Synthetases and the Evolution of the Genetic Code" was held in Berkeley, California; its patron was the Institute of Advanced Studies in Biology. The conference dealt with the development of the synthetases and that of the genetic code (see Sect. 8.2), i.e., the assignment of the various amino acids to the corresponding base triplets of the nucleic acids.

In our discussion, we will first concentrate on the activation step, i.e., the synthetase-catalysed reaction of the amino acids with ATP. The reaction occurring in today's cells can be summarized as follows:

$$AS_1 + E_1 + ATP \rightleftharpoons [AS_1 - E_1 - AMP] + PP_i \tag{5.4}$$

$$[AS_1 - E_1 - AMP] + tRNA_1 \rightleftharpoons AS_1 - tRNA_1 + E_1 + AMP \tag{5.5}$$

$$AS_1 + ATP + tRNA_1 \rightarrow AS_1 - tRNA_1 + AMP + PP_i \tag{5.6}$$

Abbreviations:

AS_1	= amino acid 1
E_1	= the aminoacyl tRNA synthetase specific for AS_1
AMP, ATP	= adenosine mono- or triphosphate
PP_i	= diphosphate (often known as pyrophosphate)

The above reactions are shown schematically in Fig. 5.1. The overall reaction is reversible but is, however, shifted to the right, since the enzyme pyrophospho-rylase hydrolyses the diphosphate formed, thus removing it from the equilibrium. Each of the 20 proteinogenic amino acids has at least one specific synthetase. In steps 1 and 2, the amino acid exists in two different forms—in step 1 as a mixed anhydride:

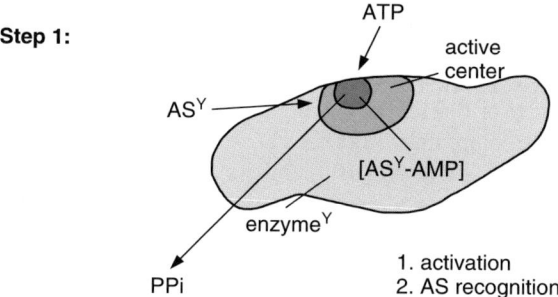

Step 1:

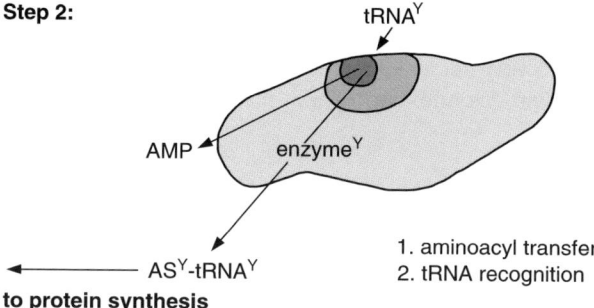

Fig. 5.1 Simplified model representation of the activation of an amino acid (AS^Y) at an amino acid-activating enzyme (i.e., an amino acid-specific aminoacyl-tRNA synthetase)

In the second step, the amino acid is bonded to the 2′ or 3′-OH function of ribose in the form of an ester, via transfer to the tRNA and the terminal adenosine residue of the tRNA. This ester formation gives a "higher-energy" bond (or to be more exact, a bond with a high group transfer potential) with a $\Delta G°$-value of -29 kJ/mol.

It was earlier considered that all the amino acid-activating synthetases were derived from a single "primeval synthetase", so that all synthetases would have similar structures. Surprisingly, however, this is not the case. When the primary sequences, and in part the secondary and tertiary structures, of all the synthetases had been determined, clear differences in their construction became obvious. The aminoacyl-tRNA synthetases consist either of one single polypeptide chain (α) or of two or four identical polypeptides (α_2 or α_4). In addition, there are heterogeneously constructed species with two sets of two identical polypeptide chains ($\alpha_2\beta_2$). This nomenclature indicates that, for each synthetase, α or β refers to a primary structure. The number of amino acids can vary from 334 to more than 1,000.

There are two classes of synthetase, each with 10 members. The amino acid sequences of these two classes have regions which are identical for all enzymes of the particular class (Eriani, 1990). The class 1 synthetases acylate the tRNA at the $2'$-hydroxyl of the terminal adenosine, while the class 2 enzymes acylate predominantly at the $3'$-function of the ribose.

The close connection of this enzyme family with the transfer of genetic information has made it a popular object of study when dealing with questions regarding the formation and evolution of the genetic code (see Sect. 8.1). It is now agreed that the aminoacyl-tRNA synthetases are a very ancient enzyme species which do not, however, arise from one single primeval enzyme, but from at least two, corresponding to the synthetase classes.

5.4 Simulation Experiments

Soon after the first successful prebiotic syntheses of amino acids by Miller and Urey, the next step, polycondensation of these monomers, was attempted. But how could the activation of the monomers have occurred on the primeval Earth without the help of special enzymes? In order to try and solve this question (in fact, there is a whole series of questions), some research groups began to work on the question using systems which were as simple as possible, in the hope of either solving it or at least coming close to an answer.

5.4.1 Prebiotic Peptides

The simplest amino acid, glycine, is almost always the one formed in highest yield in prebiotic syntheses. Thus, the first polycondensation experiments were carried out with glycine. Akabori (1955) and Hanabusa and Akabori (1959) reacted aminoacetonitrile with kaolin at 403–408 K. The starting material is readily formed as follows:

$$HCHO + NH_3 + HCN \rightarrow H_2N - CH_2 - CN + H_2O \qquad (5.7)$$

After hydrolysis of the nitrile to the aminocarboxylic acid, the authors obtained the dipeptide diglycine (Gly-Gly) and the tripeptide Gly-Gly-Gly.

Further studies were carried out mainly with glycine, since here the group R is simply hydrogen, so that no side reactions take place. In addition, glycine does not have an asymmetric carbon atom, so that chirality problems cannot occur (see Sect. 9.4). Steinman and Cole (1967) used dicyandiamide as a condensation/dehydration agent; a dipeptide was formed in about 1.2% yield.

$$(H_2N)_2C = N - C \equiv N \qquad \text{dicyandiamide}$$

As already described in Sect. 5.3, the activation of the amino acid can occur at the acid functionality:

As the phosphate anhydride
As a thioester
Using condensation agents such as dicyandiamide or its dimer

In aqueous solution, the carboxyl group of the activated amino acid reacts with nucleophilic reagents. At lower amino acid concentrations, and if no other nucleophilic compounds are present, hydrolysis takes place to give the neutral compound.

$$H_2N-CH_2-C{\overset{X}{\underset{O}{}}} \xrightarrow{+H_2O} H_2N-CH_2-C{\overset{OH}{\underset{O}{}}} + HX \qquad (5.8)$$

If no other nucleophiles are present and the amino acid concentration is high enough, the formation of a peptide bond competes with hydrolysis (Orgel, 1989).

If longer peptides are to be synthesized, the dipeptides as well as the monomers must be activated. This leads to a side reaction which can endanger the required chain-forming reaction, the formation of cyclic diketopiperazines:

$$H_2N-CH_2-\overset{\overset{\displaystyle H}{|}}{\underset{\underset{\displaystyle O}{||}}{C}}-N-CH_2-C{\overset{O}{\underset{X}{}}} \xrightarrow{-HX} \qquad (5.9)$$

The stepwise polymerisation of activated amino acids leads to the formation of activated dimers, which very often cyclise to diketopiperazines and are thus removed from the chain elongation process (Orgel, 1989).

However, in the presence of $COCl_2$, the amino acids are converted to cyclic anhydrides, which are referred to in peptide chemistry as "Leuchs anhydrides". These polymerize in aqueous media to give peptides, without the formation of larger amounts of diketopiperazines (Brack, 1982).

The formation of peptides from Leuchs anhydrides proceeds without the use of protecting groups, and no racemisation occurs. However, the use of phosgene cannot be considered a prebiotic process!

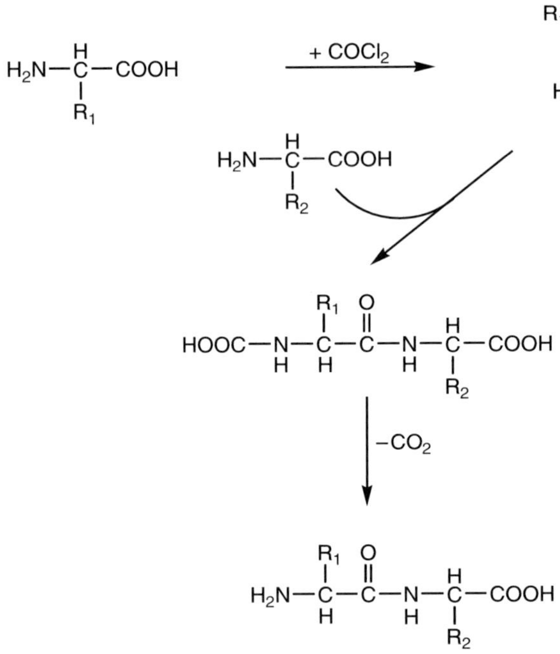

Fig. 5.2 Peptide synthesis using Leuchs anhydrides: amino acid 1 (with residue R_1) is reacted with phosgene to give the Leuchs anhydride. This reacts with amino acid 2 (residue R_2) to give the peptide carbamate. The dipeptide is obtained after cleavage of CO_2

The connection of thioamino acids to peptides was studied by Maurel and Orgel (2000). They were able to lengthen the chain of a polypeptide consisting of 10 glutamic acid residues, $(Glu)_{10}$, using the sulphur derivative of glutamic acid (GluSH), bicarbonate and $Fe(CN)_3$. $(Glu)_{11}$ was rapidly formed, and chain lengthening continued up to $(Glu)_{15}$. Details of the yields were not given, as sometimes happens: they were, however, considered to be "substantial"!

Simulation experiments of a different type were carried out by two Japanese researchers (Matsuno, 2000; Imai et al., 1999a, b). They used a simulation reactor to study processes which may occur at hydrothermal vents (see Sect. 7.2). In this case, the activation energy for the polycondensation reaction of amino acids has its origin in the Earth's interior. In the high pressure hot water reactor used, the reaction mixture circulated at around 24 MPa in a zone where the temperature was between 473 and 523 K; it was subsequently cooled to 273 K.

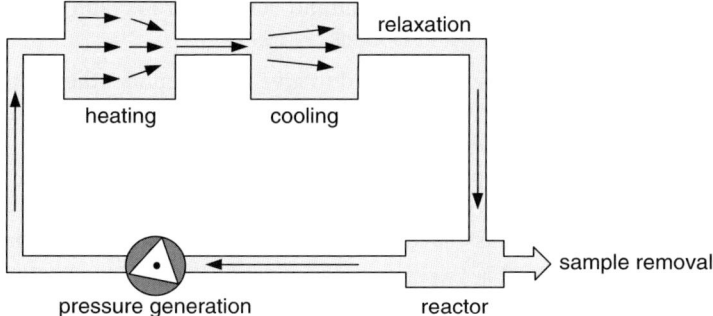

Fig. 5.3 Schematic drawing of a flow reactor for the simulation of a hydrothermal system (Imai et al., 1999)

The reaction mixture consisted only of glycine and water, without the addition of salts or other condensation catalysts. A stepwise, time-dependent synthesis of oligomers up to hexaglycine was observed. On addition of Ca^{2+} ions (pH 2.5), the oligomerisation continued further to give octaglycine.

In all simulation experiments carried out under assumed prebiotic conditions, the question of possible concentrations in a primeval ocean arises; 0.1 M solutions appear unrealistic, as this would correspond to about 12 g of amino acid per litre of seawater! Miller's "lagoons" and Darwin's "ponds" then come to mind, i.e., the concentration of dilute solutions in small localized areas due to evaporation of water. Recently, the attention of scientists has shifted towards concentration processes occurring at the surface of minerals; however, many of the problems involved remain unsolved.

New possibilities for increasing the polycondensation ability of monomers are still being sought. Experts agree that polymers will only become functional (enzymatically active) when a chain length of 30–100 building blocks (here amino acids) is reached. However, the findings of Gröger and Wilken (2001) appear to contradict this opinion; they cite an article (List et al., 2000), according to which a direct asymmetric aldol reaction was catalysed simply by the amino acid proline (see also: Movassaghi and Jacobsen, 2002; Klussmann et al., 2006; Hayashi et al., 2006).

There is no doubt that volcanic activity had various effects on the chemical processes occurring on the young Earth. Two articles deal with simulation experiments intended to study how volcanic exhalations may have affected prebiotic chemistry.

The compound carbonyl sulphide (COS) is present today in volcanic exhalations ($\sim$0.09 mol%). Since there was certainly much more volcanic activity on the young Earth than there is today, COS was then available in sufficient amounts for chemical reactions. As laboratory experiments show, amino acids can be linked under mild conditions to form peptides.

Theoretical considerations, as well as experiments, suggest the following reaction scheme:

(a) Amino acid 1 reacts with COS to give an α-amino acid thiocarbamate
(b) Intramolecular cyclisation of 2 leads to the α-amino acid N-carboxyanhydride (NCA), a Leuchs anhydride (Wieland and Bodansky, 1991)

This reacts with another molecule of the amino acid to give the dipeptide; the addition of stoichiometric amounts of Pb^{2+}, Fe^{2+} or Cd^{2+} increases the dipeptide yield, which reached 80% under optimal conditions (Leman et al., 2004).

Volcanic gases contain up to 47% sulphur dioxide (SO_2). A group from the University of Hong Kong studied the ability of SO_2 to support condensation reactions. Under weakly basic conditions, SO_2 dissolved in water forms SO_3^{2-} ions. Oxidation of these to SO_4^{2-} proceeds via SO_3. The experiments suggest the following reaction mechanism:

The amino acid, activated by uptake of SO_3, reacts with a second molecule of amino acid to form the dipeptide, which can in turn react further to form a tripeptide (and so on). This peptide synthesis model, which is supported by experimental evidence, appeals because of its simplicity; it may well correspond much more closely to conditions on the primeval Earth than do some other models (Chen and Yang, 2007).

In recent years, there has been interest in processes which may possibly have occurred three and a half billion years ago at the surfaces of rocks and minerals on the primeval Earth's surface, thus resembling solid phase syntheses in the laboratory. In a theoretical treatment, Leslie Orgel (1998) points out the problems, and also the opportunities, in synthesising long polymer chains at mineral surfaces (e.g., anion exchange minerals such as hydroxyapatite or illite). The average length of the adsorbed oligomers in the steady state (illustrated for negatively charged glutamic acid) depends on the rates of chain elongation and hydrolysis. Orgel derives an approximation formula for this process.

The complexity of the problem is shown by the same author using an example of a completely different type. Without the help of an enzyme, Chu and Orgel (1999) were able to form a peptide bond between two cysteine residues, which were bonded to a cysteine-containing peptide framework. In this case, the activation was carried out using a classical condensation agent (carbodiimidazole), which can certainly not be considered prebiotic, but the yields of dicysteine (25–60%) were amazing! The framework peptide had the structure

$$Gly\text{-}Cys\text{-}Gly_n\text{-}Cys\text{-}Glu_{10}(n = 0, 1, 2, 3)$$

The two cysteine monomers to be joined were bonded to the two cysteine residues in the framework.

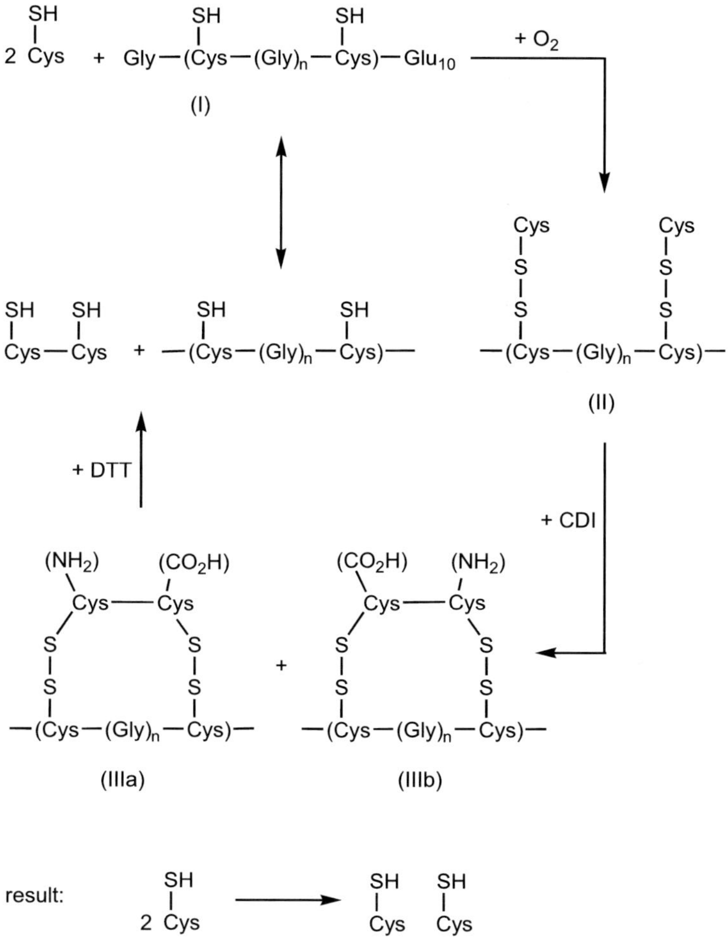

Fig. 5.4 The reaction cycle for the formation of the dipeptide cysteinylcysteine on a peptide framework (Chu and Orgel, 1999)

The two functional groups forming the peptide bond are brought close together by their positioning on the peptide framework, so that bond formation occurs readily.

It is truly amazing under what conditions, using completely different methods, amino acids can be linked up with elimination of water. Thus, it has been known for years that micelles (see Sect. 10.2) can catalyse various types of reactions in aqueous media, e.g., hydrolysis or aminolysis. Micelles are aggregates which form in aqueous solution from tensides under certain conditions (temperature, concentration). Cetyltrimethylammonium bromide $[C_{16}H_{33}N^{+}(CH_3)_3Br^{-}]$ micelles (this compound is a quaternary ammonium compound with tenside properties) are able to oligomerise relatively concentrated solutions of phosphoserine (50 mM), aspartic acid (100 mM) and glutamic acid (100 mM) after activation with carbodiimidazole (CDI). Shorter peptides are formed if the amino acid solutions are diluted by a factor of ten. The negatively charged amino acids are concentrated at the positively charged micelle surfaces to such an extent that they polymerize (Fig. 5.5).

Experiments with liposomes, which are more complex structures than micelles, point in the same direction. Liposomes consist of one (or several) lipid double layers (see Sect. 10.2), ordered concentrically around an aqueous interior. Blocher et al. (2000) carried out polycondensation reactions of amino acids and peptides using 1-palmitoyl-2-oleoyl-sn-glycero-3-phosphocholine liposomes; hydrophobically activated amino acids gave chain lengths of up to 29. The linkage of dipeptides (e.g., H–Trp-Trp–OH) to give Trp_8–OH, i.e., an octapeptide, was also carried out using liposomes. Thus, a second possibility, as well as adsorption on mineral surfaces, was available for polymer formation: via micelles and liposomes.

Bernt M. Rode and co-workers (Institute of Inorganic and Theoretical Chemistry, University of Innsbruck), have been working for more than 15 years on the formation of simple peptides under prebiotic conditions, i.e., without the addition of modern chemical condensation agents. They used only NaCl and Cu(II) ions as catalysts (Schwendiger and Rode, 1991) but needed relatively high concentrations, e.g., Gly, L-Ala, Cu^{2+} 0.49 M with an NaCl concentration about ten times higher. The reaction time was 522 h at 348–358 K, and the dipeptides Gly_2, Gly-Ala, Ala-Gly and Ala_2 were formed. Under optimum conditions, about 10% of the glycine and 5% of alanine were converted to peptides; higher peptides are also formed, but in lower yields.

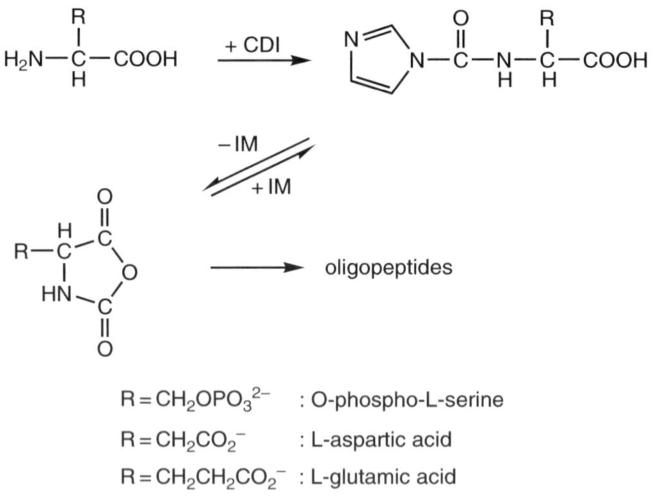

R = $CH_2OPO_3^{2-}$: O-phospho-L-serine

R = $CH_2CO_2^-$: L-aspartic acid

R = $CH_2CH_2CO_2^-$: L-glutamic acid

Fig. 5.5 The oligopeptide synthesis at cationic micelles using the condensation agent CDI leads to the intermediate (I), which is in equilibrium with an N-carboxyanhydride (II). A free primary or secondary amino acid reacts with (II) and forms an amide linkage as well as a carbamide terminus. The amino group is set free on hydrolysis. The ninhydrin method indicates that the yields are around 50% (Böhler et al., 1996)

In the same institute, experiments involving evaporation cycles were carried out, as these may have occurred periodically on Earth in the neighbourhood of volcanoes (Saetia et al., 1993); the previously described "salt-induced peptide synthesis" (SIPS) was used. Apart from the amino acids already mentioned, glutamic acid, aspartic acid, valine and proline were also used. The cycles took between 20 and 24 h at 353 or 368 K; the concentrated reaction mixtures were diluted with water at the end of each cycle. After four cycles, the diglycine yield was 6.57% of the original glycine concentration, that of triglycine 0.57%. The reaction of L-Asp with SIPS over four cycles mainly gave the βAsp-Asp dimer and the optically pure L-Asp-L-Asp dimer, but no racemic forms (L-Asp-D-Asp or D-Asp-L-Asp).

Suwannachot and Rode (1998) reported an unexpected reaction between peptides and amino acids when using SIPS: dialanine formation increased by a factor of 50 in the presence of glycine. The yield depended greatly on the glycine concentration, the optimum value being one eighth of the alanine concentration. Experimental results, as well as calculations, indicate that peptide formation is favoured by an amino acid-monochlorocuprate complex (Fig. 5.6).

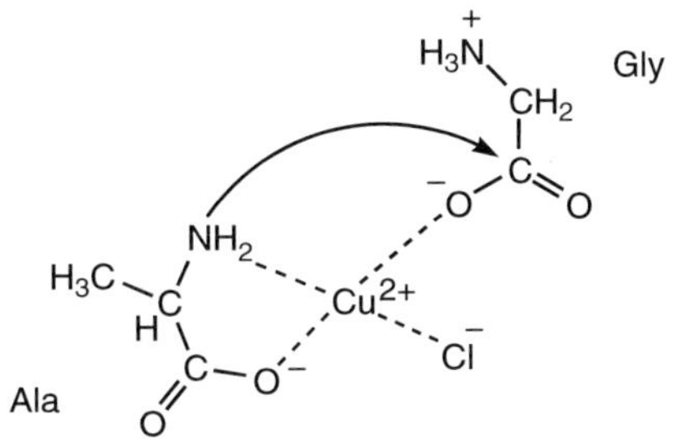

Fig. 5.6 The formation of alanine and glycine in the salt-induced peptide synthesis reaction (Suwannachot and Rode, 1998)

A combination of SIPS with the stabilising and synthesis-favouring properties of clay minerals was studied by Rode et al. (1999) in experiments involving dry/wet cycles. The simultaneous use of both SIPS and clay minerals as catalytically active surfaces led to peptides up to and including the hexamer $(Gly)_6$. The question as to whether this technique fulfils prebiotic conditions can (within certain limitations) be answered positively, since periodic evaporation phases in limited areas (lagoons, ponds) are conceivable. The "container material" could have consisted of clay minerals. Further progress in the area of peptide synthesis under conditions which could have been present on the primeval Earth can be expected.

One parameter which has so far been neglected in the discussion of the influence of physical conditions on the young Earth is the pressure in rock layers. This has been the subject of investigation by Ohara and co-workers from the Tohoku University in Sendai, Japan, who studied the pressure-dependence of the polymerisation of dry glycine at 423 K and pressures from 5 to 100 MPa. The experiments took between 1 and 32 days. Depending on the pressure, light to dark yellow products were obtained. At low pressures, the colour is probably due to the presence of melanoids.

In all cases, glycine oligomers were obtained, from the dimer up to the decamer; yields were pressure dependent. The rate of polymerisation increased during the first 8 days and then remained constant until the 31st day. The authors conclude from their experimental results that abiotic polymerisation reactions during diagenesis were more likely to have occurred in deep-lying sediments than in the primeval ocean (Ohara et al., 2007).

5.4.2 Prebiotic Proteins

About 50 years ago, shortly after publication of the Miller–Urey experiments, Sidney W. Fox and co-workers (then at the Oceanographic Institute at Florida State University) carried out work in which they heated dry amino acids under a nitrogen atmosphere in order to remove water, generated in the formation of the peptide bond, from the equilibrium. Reactions were carried out at 373–453 K for between one and several hours. The yellowish-brown pyrolysis product was dialysed to remove small molecules and then freeze-dried. After acid hydrolysis, analysis showed the presence of 60–80% of the original amino acids in the polymer mixture. Because of their protein-like character, the products were referred to as "proteinoids". Depending on the reaction conditions, some amino acids were modified to some extent: decarboxylation, deamination and (in the case of the sulphur-containing acids) partial cleavage of sulphur-containing derivatives were observed. The molecular fragments and the newly formed compounds were incorporated into the polymer chain, together with unchanged amino acids (Fox and Harada, 1958).

NMR studies on polymers containing aspartic acid showed the presence of a relatively high proportion of β-peptide bonds (Andini et al., 1975), i.e., the peptide bond involved the β-group of the acidic amino acid rather than the α-carboxyl group.

It is probable that other types of linking are present in the proteinoids: reactive molecular fragments lead to more complex compounds such as heterocycles (Heinz et al., 1979). Under other conditions (temperatures above 458 K), pteridines and flavines can be detected. The thermal polycondensation of lysine, alanine and glycine (458 K, 5 h) gave a "chromo-protenoid"; the chromophores identified were flavines (a) and diazoflavines (b) (Heinz and Ried, 1984).

A group at the Bach Institute in Moscow was able to isolate a flavine pigment (an isoalloxacine derivative) from the polymer obtained by heating a mixture of three amino acids (glutamic acid:glycine:lysine = 8:3:1); this exhibited photochemical acivity (e.g., redox reactions such as electron transfer to acceptors with lower E_0 values) under both aerobic and anaerobic conditions (Kolesnikov and Kritsky, 1999).

Proteinoids also have other properties, such as the formation of cell-like structures (microspheres), and they show weak or very weak activities such as decarboxylation (Rohlfing, 1967) or oxidoreduction (Dose and Zaki, 1971).

S. W. Fox (from 1984, director of the Institute for Molecular and Cellular Evolution of the University of Miami) made the highly controversial suggestion that the amino acid sequences in the proteinoids are not random. Nakashima prepared a thermal polymer from glutamic acid, glycine and tyrosine; the analysis showed that two tyrosine-containing tripeptides had been formed: pyr-Glu-Gly-Tyr and pyr-Glu-Tyr-Gly (Nakashima et al., 1977). The result was confirmed (Hartmann, 1981). A closer examination of the reaction mechanism showed that the formation of these two tripeptides under the reaction conditions used depends on three parameters:

> The spontaneous formation of pyr-Glu
> The slow reaction of tyrosine with itself
> The ability of glycine to form poly-Gly (Dose et al., 1982)

The experimental results, of course, only show that steric factors and the reactivity of amino acids can have a certain influence on the sequence of the polymer in thermal reactions, but that many other parameters influence the polycondensation process.

Proteinoids have gone out of fashion: it does, however, seem possible that this class of substance may return to influence the prebiotic discussion, perhaps in another form.

5.5 New Developments

New developments in chemistry can arise from the planning and testing of new synthetic methods. In contemporary chemistry, we can see developments whose roots can be found in biological processes, as the terms "signal recognition", "replication", "autocatalysis" or "self-replication" indicate. Modern ideas in these areas can be found in laboratories where work on peptide chemistry is being carried out.

The scientific world was amazed to hear that David Lee, from the laboratory of Reza Ghadiri (Scripps Research Institute, La Jolla, California), had found a self-replicating peptide (Lee et al., 1996); there are analogies to the experiments with oligonucleotides (see Sect. 6.4). Lee was able to show that a certain peptide, containing 32 amino acids, can both function as a matrix and also support its own synthesis autocatalytically. The information transfer is clearly more complex than that involved in nucleic acid replication. In the case of this particular peptide, both the amino acid sequence and the spatial structure (α-helix) determine the replication ability. The peptide is based on the sequence of the "leucine zip" region of the yeast transcription factor GCN4, a gene regulation protein.

A "superspiral" consisting of two spirals (coiled coil), known as the leucine zip, is formed in this sequence via dimerisation. The condensation reaction, carried out in the aqueous phase, involves two peptide fragments which contain 15 and 17 amino acid residues respectively. Activation takes place via thioester formation (see Sect. 5.3.1). The ligation to a complete GCN4 matrix gives a new 32 amino acid peptide, which can itself serve as a matrix. The autocatalytic reaction exhibits a parabolic increase in the peptide concentration (caused by product inhibition; see Section 6.4).

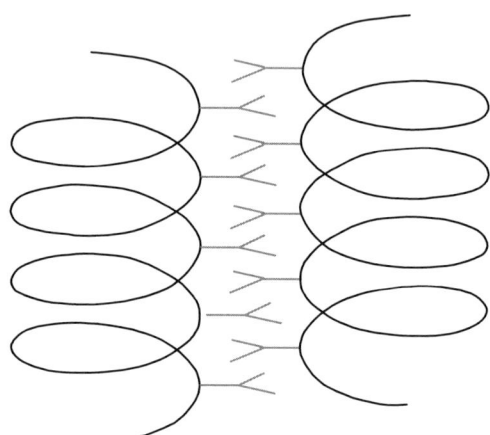

Fig. 5.7 Simplified schematic representation of a "leucine zip". The leucine residues in the two α-helices (the symbols in the centre of the diagram) interact with each other

The importance of this sensational discovery for biogenesis research only became apparent in the next few years (Lahav, 1999). It is clear that prebiotic chemistry is much more complex and versatile than was thought about 50 years ago, when Stanley Miller carried out his first successful amino acid syntheses. Experiments similar to the ones described above, as well as new ones which must first be devised, could help to slowly close the wide gap which still exists between the prebiotic and the "living" worlds (Wills and Bada, 2000).

The GCN4 system was extended by the La Jolla group, and it in turn exhibited properties which surprised the experts. The ability of organisms to survive is known to be increased by symbiotic association. Similar properties were observed by Lee et al. (1997) and Ghadiri (2000) in experiments on self-replication involving a hypercycle network. Here two—normally competing—self-replicating peptides catalysed their own formation symbiotically. These results must be seen in the context of the theories put forward by Manfred Eigen (see Sect. 8.2) and Stuart Kauffman (see Sect. 9.3). A hypercyclic peptide network system, as first described by Lee, Severin, Ghadiri and others, supports the premises of these theories that peptides played vital roles in all biogenetic processes. The newly-discovered properties of the auto-catalytic, self-replicating peptides are fascinating; they may perhaps turn out to be an important "missing link" in the transition from inanimate to living systems.

Two further important results were obtained in the Ghaziri research group. In the previously described two-peptide system (consisting of 15 and 17 amino acids), the longer peptide has electrophilic and the shorter nucleophilic character (due to the electrical charges of the amino acids from which the peptides are constructed). A dynamic error correction in autocatalytically active networks was observed using this system; apart from the two native peptides, two others were added which had a mutation (due to amino acid exchange) in their chain. During the competing condensation reactions of the peptides, both native sequences of a self-replicating peptide and mutated peptides are formed. The products are: the native sequence (no amino acid exchange), two singly mutated peptides and a doubly mutated peptide. The spontaneous errors in self-reproduction were simulated by adding starting peptides in which, in each case, one single amino acid had been replaced by another. After incubation in the aqueous phase, the clearly preferred formation of native peptide sequences was observed. It was thus assumed that the mutant peptides served as selective catalysts for the formation of native sequences. An exact analysis showed that this type of sequence selection is the result of self-organized networks, which consist of two or three (auto)catalytic systems. The preferential formation of native sequences can be seen as error correction of an autocatalytic network (Severin et al., 1998a, b).

The second report from La Jolla attempts to cast some light on the question of the homochirality of biomolecules (see Sect. 9.4). Put simply, the question is why only one of the two possible chiral forms is always found in some important classes of biomolecules.

1) $E^L + N^L \xrightarrow{\quad T^{LL} \quad} T^{LL}$

2) $E^L + N^L \xrightarrow{\quad T^{LL},\, T^{DD},\, T^{DL},\, T^{LD} \quad} T^{LL}$

3) $E^D + N^L \xrightarrow{\quad T^{LL},\, T^{DD},\, T^{DL},\, T^{LD} \quad} T^{DL}$

4) $E^D + N^L \xrightarrow{\quad \text{small matrix} \quad} T^{DL}$

5) $E^D + N^L + E^L \xrightarrow{\quad T^{LL} \quad} T^{DL} + T^{LL}$

Fig. 5.8 The reactions studied: starting materials on the left, products on the right. The matrices used are shown above the arrows (Saghatelian et al., 2001)

N^L nucleophilic peptide consisting of 15 L-amino acid residues
N^D corresponding peptide with D-amino acid residues
E^L, E^D electrophilic peptide fragments consisting of 17 D- and L-amino acid residues
T condensation product formed from E and N
T^{LL}, T^{DD} peptide containing only L- or D-amino acid sequences
T^{DL} peptide containing both amino acid species

Once again, the Ghadiri group used the peptide with the leucine zip. Starting from a racemic mixture of the peptide fragments E and N (with 15 and 17 amino acids respectively), homochiral products were preferentially formed in a homochiral selection process in the catalytic self-replication cycle. The initial mixture contained the two peptide fragments, which each consisted of D- and L-amino acids, i.e., a total of four competing molecular species (N^L, N^D, E^L and E^D). Thus, four different products could be formed in the condensation reaction: T^{LL}, T^{DD}, T^{LD} and T^{DL}. As stated above, T^{LL} and T^{DD} were formed preferentially. The experimental results can be summarized as shown in Fig. 5.8.

In accordance with the autocatalytic process, matrices are again formed. It is surprising that the autocatalysis decreases when only 1 of the 15 building blocks of the peptide has the opposite handedness, e.g., when the N-peptide fragment contains one D-amino acid as well as the 14 L-amino acids. These experimental results show that such a system is able to form homochiral products via self-replication. It can be assumed that similar mechanisms influenced the origin of homochirality on Earth (Saghatelian et al., 2001; Siegel, 2001).

However enthusiastic we may be about these discoveries and successes, we must not forget that ready-made peptides were used; we know nothing about the formation and origin of such peptides on the deserted, desolate Earth.

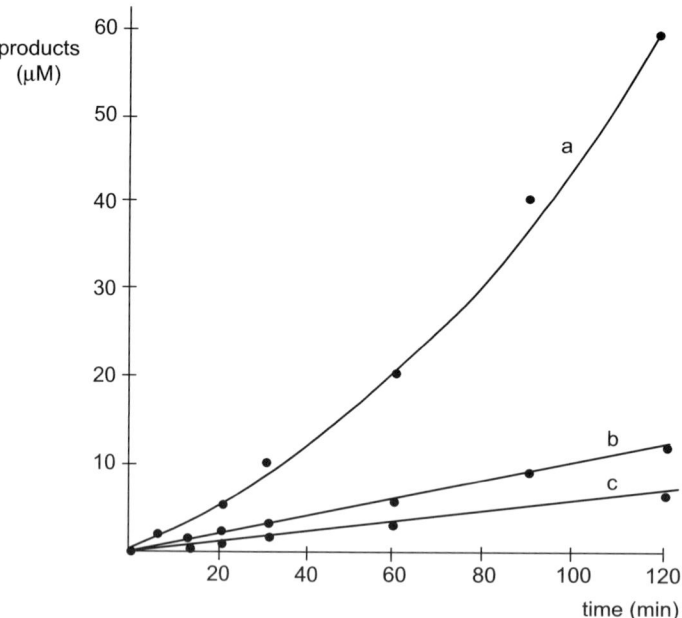

Fig. 5.9 Product formation rates as a function of time. Matrix-dependent formation **a** of the homochiral products T^{LL} and T^{DD}, **b** of the heterochiral products T^{DL} and T^{LD} and **c** using the same conditions as in **a** and **b**, except for the addition of 3 M guanidine hydrochloride (Saghatelian et al., 2001)

A detailed account of the prebiotic chemistry of the amino acids and peptides as well as a perspective on protometabolism has been provided by R. Pascal et al. (2005).

References

Akabori S (1955) Kagaku 25:54
Andini S, Benedetti E, Ferrara L, Paolillo L, Temussi PA (1975) Origin of Life 6:147
Blocher M. Liu O, Luisi PL (2000) Orig Life Evol Biosphere 30:123
Böhler C, Aubrey R, Hill JR, Orgel LE (1996) Orig Life Evol Biosphere 26:1
Brack A (1982) Biosystems 15:201
Brack A (1989) Orig Life Evol Biosphere 19:77
Chen F, Yang D (2007) Orig Life Evol Biosphere 37:47
Chu BFC, Orgel LE (1999) Orig Life Evol Biosphere 29:441
de Duve C (1991) Blueprint for a Cell: The Nature and Origin of Life, Neil Pattersons Publishers, Carolina Biological Supply Company, Burlington
de Duve C (1995) Vital Dust. Life as a Cosmic Imperative. Basic Books, Harper Collins Publ. Inc.
Dose K, Zaki L (1971) Z Naturforsch 26b:2:144
Dose K, Rauchfuss H (1972) On the electrophoretic behavior of thermal polymers of amino acids. In: Rohlfing DL, Oparin AI (Eds.) Molecular Evolution: Prebiological and Biological. Plenum, New York, pp 199-217
Dose K, Hartmann J, Brand CM (1982) Biosystems 15:195
Eriani G, Delarue M, Pock O, Gangloff J, Moras D (1990) Nature 347:203
Ferris JP (1989) Nature 337:609
Fox SW, Harada K (1958) Science 128:1214
Ghadiri MR (2000) Orig Life Evol Biosphere 30:247
Gröger H, Wilken J (2001) Angew Chem 113:54
Hanabusa H, Akabori S (1955) Bull Chem Soc Japan 32:626
Hartmann J, Brand CM, Dose K (1981) Biosystems 13:141
Hayashi Y, Matsuzawa M, Yamaguchi J, Yonehara S, Matsumoto Y, Shoji M, Hashizume D, Koshino H (2006) Angew Chem 118:4709, Int Ed 45:4593
Heinz B, Ried W, Dose K (1979) Angew Chem 91:510, Int Ed 18:478
Heinz B, Ried W (1984) Origins of Life 14:279
Hoagland MB (1955) Biochim Biophys Acta 16:288
Imai E-I, Honda H, Hatori K, Matsuno K (1999a) Orig Life Evol Biosphere 29:249
Imai E-I, Honda H, Hatori K, Brack A, Matsuno K (1999b) Science 283:831
Klussmann M, Iwamura H, Mathew SP, Wells Jr DH, Pandya U, Armstrong A, Blackmond DG (2006) Nature 441:621
Kolesnikov MP, Kritsky MS (1999) Book of Program and Abstracts, ISSOL-99. Mandeville Center, Univ. California, P1.31, p 73
Lahav N (1999) Biogenesis. Oxford University Press, New York, Oxford, p 230
Lee DH, Cranja JR, Martinez JA, Severin K, Ghadiri MR (1996) Nature 382:525
Lee DH, Severin K, Yokobayashi Y, Ghadiri MR (1997) Nature 390:591
Leman L, Orgel LE, Ghadiri MR (2004) Science 306:283
List B, Lerner RA, Barabas III CF (2000) J Amer Chem Soc 122:2395
Matsumu K (2000) Orig Life Evol Biosphere 30:203
Maurel MC, Orgel LE (2000) Orig Life Evol Biosphere 30:165
Movassaghi M, Jacobsen EN (2002) Science 298:1904
Nakashima T, Jungck JR, Lederer E, Fox SW (1977) Int J Quantum Chem 4:65
Orgel LE (1989) J Mol Evol 29:465

Orgel LE (1998) Orig Life Evol Biosphere 28:217
Pascal R, Boiteau L, Commeyras A (2005) Top Corr Chem 259:69
Rhodes GW, Flurkey WH, Shipley RM (1975) J Chem Education 52:197
Rode BM, Son HL, Suwannachot Y, Bujdak J, (1999) Orig Life Evol Biosphere 29:273
Rohlfing D, Fox SW (1967) Arch Biochem Biophys 118:122
Saghatelian A, Yokobayashi Y, Soltani K, Ghadiri MR (2001) Nature 409:797
Saetia S, Liedl KR, Eder AH, Rode BM (1993) Orig Life Evol Biosphere 23:167
Schwendinger M, Rode BM (1991) Inorganica Chemica Acta 186:247
Severin K, Lee DH, Martinez JA, Vieth M, Ghadiri MR (1998b) Angew Chem Int Ed 37:126
Siegel JS (2001) Nature 409:777
Steinman G, Cole MN (1967) Proc Natl Acad Sci USA 58:735
Suwannachot Y, Rode BM (1998) Orig Life Evol Biosphere 28:79
Synge RCM (1945) J Biochem 39:351
Wieland T, Schäfer W (1951) Angew Chem 63:146
Wieland T, Pfleiderer G (1957) Advances in Enzymology 19:235
Wills C, Bada Y (2000) The Spark of Life. Perseus, Cambridge Mass, p 13

Chapter 6
The "RNA World"

6.1 Introduction

The term "RNA world" appears in many publications on biogenesis and has become almost a catchword, like the term "primordial soup". The RNA world theory—or more correctly, hypothesis—was put forward about 20 years ago. The expression was coined by the Nobel Prize winner Walter Gilbert from Harvard University (Gilbert, 1986). However, the predominant role of this particular nucleic acid was recognized 40 years ago by Carl Woese (1967), Francis Crick (1968) and Leslie Orgel (1968) in connection with processes which could lead to biogenesis. In Germany, Manfred Eigen (1971) published a comprehensive article on "Self-organisation of matter and the evolution of biological molecules", while Hans Kuhn (1972) published one called "Self-organisation of molecular systems and the evolution of the genetic apparatus"; the exceptional role played by RNA in the origin of life was stressed in both articles (see Sects. 8.2 and 8.3). In recent years, the RNA world hypothesis has changed the direction of research in many institutes where work on biogenesis is being done.

Almost 20 years ago, in a review article in *Nature*, Gerald Joyce (1989) explained why RNA fulfils almost all the conditions which a type of substance requires in order to be able to transfer information:

The matrix properties of RNA make the self-replication process easier. RNA matrices are able to control the synthesis of complementary oligonucleotides. The double function of RNA comprises:

Its ability to transfer information (see Sect. 6.4) and
Catalytic properties (see Sect. 6.5)

In all of today's living systems, RNA is involved in processes which are very old from an evolutionary point of view. RNA occurs in protein biosynthesis in three different forms, and thus with three different functions: tRNA, mRNA and rRNA.

Most biochemically active coenzymes are either nucleotides or compounds which could be derived from nucleotides.

H. Rauchfuss, *Chemical Evolution and the Origin of Life*,
© Springer-Verlag Berlin Heidelberg 2008

The amino acid histidine has important functions in the active centres of several enzymes. Its biosynthesis involves intermediates which are possibly related to a precursor molecule (purine base) of a ribozyme.

Thymine is found only in DNA but not in RNA. Deoxythymidylic acid is formed by methylation of deoxyuridylic acid; this may be an indication that ribonucleosides were present on the young Earth before the deoxynucleosides.

Even its proponents do not always discuss the RNA world in the same terms. However, there are three basic assumptions on which there is consensus (Joyce and Orgel, 1999):

After a certain stage in the development and evolution of life, genetic continuity was ensured by RNA replication.

The basis of the replication process is Watson–Crick base pairing.

During the development phase of the RNA world, there were no functioning processes which were influenced by protein catalysts.

The differences between the different versions of the RNA world mainly concern the questions:

What came *before* the RNA world?

Which complex mechanisms were active in the RNA world?

What was the function of the cofactors?

In spite of all the enthusiasm for the RNA world, critical observers point out (as do Joyce and Orgel in 1989) that the RNA world is only a hypothesis! It has, however, led to the development of a series of interesting biogenesis models. Some points have been verified by means of successful experiments, but the hypothesis poses questions to which we have no answers at present (see Sect. 6.6).

6.2 The Synthesis of Nucleosides

Expressed simply, nucleosides are formed by linking an organic base (here guanine, adenine, uracil or cytosine) to a sugar (here D-ribose) with the elimination of water:

$$base + sugar \rightarrow nucleoside + water$$

This reaction looks simple, but an enzyme-free prebiotic synthesis, in particular involving pyrimidine bases, is still very difficult. It is possible that completely new synthetic strategies will need to be devised.

The problems in the nucleoside synthesis arise in the linkage of the 3-N atoms of the pyrimidines and the 9-N atoms of the purines with the $1'$-C atom of ribose, not only without enzyme control, but also under conditions extant on the primordial Earth. How might such reactions occur? There have naturally been many attempts to solve the problem. In 1963, Ponnamperuma et al. reported the detection of small amounts (0.01%) of adenosine when a 10^{-3} M solution of adenine, ribose and phosphate was irradiated with UV light. Several years later, Fuller et al. (1972a, b) were able to carry out nucleotide synthesis in low yield by heating a mixture of purines (adenine, hypoxanthine and guanine) with ribose and Mg salts; however, similar experiments using pyrimidines were unsuccessful. This fact highlights a dangerous weakness in the whole RNA world hypothesis. The solution of the tiresome nucleoside problem thus requires some quite unorthodox, imaginative ideas for its solution. Perhaps the route presently used in living cells could be followed. Nature uses 5-phosphoribosyl-1-pyrophosphate (PRPP), i.e., ribose which is activated prior to the coupling with the base (Zubay, 2000).

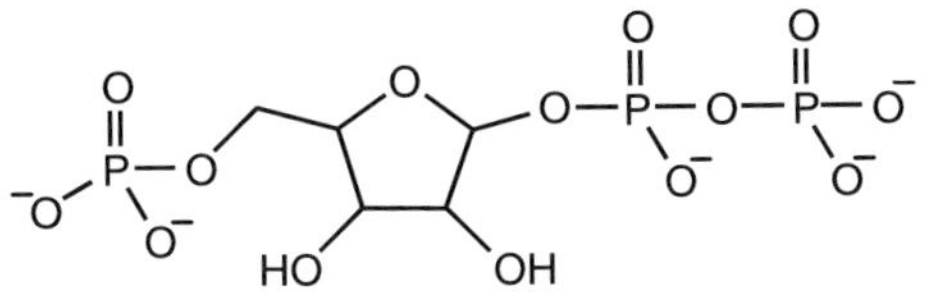

6.3 Nucleotide Synthesis

The actual building blocks for the nucleic acids are the nucleotides, which are formed in an esterification reaction between nucleosides and phosphate; three OH functions of ribose, and two of deoxyribose, can undergo esterification:

$$\text{nucleoside} + H_3PO_4 \rightarrow \text{nucleotide} + H_2O$$

The reaction, using the example of adenosine-5′-phosphate, is shown in Fig. 6.1.

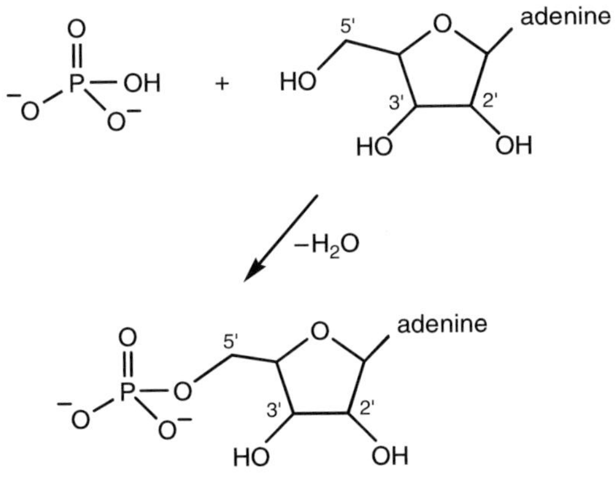

Fig. 6.1 Phosphate and adenosine give adenosine monophosphate (AMP)

Although the results of all experiments so far carried out on nucleoside synthesis under prebiotic conditions have been disappointing, the next step, to give the nucleotides, has been carried out using nucleosides synthesized in today's laboratories. There are two preconditions for nucleotide syntheses:

A relatively readily available source of phosphate (see Sect. 4.7)
An activation mechanism which permits the phosphate residue to form a phosphate ester

As early as 45 years ago, G. Schramm et al. (1962) carried out the synthesis of AMP, ADP and ATP using ethyl metaphosphate as the phosphorylating agent. These successful syntheses led to the formation of longer nucleotide chains; however, they by no means correspond to the conditions present on the primordial Earth. Thus, the question as to the source of the phosphates remains paramount. According to Schwartz (1998), the following phosphate sources deserve consideration:

Fluoroapatite
Schreibersite
Alkyl phosphonic acids

The RNA world hypothesis caused prebiotic phosphate chemistry to become an attractive research area again; unfortunately, no clear evidence for a realistic nucleotide synthesis under the simplified conditions of a primitive Earth has yet appeared. Important work on nucleoside phosphorylation has, however, been done. It is important to distinguish between:

Phosphorylation of a sugar to give monophosphate and
Further reactions of monophosphate to give di- or triphosphate.

The search for routes to phosphorylated nucleosides or phosphorylated ribose started, as mentioned in Sect. 4.7.3, with work done by Ponnamperuma (1963). The synthesis of ATP was carried out by phosphorylation of ADP using carbamoyl phosphate in the presence of Ca^{2+} ions as catalyst; yields of up to about 20% were obtained (Saygin and Ellmauerer, 1984).

G. Zubay and his co-workers have been searching for some years for synthetic routes to nucleotides which could have been realized under prebiotic conditions, starting from work done by Lohrmann and Orgel in 1971 and developing it further. Reimann and Zubay (1999) improved the synthesis of 5'-nucleoside monophosphates (5'-NMP) using the thin layer method introduced by Lohrmann and Orgel; these layers are obtained by evaporation and heating of solutions and consist of nucleosides (uridine, cytidine, inosine and adenosine), inorganic phosphate and urea. The four nucleosides react similarly, but there are differences in their reaction behaviour which lead to differing yields. The products are mixtures of 5'-NMPs and 2'(3')-NMPs in a ratio of 2:1. Variation of pH between 4.2 and 7.4 has little effect, but the yield is considerably lower at pH 9.0. Monoionic phosphate is assumed to be the active phosphate species; no phosphorylation occurs in the absence of urea, while other nitrogen compounds such as ammonium formate, arginine, asparagine or imidazole have no effect. The next problem to be solved is product separation under prebiotic conditions.

Reimann and Zubay found that 5'-AMP is selectively converted to adenosine-5'-polyphosphate when a solution of nucleotides, trimetaphosphate and MgCl$_2$ is evaporated down. These are the same conditions under which 2'(3')-AMP is converted to 2',3'-cyclic AMP.

If the assumption that all living things had their origin in water is correct, a synthesis of nucleic acid building blocks in an aqueous phase must also be feasible. Yukio Yamagata from the Institute for Chemical Evolution in Kanazawa, Japan, has published work on this problem (Yamagata 1999, 2000). He was able to carry out the phosphorylation of ADP to ATP in aqueous solution with the help of cyanate as the condensation agent, in the presence of calcium phosphate. The formation of ADP from AMP took place under similar conditions. Yields of ADP and ATP from AMP of 19% and 7% respectively were obtained at 277 K and a pH of 5.75. Other nucleoside triphosphates could also be obtained from their diphosphates in the same way, although in lower yields.

The thermal synthesis of nucleoside-5'-phosphite monoester using $(NH_4)_2HPO_3$ was carried out under relatively mild conditions (60°C, reaction time about 24 h) by A. W. Schwartz's group in Nijmegen, Holland; in the case of uridine, the yield was 20%. Ammonium phosphate, however, cannot be used: it gave yields of only $\sim$0.15% after very long reaction times (46 days). This confirms earlier suggestions that nucleoside-H-phosphonates, and condensation products possibly derived from them, would have been formed more readily on the primeval Earth than nucleotides (de Graaf and Schwartz, 2005).

In summary, it has been established that all NMPs and NDPs are phosphorylated in reactions where calcium phosphate is present. However, no phosphorylation products are obtained when diphosphate (PP$_i$) or tripolyphosphate (PPP$_i$) is used instead of phosphate (P$_i$). Neither cyanamide nor dicyandiamide can be used instead of cyanate.

It is not clear whether the conditions used in the above experiments correspond to those present on the young Earth. Cyanate has been detected in cosmic nebulae (Yamagata, 1999), while water-soluble phosphates and diphosphates can be formed during volcanic activity, as Yamagata showed as early as 1991 (see Sect. 4.7.3).

The catalytically active RNA species (ribozymes) have been shown in recent years to undergo an unbelievable range of reactions. This led to the question as to whether they are also involved in nucleotide syntheses. Unrau and Bartel (1998) have reported successful nucleotide syntheses carried out using ribozymes (see Sect. 6.5). It was possible to isolate in vitro selected RNA which acted catalytically in the synthesis of a pyrimidine nucleotide; it remains unclear whether these results are important for biogenesis.

Zubay and Mui (2001) have pointed out an important, but as yet unsolved, problem of prebiotic nucleotide synthesis: the nitrogen-containing nucleobases very probably have HCN as the main precursor for their construction. In the case of ribose, formaldehyde is assumed to be a building block (see Sect. 4.4). Both species could, however, hardly have been present at the same time and place, as they would have immediately reacted with each other. Thus, their availability for the formation of bases or sugars would have been greatly diminished; perhaps as yet unknown reaction mechanisms may have led to other unknown prenucleotides.

6.4 The Synthesis of Oligonucleotides

The polycondensation of several nucleoside monophosphates gives oligonucleotides (up to 40–50 units). If the chain is even longer, the polymer is referred to as a polynucleotide. Initial experiments on the polycondensation of nucleotides to give longer chains were carried out about ten years after the discovery of the DNA double helix (G. Schramm, Sect. 6.3).

Naylor and Gilham (1966) took another route: they were able to link short DNA fragments to a complementary matrix without using an enzyme. The reactions were carried out in aqueous solution, and the molecules first had to be converted into a reactive state by chemical activation; the activation agent used was a water-soluble carbodiimide.

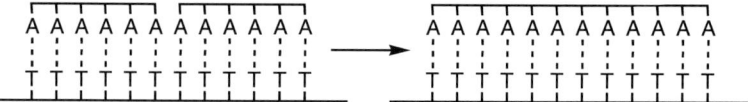

Fig. 6.2 Schematic representation of the coupling of two adenylhexanucleotides on a T-matrix to give the dodecaoligonucleotide

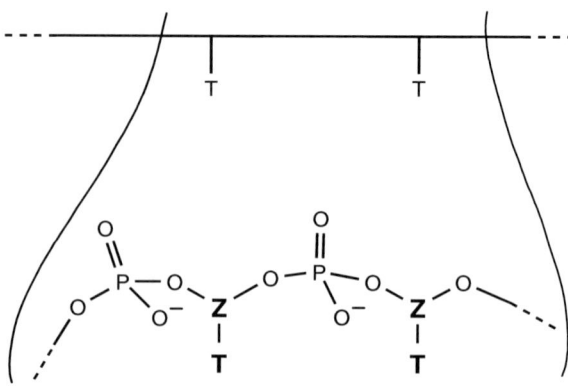

Fig. 6.3 Detail of the scheme shown in Fig. 6.2

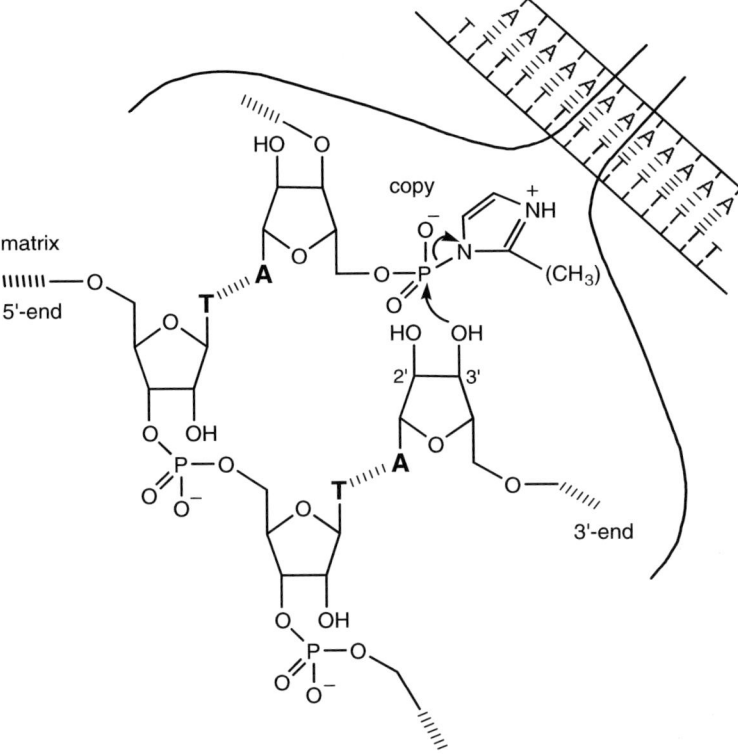

Fig. 6.4 The exact chemical mechanism of the ligation of two oligonucleotides at a matrix (see also Figs. 6.2 and 6.3)

Two molecules of hexathymidylic acid-5′-phosphate were linked to give one of dodecadeoxythymidylic acid-5′-phosphate. This reaction required the presence of a complementary matrix made up of polydeoxyadenylic acid (poly-dA). The coupling reaction, also known as "ligation" of the two hexamers, is an esterification reaction. The OH group at the chain end of the first oligonucleotide carries out a nucleophilic attack at the activated phosphate group of the second hexamer; a phosphodiester bond is formed between the two hexamers, and the activating agent is split off as a urea derivative.

The experimental result shows that two conditions must be fulfilled in order for a ligation to take place:

One building block must be activated, and
A complementary matrix must be present.

Even if the Naylor-Gilham experiment was successful, the question as to how the hexamers were formed still remains.

Leslie Orgel and co-workers took up this problem and studied the non-enzymatic polymerisation of mononucleotides, i.e., the question as to whether single nucleic acid building blocks can undergo polycondensation on a corresponding complementary matrix. The substrates used were the 5'-phosphoimidazolides of adenosine (ImpA) and guanosine (ImpG), the matrices poly(U) and poly(C).

ImpA (G)

The result was quite disappointing, as instead of the required 3'-5'-phosphodiester linkage, which is found in nucleic acids today, the main products obtained were those with the unnatural 2'-5'-bond between the nucleotides. Further experiments showed that the presence of divalent metal ions had a clear positive effect on the matrix-dependent polycondensation. The addition of $1-10\,mM\,Pb^{2+}$ to $100\,mM$ of poly(U) as the matrix and $50\,mM$ of ImpA monomer caused the yield of oligomeric product (pentamers and longer) to increase by a factor of four (Sleeper et al., 1979).

Apart from Pb^{2+}, Zn^{2+} ions also catalyse the poly(C)-supported polymerisation of the activated guanylic acid derivative, guanosine-5'-phosphoimidazolide; polymers with 30–40 units were obtained, the yield dropping considerably as the chain length increased. 2'-5'-linking of the G oligomer predominates when Pb^{2+} is added, while the addition of Zn^{2+} favours the natural 3'-5'-bonding. The regioselective behaviour of metal ions shows the sensitivity of these reactions, even under these (not particularly prebiotic) conditions (Lohrmann et al., 1980). It is interesting to note that some of today's nucleic acid polymerases contain bonded Zn^{2+} ions as well as Mg^{2+} as a cofactor for their enzymatic activity.

Inoue and Orgel (1981) found that the oligomerisation of guanosine-5'-phosphoimidazolide on a poly(C) matrix can take place in the absence of Pb^{2+} or Zn^{2+}, if guanosine-5'-phospho-2-methylimidazolide is used for the oligo-G synthesis.

G = guanosine
Guanosine-5'-phosphoimidazolide
In the case previously described, $R_1 = Me, R_2 = H$; otherwise $R_1 = R_2 = R_3 = H$.

A further unusual feature of the matrix-dependent polycondensation lies in the character of the nucleobases themselves. Purine mononucleotides undergo poly-condensation, in good yields, at complementary matrices consisting of pyrimi-dine polymers. However, the synthesis of pyrimidine oligonucleotides from their mononucleotides at purine matrices is not effective. This important fact means that a pyrimidine-rich matrix leads to a purine-rich nucleic acid, which is itself not suitable to act as a matrix. This phenomenon also occurs when matrices are used which contain both basic species, i.e., purines *and* pyrimidines. An increase in the amount of purine in a matrix leads to a clear decrease in its effectiveness (Inoue and Orgel, 1983). However, the authors note self-critically that the condensation agent used cannot be considered to be prebiotic in nature.

Exactly this problem was the subject of synthetic experiments carried out by J. Oró et al. (1984), which were intended to clarify the possible formation of these condensation agents. They used simple compounds, such as formaldehyde, acetalde-hyde, glyoxal and ammonia as starting materials, and were able to synthesize imi-dazole as well as its 2- and 4-methyl derivatives.

The dilemma described above, that cytosine-rich matrices lead to (complemen-tary) sequences which are low in cytosine and are themselves ineffective matrices, makes the synthesis of nucleic acids in the absence of enzymes almost impossible. Thus, other models and model experiments must be looked for.

The first successful oligonucleotide synthesis at an oligomeric matrix with a de-fined nucleoside sequence was carried out in 1983 (Inoue et al., 1984). A mixture of 2-MeImpC and 2-MeImpG was allowed to react at a CCGCC* matrix, giving the corresponding complementary pentanucleotide, with the sequence pGGCGG, in 19%; other products were naturally also formed, such as the tetramer pGGCG. No products were formed in the absence of a matrix. However, the nucleotide sequence would need to be multiplied in further cycles in order to obtain an effective infor-mation transfer. This has not yet been possible using the known systems for linking nucleotides at oligomeric matrices.

Thus it was necessary to develop models which would make self-replication of nucleic acids possible. Such models would need to be simple, since they should sim-ulate prebiotic processes (von Kiedrowski, 1999). The realisation of such a scheme would require the fulfilment of several conditions:

> The system must contain a self-replicating matrix (C), i.e., the copy generated at the matrix must have the same nucleotide sequence as the matrix itself.
>
> The system should consist of only two building blocks A and B. These monomeric building blocks give a dimeric matrix, two dimeric building blocks giving a tetrameric matrix, and so on. Analysis of kinetic and thermo-dynamic data have indicated that a hexameric matrix, i.e., a matrix consisting of six nucleotide building blocks (and containing only C and G[1]), will suffice to bond trimeric oligonucleotides strongly enough.

---

[1] In the simplified notation used here and in the following, the letters C, G, A, U and T do not refer to the nucleobases themselves, but to the nucleotides; thus, C corresponds to cytosine monophos-phate in the oligonucleotide.

In order to simplify the analysis of the products, the system should react without complications. Thus, the more easily handled DNA analogues are used instead of RNA oligomers, as they lack one functional group (the 2′-OH group of the pentose).

The minimal system would therefore consist of the following reaction sequence:

$$A + B + C \leftrightharpoons ABC \rightarrow C_2 \leftrightharpoons 2C$$

A and B:	The two building blocks to be linked (mono-, di-, trinucleotides or oligonucleotides)
C:	The matrix
ABC:	The two building blocks A and B bonded to the matrix by hydrogen bonds and positioned in such a way that ligation can take place
C_2:	The duplicate
2 C:	The two matrices C

Günter von Kiedrowski (Ruhr-University, Bochum, Germany) used two trideoxynucleotide substrates in experiments to study the possibility of a self-replicating, enzyme-free nucleotide system. Chemically speaking, the three symbols A, B and C represent:

A: d(5′-CCGp)—the 5′ end of the trinucleotide is protected as the methyl ester, while the 3′ end carries a phosphate group activated by the condensation agent (a carbodiimide).

B: d(5′-CCGp-o-PhCl-3′)—the 5′-OH function is free, and in the ligation step, it attacks the activated phosphate group of A in a nucleophilic manner, thus forming the required 3′-5′-phosphodiester bond between A and B. The 3′ end of B is inactivated by means of an o-chlorophenylphosphate protecting group.

C: The matrix of the hexanucleotide d(5′-MeCCGCGCp-o-PhCl-3′).

The attachment of the two trinucleotides A and B to the matrix takes place via Watson–Crick base pairing: they approach each other in such a manner that the formation of the 3′-5′-phosphodiester bond can occur. The nature of the trinucleotide sequence chosen means that the new hexanucleotide formed at the matrix has the same sequence as the matrix itself. After its separation from the original matrix, the newly formed hexamer can itself function as a matrix.

The term "self-replication" as used below involves two separate functions:

Autocatalysis and
Information transfer.

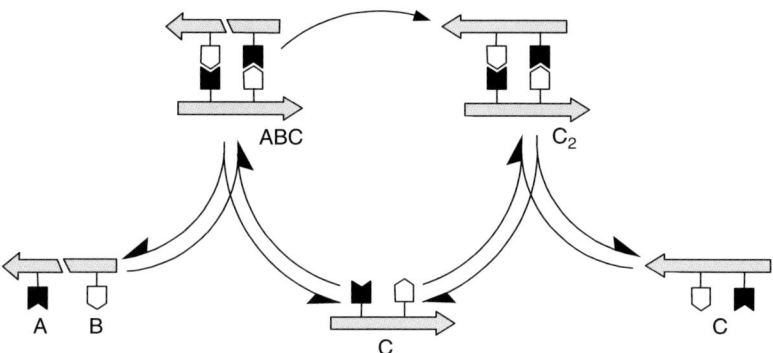

Fig. 6.5 Scheme necessary for a minimal self-replicating system with an autocatalytic reaction cycle (von Kiedrowski, 1999)

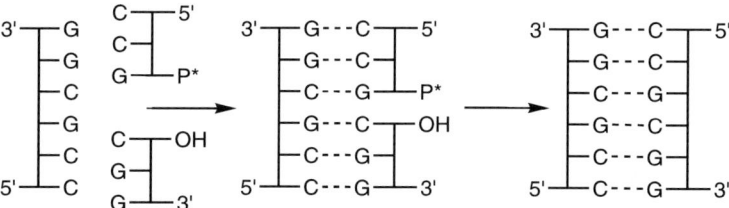

Fig. 6.6 Schematic representation of the first minimal self-replicating system. The hexameric product obtained from linking the two trideoxynucleotides can itself function as a matrix (von Kiedrowski et al., 1992)

We can speak of self-replication only when these two functions occur together.

The experiment described above was carried out on a microlitre scale, the analysis of the products being carried out by HPLC. However, as so often happens when working with complex systems, the hoped-for success was not immediate; the necessary experience first had to be accumulated.

The initial rate at which the matrix C is formed in these matrix-dependent experiments is related to the initial concentration c by a square-root dependence. This "square root law of autocatalysis" is found in most self-replicating systems:

$$v_c = \frac{dc}{dt} = \alpha \sqrt{c} + \beta$$

where v_c is the initial rate, and c is the initial concentration of the matrix C. α and β result from a plot of the measured initial rate v_c against the given initial concentration c; α is the gradient, β the intercept on the y-axis. α is a measure of the effectiveness of the autocatalysis, β of the rate of matrix synthesis when no matrix molecules are present in the reaction mixture (i.e., $c = 0$).

More information on the square root law and on kinetic and thermodynamic aspects of the minimal replicator theory can be found in the literature (von Kiedrowski, 1999 and 1993). The square root law has its origin in the product inhibition involved in the mechanism of self-replication. The more C units are formed, the greater is the tendency of the molecules of C to dimerize to give C_2; this species, however, cannot function as a catalyst.

The square root law was confirmed experimentally by Zielinski and Orgel (1987); they replaced the 3′-OH group of ribose by a 3′-NH$_2$ group, thus increasing the nucleophilic character of the 3′ terminus of the sugar. In this synthesis, a tetranucleotide phosphoramidate ($G_{NHp}C_{NHp}G_{NHp}C_{N_3}$) was formed from the two nucleotide analogs ($G_{NHp}C_{NH_2}$) and ($pG_{NHp}C_{N_3}$)[2] after reaction with the condensation agent 1-ethyl-3-(3-dimethylaminopropyl)-carbodiimide (EDC).

Let us return to the hexanucleotide synthesis described above: the yields of new matrix were not as high as had been hoped. There are various reasons for this; for example, the condensation agent was inactivated by hydrolysis, and higher amounts of carbodiimide had to be used. In order to obtain higher yields of C, it was necessary to modify the building block B (for example by an amino modification). These changes led to an increase of the reaction rate by a factor of 10^4.

The experimentally detected square root law found by von Kiedrowski does not lead to the exponential growth shown by biological systems, but to a "parabolic growth" which lies between linear and exponential growth. As Eörs Szathmary from the University of Budapest demonstrated from theoretical principles, parabolic growth (which he described as "sub-exponential and supra-linear") does not lead to selection in the Darwinian sense but to a "more peaceful" state, i.e., coexistence in the competition of the matrices for the building blocks in a system (Szathmary and Gladkih, 1989). Parabolic replicators were also discovered in other laboratories, by Julius Rebek, L. Luisi (see Sect. 10.2) and R. Ghadiri (see Sect. 5.5). Thus, the range of this popular area of chemistry covers nucleic acids, membranes and peptides.

---

[2] Nomenclature: in the first nucleotide ($G_{NHp}C_{NH_2}$), G (guanosine) is linked to C (cytidine) via an NH–O–PO$_2$–O– bridge. The latter has a free amino group at the 3′ position of the ribose. In the second dinucleotide ($pG_{HNp}C_{N_3}$), a phosphate residue is present at the 5′ position of the sugar. The ligation of the two nucleosides takes place as described previously. The 3′ position is blocked by the azide residue (N_3).

In spite of all these successes, critics point out that from the perspective of bio-genesis and the self-replication methods reported, nature uses matrices which are almost solely complementary, and not self-complementary. These concerns also in-clude Orgel's result that a C-rich oligonucleotide leads, after replication, to a G-rich polymer which itself has only very limited use as a matrix. All these results refer to experiments on mononucleotides, so that the question arises as to whether the behaviour previously observed would remain unchanged if oligonucleotides were used instead of monomers.

We still need to clear up one or two points of nomenclature: in normal repli-cation of nucleic acids, the matrix (the + strand) and the newly formed daughter strand (− strand) are held together by Watson–Crick hydrogen bonding. This pro-cess is also referred to as "cross-catalytic". Normal autocatalysis is different: it leads to a product which corresponds in structure to the matrix, so that there is no differ-ence between the + and − strands. Such self-complementary sequences are called palindromes.

Sievers and von Kiedrowski (1994) showed that the fear that G-rich strands would be useless as matrices was unfounded. The system they used consisted of two classes of trimers: CCG and CGG. One class contains 3′-phosphate groups (CCGp and CGGp), the other 5′-amino residues (nCCG and nGGC). They were able to synthesize four hexanuncleotides, two of which were self-complementary: CCGpnCGG and CGGpnCCG. The others were complementary: CCGpnCCG and CGGpnCCG. The two classes of hexamer differ in their C and G content; while the first two contain the same number of C and G units, the distribution of C and G in the other two is unequal. In further experiments the authors found that self-complementary sequences are formed more quickly than complementary, and that there is no difference between C- and G-rich units. Temperature stud-ies suggest interactions in base stacking. Complementary sequences need cross-catalysis for their formation, i.e., reaction between + and − strands (von Kiedrowski, 1999).

Li and Nicolaou (1994) from the Scripps Institute, La Jolla, chose a differ-ent way of solving the problem: they used palindromic, double-strand deoxyri-bonucleotides consisting of 24 monomers. Each single strand consisted of only one base species, i.e., one contained only purines, the other only pyrimidines. Such a structure is able to bind complementary dodecameric pyrimidine-containing oligomers. Bonding occurs via the so-called Hoogsteen base pairing (a non-Watson–Crick base pairing of homogeneous double strands). The two dodecameric deoxy-oligonucleotides require a condensation agent for the activation of the 5′-phosphate chain end.

The liberation of the third, newly formed strand is of great importance in this process; it is made possible by adding free dodecameric purine oligonu-cleotides, which can bond to the newly-formed pyrimidine matrix by Watson–Crick pairing.

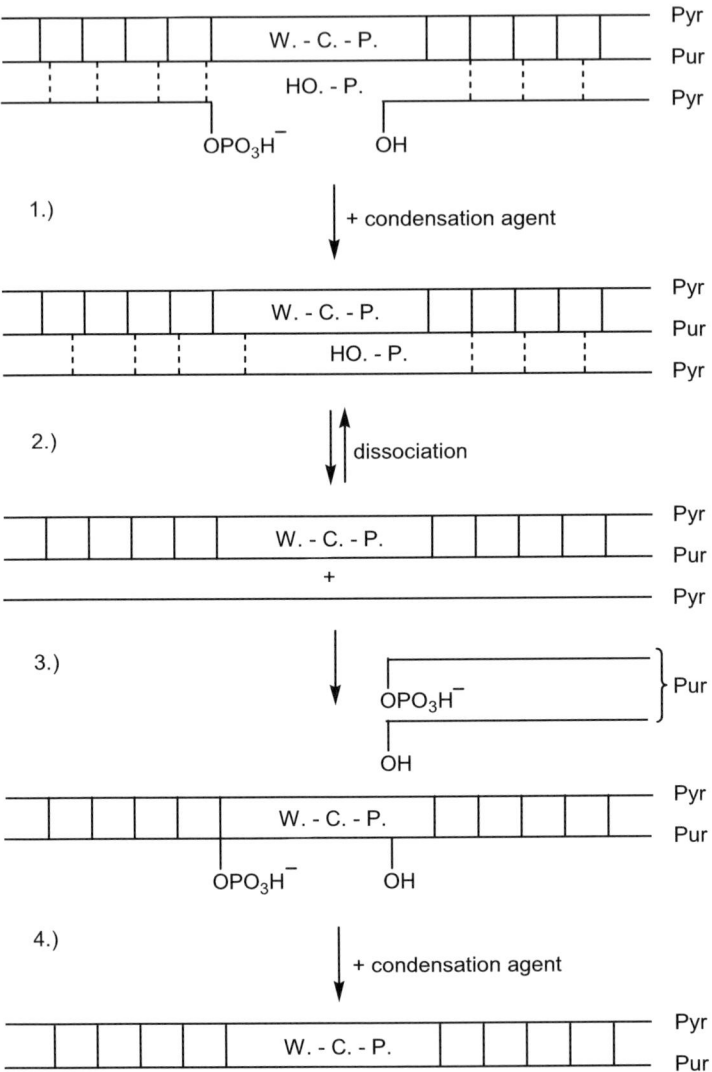

Fig. 6.7 Simplified scheme of the copying process in DNA replication via triple helices, as described by Li and Nicolaou (Ferris, 1994)

The following pairing rules apply to triple strands:

$$C \cdot G - CH^+ \text{ and } T \cdot A - T$$

The symbol $\cdot$ denotes Watson–Crick pairing, while $-$ means Hoogsteen pairing; CH^+ is protonated C (cytosine).

The mechanism of the Li–Nicolaou experiment shows the manifold possibilities for oligonucleotide synthesis. However, this replication system is limited to special cases. The palindromic sequences present must contain only a single species of base; these preconditions greatly restrict the importance of this approach as a biogenesis model.

The use of reactive surfaces for the specific synthesis of biomolecules, or as a model for replication processes, was first reported by Cairns-Smith and Weiss (see Sect. 7.1) and continued by G. Wächtershäuser (see Sect. 7.3), as well as J. Ferris and L. Orgel. It was thus appropriate to study the stabilisation of the reaction partners in enzyme-free self-replication at surfaces with reactive properties. As early as 1995, the group of G. von Kiedrowski (then at Freiburg, Germany) bonded reacting molecules at surfaces and then added the other required reaction components to the system in a stepwise manner (the latter process is referred to as "feeding").

The process known as SPREAD (Surface Promoted Replication and Exponential Amplification of DNA Analogues) attempts to reach the target, striven for by many researchers, of an exponential proliferation of biomolecules in model systems. As already mentioned, product inhibition (e.g., by dimerisation of the new matrices to give C_2) only allowed parabolic growth. In the SPREAD process, both solid phase chemistry and "feeding" have a positive effect on the synthesis. Thus, no separation processes are required, as excess reagents can be removed just by washing. The synthetic process consists of four steps:

Immobilisation of the matrix: in the SPREAD process, this takes place not at the surface of minerals, but at thiosepharose, a polymeric carbohydrate which carries an SH group.

Hybridisation: the matrix bonds the complementary units, one of which contains the function necessary for immobilisation.

Ligation: the two fragments are linked to give the complementary copy.

Separation (dissociation): separation of the copy from the matrix (this is carried out by washing with dilute sodium hydroxide).

The copy itself can now bond to the surface via the adhesive group and function as a matrix (provided that enough fragments are present).

What could be achieved by using the SPREAD process? The re-association of the newly formed copy and the matrix, and thus the product inhibition which causes parabolic growth, are prevented by bonding the matrices to a surface. The critical reader might remark that this process could not have been realized on the primordial Earth. We will return to this point at the end of the section.

"Feeding" can be avoided in the previous experiments if the molecular replication process at a surface is coupled with chromatographic separation. The two building blocks A and B are added to the mobile phase, while the matrices C are formed at the surface of the stationary phase. The main difference from the process described previously is that the matrices are not bonded covalently to the surface, but reversibly. During elution (which supplies new building block molecules to the system) a certain amount of matrix molecules is washed off and must be replaced by replication (von Kiedrowski, 1999).

Fig. 6.8 The coupling of
replication and
chromatography: a model
system on a chromatographic
column (von
Kiedrowski, 1999)

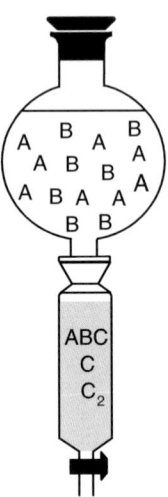

New developments point to future research directions: the connection between information processing and the replication processes of oligonucleotides. Branched oligonucleotides have been synthesized in which three molecules (either of the same type or of different types) are bonded at their 3′ ends to trifunctional linker molecules; the authors call these "trisoligonucleotidyles" (von Kiedrowski, 1999; Scheffler et al., 1999a, b). Under certain conditions, these polymers can associate to give nanocomplexes which, if a DNA double helix is considered as a single chemical bond and this way of thinking extended to the newly-formed product, can be given the name "nano-acetylene" or "nano-cyclobutadiene".

Fig. 6.9 Possible topologies
of a biomolecular complex
constructed from two
complementary
trisoligonucleotidyles. Each
DNA double strand is
represented as a single bond,
each unpaired
oligonucleotide as an
unpaired electron. Slightly
simplified from Scheffler
et al. (1999a, b)

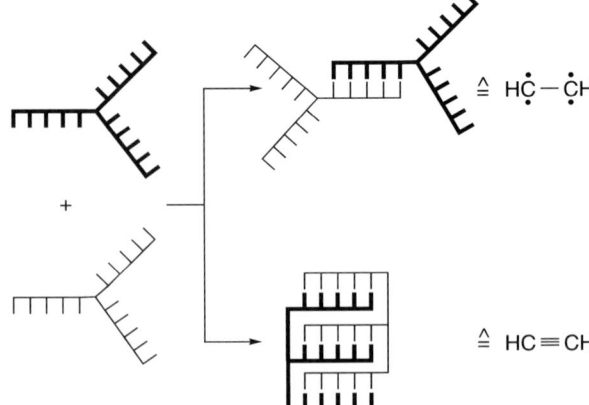

These experimental results are astounding; a few years ago, the reaction mechanisms for the self-replication of oligonucleotides which have been discovered and established would have been considered impossible!

However, the question must always be asked as to whether these processes could have taken place on the primordial Earth in its archaic state. The answer requires considerable fundamental consideration. Strictly speaking, most of the experiments carried out on prebiotic chemistry cannot be carried out under "prebiotic conditions", since we do not know exactly what these were. In spite of the large amount of work done, physical parameters such as temperature, composition and pressure of the primeval atmosphere, extent and results of asteroid impacts, the nature of the Earth's surface, the state of the primeval ocean etc. have not so far been established or even extrapolated. It is not even sure that this will be possible in the future. In spite of these difficulties, attempts are being made to define and study the synthetic possibilities, on the basis of the assumed scenario on the primeval Earth. Thus, for example, in the case of the SPREAD process, we can assume that the surface at which the reactions occur could not have been an SH-containing thiosepharose, but a mineral structure of similar activity which could have carried out the necessary functions just as well. The separation of the copy of the matrix could have been driven by a periodic temperature change (e.g., diurnal variation). For his models, H. Kuhn has assumed that similar periodic processes are the driving force for some prebiotic reactions (see Sect. 8.3).

Replication—and in particular, self-replication—phenomena are closely linked to the problem of "molecular recognition". Julius Rebek Jr. has worked on this important question (Rebek, 1990). As early as 1990 (when he was at MIT), he described a compound, synthesized in the laboratory, which was capable of replication (Tjivikua et al., 1990). It consisted of a nucleic acid component (an adenosine derivative) and a component with a recognition structure with which adenine could interact; the two components were joined by an amide bond of the type present in peptides (see Sect. 5.2); the system obeyed the square root law already described. The autocatalytic reaction observed led to daring suggestions of possible "initial primitive expressions of life". It was unavoidable that critical voices would respond to this highly optimistic interpretation of the successful experiments: such models do not deal with the formation of a living system, but with the study of mechanisms which could possibly have initiated a replication process. Thus, for example, Rebek used the highly reactive pentafluorophenyl group (which certainly never occurred on the primeval Earth) for the activation of one half of the molecule in preparation for linking the two components (Nowick et al., 1991).

When these reactions are considered, the question arises as to the forces which make complex structures possible, and which maintain them, leading to strong covalent bonds. There are three types of bond which are much weaker than covalent bonds, but are vital for their formation:

Hydrogen bonds
van der Waals attractive forces
Aromatic stacking

These (weak) interactions in particular lead to molecular recognition and thus to replication and self-replication; while they lead relatively easily to bonding, the bonds formed can, however, just as easily be broken again (Rebek, 1990, 1994).

6.5 Ribozymes

The following two important statements were considered at one time to be established scientific "dogma" in molecular biology:

The information stored in DNA is translated in the living cell to RNA, which leads to proteins:

$$DNA \rightarrow RNA \rightarrow proteins$$

The two nucleic acids serve only as information carriers, while proteins (as enzymes) are only responsible for functions in the cell:

$$DNA, RNA \rightarrow information\ carriers$$
$$proteins\ (enzymes) \rightarrow functions\ (catalysis)$$

Both dogmas had to be revised or expanded, the first because of the discovery of reverse transcriptase (an RNA-dependent DNA-polymerase) by D. Baltimore (MIT) and H. M. Temin (University of Wisconsin). This is an enzyme which catalyses the synthesis of DNA from single-strand RNA:

$$DNA \leftrightharpoons RNA \rightarrow proteins$$

The second dogma had to undergo drastic correction about ten years after the correction of the first: Sidney Altman from Yale University and Thomas Cech from the University of Colorado at Boulder independently discovered enzymatically active ribonucleic acids in different RNA species. This new class of RNA was called ribozyme (from *ribo*nuclease and en*zyme*). It was now necessary to modify the second dogma as follows:

$$Execution\ of\ a\ function \rightarrow protein\ \ +\ \ RNA$$
$$(catalysis)\qquad\quad (enzyme)\quad (ribozyme)$$

All four scientists whose work led to modification of the dogmas received the Nobel Prize. Thomas Cech (1987) was the first to observe enzyme-like reactions taking place at the same RNA strand, in ribosomal RNA (rRNA) from the protozoon *Tetrahymena thermophila*. The RNA produced, which is completely viable, is formed in a process in which certain sections (introns) of the primary copy (the transcription of DNA to mRNA) are cut out, the two remaining ends of the exon then being rejoined (spliced).

Sidney Altman discovered this property of RNA in the course of studies on precursor transfer RNA. It was realized that the catalytic properties of RNA are not exactly the same as those of protein enzymes, since the ribozyme is itself active and thus undergoes change during the catalytic reaction. This does not correspond to the generally accepted definition of an enzyme. Later studies, however, showed that some ribozymes are capable of acting catalytically at other RNA molecules. The ribozymes remain completely unchanged in this process, and thus fulfil the definition of a "real" enzyme.

Cech and co-workers obtained a ribozyme which had been shortened via splicing, the L-19 RNA, from the system described above: in the presence of guanosine or guanosine nucleotide as a cofactor, an intron with 414 nucleotides is cut out of the precursor rRNA, the result of splicing being L-19 RNA, which contains 395 nucleotides; Cech was able to show that L-19 RNA is able to shorten and lengthen other oligonucleotide chains.

The question arises as to whether comparisons with protein enzymes are justified. In other words, what can ribozymes really do? An important parameter for measuring the efficiency of enzymes is the value of k_{cat}/K_M. This quotient is derived from the values of two important kinetic parameters: k_{cat} is a rate constant, also called turnover number, and measures the number of substrate molecules which are converted by one enzyme molecule per unit time (at substrate saturation of the enzyme). K_M is the Michaelis–Menten constant; it corresponds to the substrate concentration at which the rate of reaction is half its maximum.

In many enzymes, the value of k_{cat}/K_M lies between 10^8 and 10^9 $M^{-1}s^{-1}$. The value for L 19 RNA is 10^3 $M^{-1}s^{-1}$, i.e., five orders of magnitude lower than for protein enzymes with high catalytic activity. However, L 19 RNA does compare in its efficiency to the enzyme ribonuclease A. The capabilities of ribozymes referred to above dealt solely with interactions of RNA (i.e., ribozymes) with RNA molecules. In a (hypothetical) RNA world, they would, however, need to be capable of doing much more, e.g., carrying out reactions at the carbon skeletons of biomolecules.

Cech's group was the first to have success in this direction (Piccirilli, 1992). Using a genetically modified Tetrahymena ribozyme, they were able to hydrolyse an ester bond between the amino acid N-formylmethionine and the corresponding tRNA$^{f\text{-Met}}$. The reaction was, however, very slow, only about 5 to 15 times faster than the uncatalysed reaction. The authors ventured to suggest that these ribozymes could have functioned as the first aminoacyl tRNA synthetases.

Soon after this report, the group of M. Yaros, also working in Boulder, was able to demonstrate ribozyme activity with a much higher performance (Illangsekare, 1995). Using a random mixture of many billions of RNA sequences, they selected one species which was able to catalyse the aminoacyl synthesis. In other words, the selected ribozyme aminoacylated its $2'(3')$ end when offered phenylalanyl-AMP; the addition of Mg^{2+} and Ca^{2+} was necessary. The catalysed reaction was about 10^5 times faster than in the absence of ribozyme. Thus the group was able to show that a fundamental reaction of contemporary protein biosynthesis can also be catalysed by a ribozyme (see Sect. 5.3.2). The assiduous search for further activities continues.

The importance of water molecules for the structural dynamics and the functioning of ribozymes was investigated by Rhodes and co-workers. They studied non-coded RNA using a combination of explicit solvent molecular dynamics and single molecule fluorescence spectroscopy approaches (Rhodes et al., 2006).

Jeff Rogers and Gerald F. Joyce (1999) studied the question as to whether a ribozyme containing only three different types of nucleotide rather than the normal four can also be catalytically active. They decided that the nucleoside cytidine (with the base cytosine) should be eliminated, for two reasons:

> Cytidine is the most labile of the four nucleosides (see Sect. 6.2). It deaminates spontaneously to uridine ($t^1/_2 = 340$ years at pH 7 and 298 K). This was possibly the reason that cytidine was not present in the original genetic material on the primeval Earth.
>
> Nucleic acids which contain only adenosine, guanosine and uridine are able to form A-U Watson–Crick pairs and G-U wobble pairs. They should be able to build up complex secondary and tertiary structures.

The authors obtained an RNA ligase ribozyme using the method of "in vitro evolution". Here, macromolecules are allowed to go through a series of synthetic cycles, which are followed by a proliferation phase, mutation and selection. As in Darwinian evolution, the goal is to carry out laboratory selection of molecules with certain required properties.

Fig. 6.10 Schematic representation of the principle of the evolution of a ribozyme in a test tube. Several mutants are selected in each cycle and proliferate in the next step. Slightly modified after Culotta (1992)

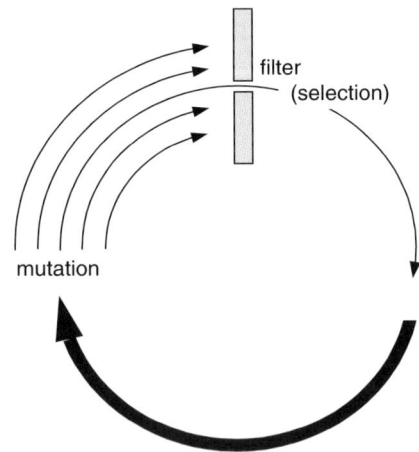

An introduction to the method of in vitro evolution is given by Wilson and Stoszak (1999). The RNA lipase ribozyme, with about 140 nucleotides (but without the pyrimidine base cytosine), folded in a defined structure and was able to reach a reaction rate 10^5 times higher than in the uncatalysed reaction. This result certainly surprised those biogenesis researchers who were critical of the RNA world; but we do not know whether the result changed their attitude to it!

"The ribosome is a ribozyme": this is how Cech (2000) commented on the report by Nissen et al. (2000) in *Science* on the successful proof of ribozyme action in the formation of the peptide bond at the ribosome. It has been known for more than 30 years that in the living cell, the peptidyl transferase activity of the ribosome is responsible for the formation of the peptide bond. This process, which takes place at the large ribosome subunit, is the most important reaction of protein biosynthesis. The determination of the molecular mechanism required more than 20 years of intensive work in several research laboratories. The key components in the ribosomes of all life forms on Earth are almost the same. It thus seems justified to assume that protein synthesis in a (still unknown) common ancestor of all living systems was catalysed by a similarly structured unit. For example, in the case of the bacterium *E. coli,* the two subunits which form the ribosome consist of 3 rRNA strands and 57 polypeptides. Until the beginning of the 1980s it was considered certain that the formation of the peptide bond at the ribozyme could only be carried out by ribosomal proteins. However, doubts were expressed soon after the discovery of the ribozymes, and the possibility of the participation of ribozymes in peptide formation was discussed.

6.6 Criticism and Discussion of the "RNA World"

The RNA world has its critics as well as its proponents. A third group is formed by the "soft critics", who warn of too much optimism; one of these is Leslie Orgel, an expert on the problems involved, who made it quite clear that the RNA world is a hypothesis and nothing more.

R. Shapiro is naturally one of the critics, and his voice cannot be ignored. In his book *Planetary Dreams* (1999) he lists the various arguments which cause him and others to favour alternative models of biogenesis over the RNA world. Shapiro is happy to report that scientists from a well-known laboratory gave him the nickname "Dr. No". Critical analysis of the experimental results on synthesis of nucleobases makes it clear that some of them were "forced" to proceed under very drastic conditions (Shapiro, 1996). Thus, for example, one guanine synthesis required cyanide concentrations of 2.2 M and more in order to obtain products! A guanine synthesis involving polymerisation of NH_4CN has been reported in Sect. 4.3. This dilemma can only be solved by means of new synthetic methods using highly active but as yet unknown catalysts.

Shapiro (2000) draws our attention to another type of problem: the homopolymer problem in biogenesis. Many biogenesis hypotheses presuppose the spontaneous formation of polymeric organic replicators which were formed from a mixture of inorganic compounds; these replicators consist of subunits of a known chemical species. Their structure involves a backbone which is bonded to information-transmitting residues. As we shall discuss in Sect. 6.7, not only RNA and DNA, but also proteins and peptide nucleic acids (PNA) have been suggested as possible information transmitters. Shapiro rightly considers that until now, not enough attention has been paid to the difficulties of obtaining such polymers. The prebiotic mixtures of chemicals present on the primeval Earth probably contained an immense number of compounds which competed with each other for incorporation in (hypothetical) polymers. He takes as an example the probability of obtaining a polypeptide consisting only of L-amino acids. If we include the components found in the Murchison meteorite (see Sect. 3.3.2.2), L-amino acids compete with D-amino acids, β-amino acids and also hydroxyacids for incorporation in the molecular chain. Some reactions lead to chain termination. Trifunctional substances such as aspartic acid, glutamic acid and serine can lead to chain branching. Other substances which are capable of replication probably have their own difficulties.

In prebiotic simulation experiments, the problems referred to were avoided by keeping competing substances out of the polymerisation mixtures, i.e., the conditions were idealized. The formation of an information-transmitting homopolymer from a complex mixture of substances cannot be excluded, but it is extremely unlikely!

How could such difficulties be avoided? Shapiro lists minerals which could have served either as the first replicators or as highly selective polymerases. He also considers a further possibility: life may have begun as a metabolic network of reactions which involved monomers; the replicators may have evolved in a later evolutionary phase. The misgivings mentioned, and the open questions referred to in earlier chapters, indicate that a de novo synthesis of RNA under the conditions present on the young Earth was almost impossible. Thus, models were and are being looked for which could bypass as many as possible of the problems referred to.

Shapiro remained true to his role of critical observer at the ISSOL conference in 2002 in Mexico; there he expressed the opinion that the beginnings of life did not involve polymers at all (be they nucleic acids or proteins, or their hypothetical precursors pre-nucleic acids or pre-proteins), but initially involved interactions between monomers, the polymeric biomolecules being formed in later phases of molecular evolution. In this "monomer world", reactions were supported by small biocatalysts (Shapiro, 2002).

The still open question, Information or metabolism first? has again been discussed by Robert Shapiro. In an article with the title "Did This Molecule Start Life? A Simple Origin for Life", he again stresses that it is improbable that life could have begun in an RNA world (referred to here as "RNA-first"). Shapiro offers his own suggestion in the metabolism debate; he assumes that cyclical processes, occurring in small compartments, lead from small molecules to systems of higher complexity. The Shapiro model takes into account aspects of the approaches and hypotheses proposed by Wächtershäuser (see Sect. 7.3), de Duve (see Sect. 7.4) and Kauffmann (see Sect. 9.3). In order to avoid one-sidedness, Shapiro's article is accompanied by a short reply "An RNA-First Researcher Replies". In this way, the reader is shown in a clear and understandable manner what the differences between the two approaches are (Shapiro, 2007).

An article by Dworkin, Lazcano and S. Miller in the *Journal of Theoretical Biology* shows how much is still unclear in the search for the first information-transferring species on Earth. The authors assume the RNA world to be a generally accepted hypothesis; however, there are still no acceptable models for developments before and after the RNA world. For example, did DNA dominate evolutionary events prior to or subsequent to the proteins? Metabolic arguments suggest that RNA genetic material preceded DNA, but the opposite can also be justified: 2'-deoxyribose is more stable, more reactive and more soluble than ribose. Although such discussions are only speculative, they offer strong incentives for carrying out further experiments (Dworkin et al., 2003).

The wide gap between the two opposing theories, "replication first" and "metabolism first", was analysed by Pross from the Ben Gurion University of the Negev (Israel). Pross concludes that replication came first! He is convinced that a causality between the two theories can only be established if it is assumed that the replication-first thesis is correct. His analysis also shows that more of the experimental results and theoretical rationales favour the replication thesis. The author finds his assumption justified that life processes are strongly kinetically controlled and that the development of metabolic pathways can only be understood if life is considered as a manifestation of "replicative chemistry" (Pross, 2004).

6.7 The "Pre-RNA World"

The considerable problems associated with the RNA world led to a search for simpler systems. Nelson et al. (2000a, b) sum up the greatest problems of the theory as follows:

> The instability of ribose and other sugars
> The difficulties of implementing glycosidic bonds between the nucleotides under the conditions present on the primeval Earth
> The inability to achieve double-sided, non-enzymatic matrix polymerisation

Efforts were concentrated mainly on the "backbone" of RNA, i.e., the sequence phosphate – D-ribose – phosphate – D-ribose – phosphate, which needed to be simplified without losing the most important RNA functions, such as base pairing and information transfer.

An extremely interesting polymeric substance which fulfilled the above conditions was discovered by researchers working on a quite different problem. P. E. Nielsen et al. from the Research Centre for Biotechnology in Copenhagen (1991) and Egholm et al. (1992) were looking for antisense agents[3] during work on computer models when they came upon a polymeric substance which they described as a "peptide nucleic acid (PNA)". PNA was intended to interact with complementary DNA or RNA via the normal Watson–Crick base pairing in order to block certain regions of, for example, virus nucleic acids.

---

[3] Antisense agents are substances (in this case oligonucleotides) which are intended to prevent illnesses, for example by blocking certain mRNA sequences.

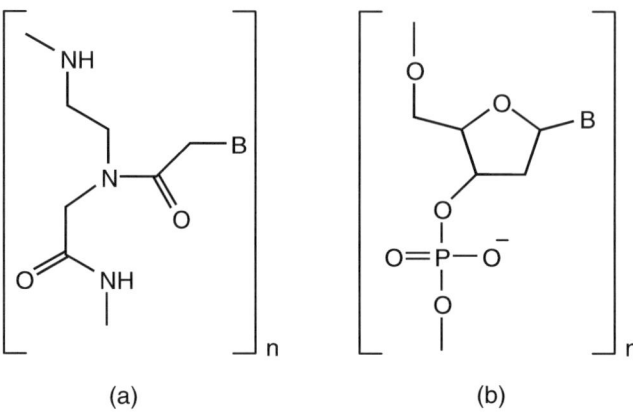

Fig. 6.11 Comparison of the structures of PNA (**a**) and DNA (**b**)

As Fig. 6.11 shows, the normal ribose phosphate chain of native RNA is replaced by the simpler backbone of polyamide polymers. The laboratory synthesis of these new polymers has been carried out.

The term "peptide nucleic acids" was chosen because of the peptide bond in the polymer (see Sect. 5.2). The bond between the polyamide strand and the organic bases involves an acetyl group. The formation of DNA-like double helix structures by PNAs was described by Pernilla Wittung et al. (1994). The question arises as to whether peptide nucleic acids can in fact be synthesized under prebiotic conditions.

One of the founding fathers of prebiotic chemistry, Stanley Miller, took up this question. The PNA polymer consists of ethylenediamine monoacetic acid (EDMA) units, which can also be called N-(2-aminoethyl)-glycine (AEG).

$$H_2N \diagdown \diagup \underset{H}{N} \diagup COOH$$

As early as 1996, Nelson and Miller (1996) reported initial successful experiments. They synthesized AEG and small amounts of ethylenediamine diacetic acid from mixtures containing ethylenediamine, formaldehyde and HCN; thus the formation of AEG (Fig. 6.12) is possible via the Strecker synthesis (see Sect. 4.1).

The yields of AEG were between 11 and 79%, at starting concentrations of 10^{-1}–10^{-4} M. AEG cyclizes to the lactam, monoketopiperazine, at neutral pH and around 373 K. The synthesis of the monomeric PNA units was carried out three years later (Nelson et al., 2000a, b). The synthesis of pyrimidine-N'-acetic acid can be achieved if the reaction of cyanoacetaldehyde with high concentrations of urea mentioned above is modified slightly, with hydantoic acid replacing urea (Robertson and Miller, 1995).

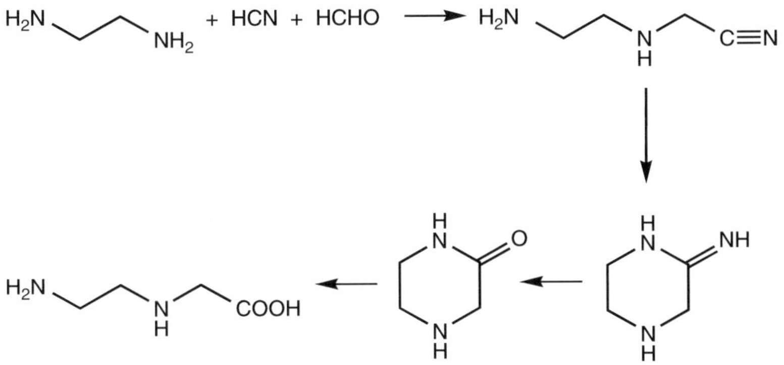

Fig. 6.12 The synthesis of *N*-(2-aminoethyl)-glycine, which is required for the construction of peptide nucleic acids. The reaction corresponds to the Strecker synthesis

Hydantoin, the cyclic form of hydantoic acid, was detected in the Murchison meteorite and also as a product of polymerisation of HCN (Ferris et al., 1974). The yields obtained (based on cyanoacetaldehyde) were 18% for cytosine-*N'*-acetic acid, but only 1.8% for the corresponding uracil derivative (1 mM cyanoacetaldehyde and 2 M hydantoic acid were allowed to react at around 373 K).

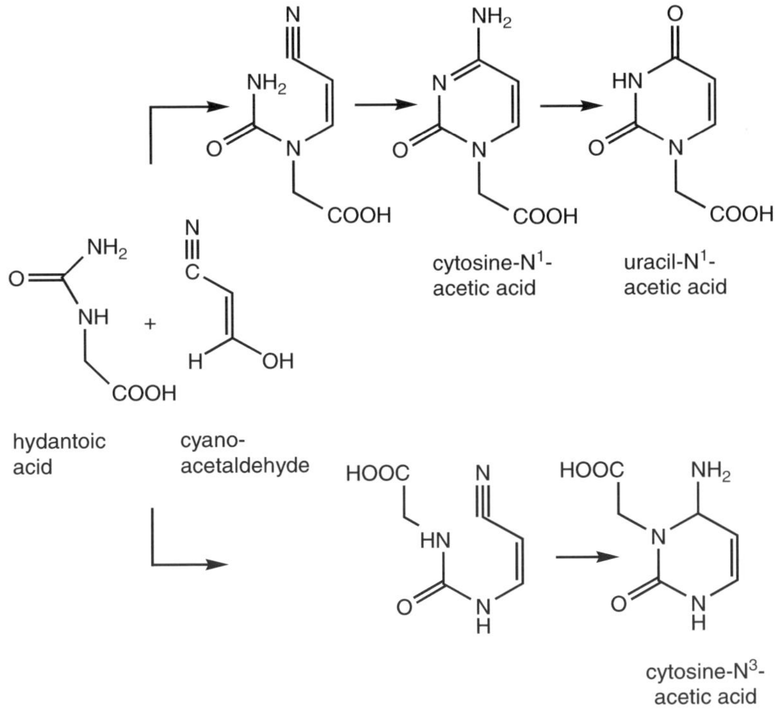

Fig. 6.13 Reaction routes for the synthesis of other components of PNA, such as cytosine- and uracilacetic acid (see Fig. 6.12) from hydantoic acid and cyanoacetaldehyde (Nelson et al., 2000b)

Although many questions are still open, peptide nucleic acids are easier to synthesize via simple reaction routes than is natural RNA. The PNAs have another important advantage: they are achiral and uncharged, i.e., they contain no chiral centres in the polymeric backbone (see Sect. 9.4). Unfortunately, however, they do not fulfil all the necessary conditions for molecular information storage and transfer. Thus, the search for other possible candidates for a pre-RNA world continues.

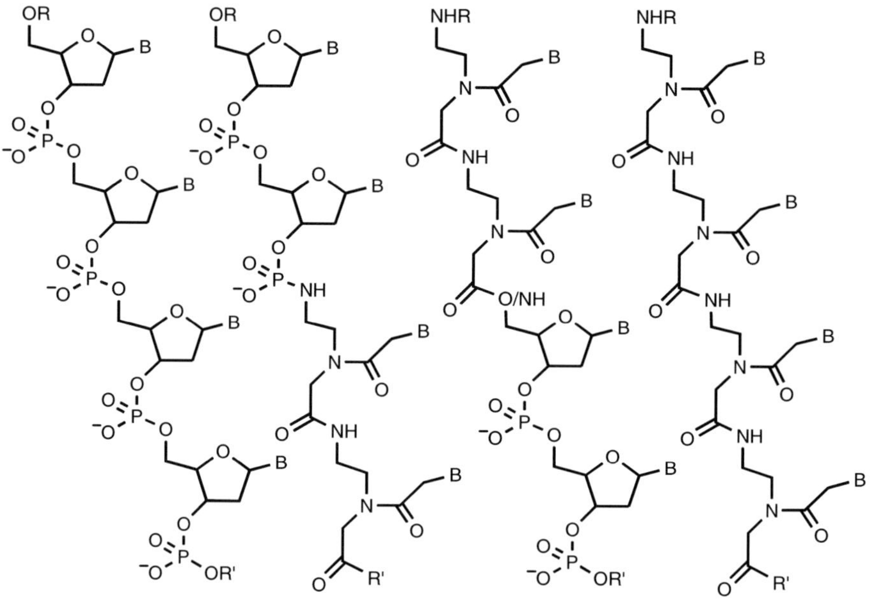

(a)   (b)   (C)   (d)

Fig. 6.14 The structures of DNA (**a**), the two chimera linkage products (**b** and **c**), and PNA (**d**) (Koppitz et al., 1998)

We cannot exclude that not just one but two, or even several, types of molecules predated the RNA phase. In this case, an evolutionary transition from one polymeric species to another without loss of functionality (i.e., information transfer and replication properties) must have been possible. Such a process would be facilitated by the formation of "chimeras", i.e., of polymers which contained both types of species. This problem, the formation of oligonucleotide-PNA-chimeras in a matrix-directed reaction, has been stressed by Koppitz et al. (1998). The term "chimera" comes from Greek mythology; a chimera was a creature consisting of parts of a lion, a goat and a snake. Prebiotic chimeras contain PNA and normal DNA in the form of oligomers which contain only a few monomers. Linkage via the phosphoamidate and the 5'-ester bond between DNA and PNA was favoured. In both cases, both PNA and DNA could function as a matrix for the linkage of the two monomers. Ligation via a 5'-phosphoamidate bond was unfavourable (Fig. 6.14). Studies on the olefinic peptide nucleic acids (OPAs) introduced by Schütz et al. (2000) are mainly concerned with antisense research.

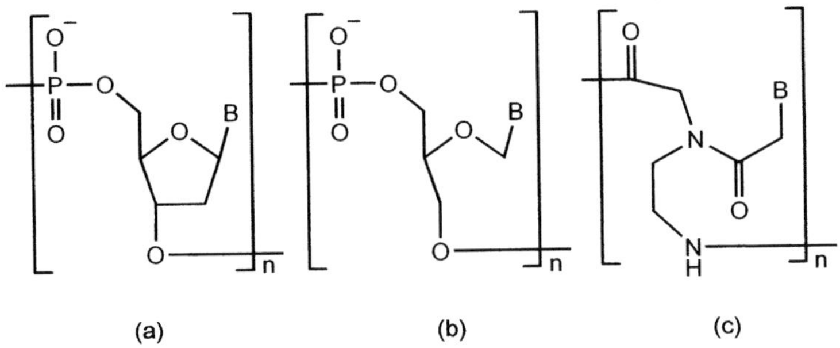

Fig. 6.15 Structural comparison DNA (**a**), Gly-NA (**b**), and PNA (**c**), (Merle and Merle, 1996)

The search for other powerful backbone structures led to the poly(glycerotides). Acyclic oligonucleotides are referred to as glycerine nucleic acids (Gly-NA).[4] They consist of a glycerine backbone in which the C2′ atom of ribose is missing. This class of molecules was first described by Schwartz and Orgel (1985). The molecules have one asymmetric carbon atom per monomer unit (Merle et al., 1993). Some Gly-NAs form complexes with natural polynucleotides (J. and M. Merle, 1996), e.g., dodecaglycerine-adenine $(Gly-A)_{12}$ complexes with dodecathymidylate $(dT)_{12}$ at temperatures below 298 K. However, this complex is less stable than a corresponding oligonucleotide duplex. The thermal stability of various oligonucleotides and their analogues in double strands falls in the following sequence:

$$PNA > DNA > RNA > Gly-NA$$

Thus, the PNAs seem more likely to be RNA precursors than do Gly-NAs (J. and M. Merle, 1996).

Some biogenesis researchers are of the opinion that the model substances considered until now for a pre-RNA are still too complex. Prebiotic chemistry on the primeval Earth was certainly much simpler than some models suggest. In an intensive search for new classes of compound which could replace the phosphate backbone of RNA, chemists from Atlanta, Los Angeles and London found a known, simple organic compound which could do this job in an excellent manner: glyoxylate, the ionized form of the C_2 compound glyoxylic acid, the simplest aldehydecarboxylic acid. In the pre-RNA world phase, this could have replaced the phosphate group in polymers. The authors use the term gaRNA (from glyoxylate-acetate). The yields obtained until now under neutral conditions are still very low (about 1%), but increases appear possible. However, two conditions must be fulfilled: the reaction requires the presence of Mg^{2+} ions, and water has to be removed from the reaction mixture (for example by evaporation). More detailed studies show that the structure of the polymer as well as its electrostatic charge correspond to "normal" RNA (Bean et al., 2006).

---

[4] Unfortunately, the literature cited here uses the abbreviation Gly, which in biochemistry is generally reserved for the amino acid glycine, for glycerine.

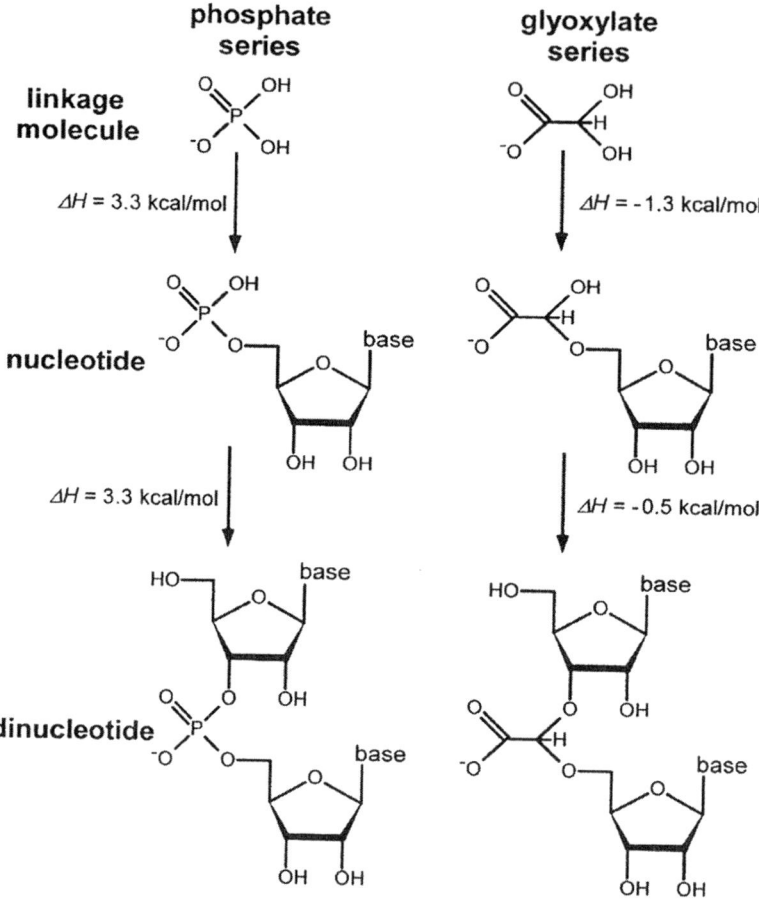

Fig. 6.16 Comparison of phosphate and glyoxylate chemical structures, and their associated nucleotides and dinucleotides (From H.D. Bean et al., 2006)

Oligonucleotides with modified sugar components are another alternative to PNAs; work in this direction was begun by Albert Eschenmoser, a famous synthetic chemist who was interested in the question as to why nature chose certain biomolecules for the processes of life and not others (Eschenmoser, 1991). This group carried out studies on the sugar components of the nucleic acids, in order to find out why D-ribose was used rather than another sugar.

In successful sugar syntheses (see Sect. 4.4), for example, in the chemistry of glycolaldehyde phosphate, the problem arose as to why only pentose nucleic acids are formed, but no hexose nucleic acids. The hexose D-allose occurs in two forms: as the five-membered furanose ring and the six-membered pyranose ring. The pyranose is more stable, as it is able to exist in the more thermodynamically stable chair form. For technical reasons, the Eschenmoser group first used the sugar in a form in which the two OH groups in 2′ and 3′ position are missing, i.e., the 2′,3′-dideoxyallopyranose. Since the sugar component contains one more CH_2 group than does the deoxyribose in normal DNA, this derivative is known as "homo-DNA"; it forms antiparallel double strands. The melting points (T_m values) of homo-DNA lie 30–40 K higher than those of analogous oligonucleotides in normal DNA, i.e., the double strand is separated into single strands at higher temperatures. Homo-DNA does not interact with normal DNA. The intensive study of sugar analogs in oligonucleotides led to the insight that the Watson–Crick base pairing rules are not only a result of the hydrogen bonding properties of the bases involved, but also of the furanose structure of the sugars in the nucleic acid chain (Eschenmoser, 1991).

As already shown in Sect. 4.4, ribose-2,4-diphosphate is obtained in a base-catalysed condensation of glycolaldehyde phosphate in the presence of formaldehyde (Müller et al., 1990). The phosphate group in the 4 position of the sugar prevents the formation of a 5-membered furanose ring, but a 6-membered pyranose structure can be formed.

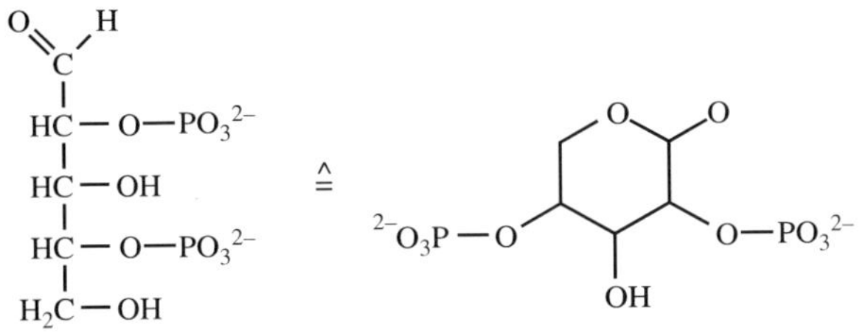

The oligonucleotides formed with the 6-membered ring sugar are referred to as "pyranosyl RNA (pRNA)"; in this derivative, the nucleotides are bonded via a $2'$-$4'$ linkage (Eschenmoser, 1994).

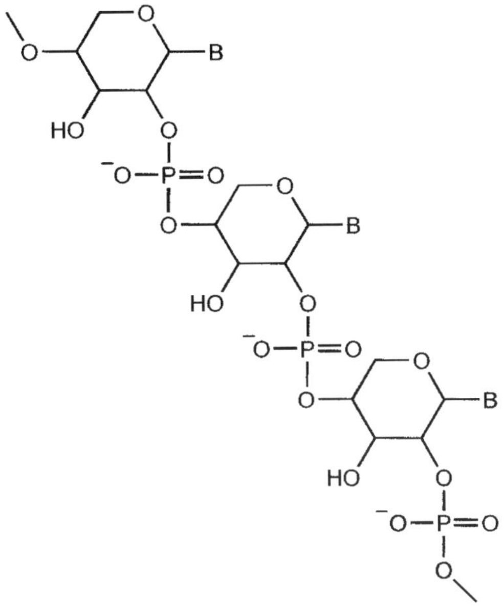

Fig. 6.17 Constitution of pyranosyl-RNA (pRNA) (Eschenmoser 1994)

Complementary pRNA strands form double strands, and they interact more strongly and selectively than do DNA or RNA strands (Schwartz, 1998). It does, however, remain unclear as to whether pRNA can be considered a candidate for a pre-RNA species. Eschenmoser has provided a comprehensive account of the aetiology of the nucleic acid structures in *Science* (Eschenmoser, 1999). pRNA is probably not a suitable RNA precursor, as it cannot undergo Watson–Crick base pairing with normal RNA; this deficit precludes information transfer from precursor to successor. However, since this "genetic takeover" is considered to be the most important property of prenucleic acids, the question remains unsolved. Eschenmoser's group was also successful here; it had previously been considered that the backbone of the nucleic acids must contain at least six atoms between the repeating units in order to make normal base pairing possible, but the group showed that this assumption is not correct (Orgel, 2000).

When working as a guest at the famous Salk Institute for Biomedical Research in La Jolla (California), Eschenmoser (now a professor emeritus) was able to study a series of sugar derivatives as possible candidates for pre-RNA models. It turned out that five atoms in the nucleic acid chain suffice, if they are present in an optimal extended form. This was shown using a polymer containing (L)-α-threofuranosyl-$(3' \rightarrow 2')$-oligonucleotides (TNA) (Fig. 6.18) (Schöning et al., 2000).

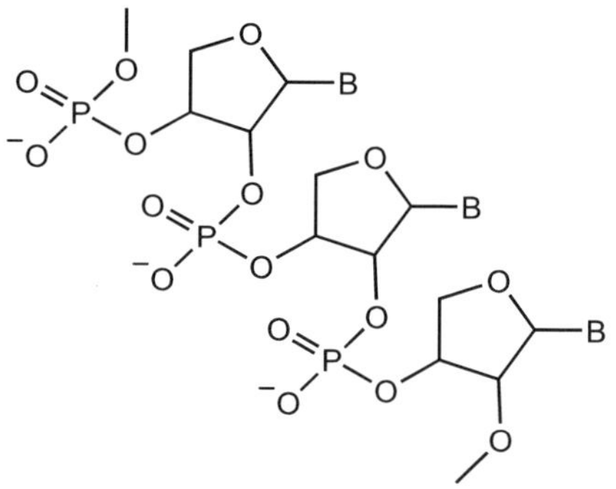

Fig. 6.18 Constitution of the "TNA", (L)-α-threofuranosyl-$(3' \rightarrow 2')$-oligonucleotide, strand (Schöning et al., 2000)

The 5-membered ring sugar contains only 4 carbon atoms; there are only 2 carbon atoms between the oxygen atoms linking the sugar moieties, and not 3 as in ribose or deoxyribose units. The most important reason why tetroses have advantages as prenucleic acid sugars is that they are easier to synthesize. Sugars containing 4 carbon atoms would have probably been formed more easily on the primeval Earth than pentose sugars with five. Thus tetroses can be formed directly from two identical C_2 units, for example two glycolaldehyde molecules. Pentose sugars are formed in a complex synthesis from a C_2 and a C_3 building block; the synthesis leads to mixtures of sugars with four, five and six carbon atoms, which are difficult to separate. TNAs can form stable Watson–Crick double helices and—to the surprise of the experts— they are also capable of forming stable double helix bonds with complementary RNAs and DNAs (Eschenmoser, 2004).

Although the pre-RNA world is now much more the centre of scientific attention
in prebiotic chemistry, there have been several attempts in recent years to understand
the synthesis of oligonucleotides from the normal nucleotides by using simulation
experiments (Ferris, 1998). In condensation reactions in aqueous media, there is
always competition between synthesis and hydrolysis; synthesis is generally only
successful when supported by catalysts.

A new scenario has been provided by polymer formation catalysed by mineral
surfaces; the most important feature here is probably the concentration of the starting molecules at the surface.

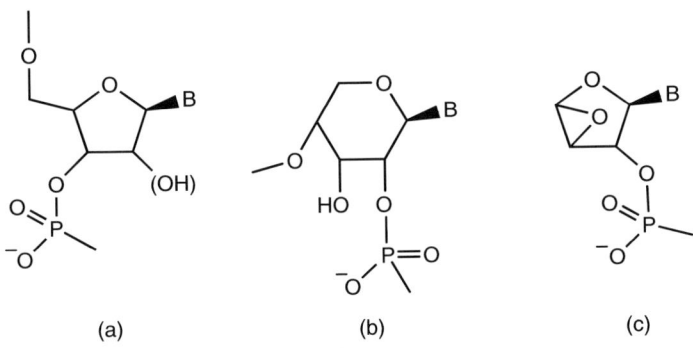

(a)   (b)   (c)

Fig. 6.19 Three different units for the synthesis of oligonucleotides (Herdewijn, 2001):
(**a**) RNA ($5'{\rightarrow}3'$)-β-D-ribofuranosyl
(**b**) p-RNA ($4'{\rightarrow}2'$)-β-D-ribopyranosyl
(**c**) TNA ($3'{\rightarrow}2'$)-β-L-threofuranosyl

Certain minerals, such as montmorillonite (see Sect. 7.1), have been used successfully in such experiments for many years. The polycondensation of phospho-
imidazolides of adenosine (ImpA) using montmorillonite led to chains of up to 55
nucleotides. The reaction requires the presence of Mg^{2+} ions and an oligo-A-primer.
Since montmorillonite layers carry negative charges, the doubly positively charged
magnesium ions function as intermediaries between the negative charges of the nucleotide phosphate esters and the mineral surface. A chain length of 50 nucleotides
is not reached "in one go", but by "feeding" the reaction mixture, i.e., stepwise
addition of the activated nucleotides over the course of about 14 days (Ferris, 1996).

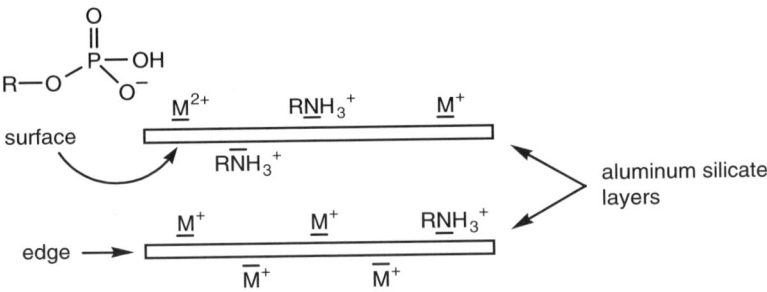

Fig. 6.20 Simplified diagram of montmorillonite clay with anionic and cationic organic compounds bonded to it (Ferris, 1998)

A chain length of 50–60 nucleotides is generally considered to be sufficient to allow such an oligonucleotide to act as a catalyst (as does a ribozyme). The groups of Ferris and Orgel, working together, were able to synthesize homo-oligomers with the two amino acids aspartic and glutamic acid, although at the surface of other minerals (illite and hydroxyapatite). Physicochemical processes, as well as studies of the action of inhibitors on RNA formation on montmorillonite surfaces, were used to elucidate the reaction mechanism: an upper limit of $5–10 \times 10^{15}$ catalytically active regions on around 50 mg of montmorillonite was determined (Wang and Ferris, 2001).

In a closer examination of the formation of oligomers, Ferris and co-workers found that the reaction is favoured by relatively high salt concentrations (e.g. 1M NaCl), while the presence of divalent cations is not necessary. The formation of RNA oligomers was found to be temperature sensitive: the yields decrease when the temperature is raised from 4°C to 50°C. The addition of meteorite material (from 3 meteorites) does not catalyse the polymerisation reaction; only galena (PbS) can do this. The authors thus assume that RNA oligomers could have been formed on the young Earth in solutions of alkali metal salts in the presence of montmorillonite and a pH value of 7–9 (Miyakawa et al., 2006). Ferris (2006) has provided a short but up-to-date survey of montmorillonite-catalysed RNA oligomerisation.

Numerous studies on prebiotic syntheses carried out in recent years make the assumption that adsorption processes of micromolecules on clay minerals must have been of decisive importance: on the one side for the concentration of aqueous solutions, and on the other because the reactions were only possible due to catalytic effects at the mineral surface. Recent experiments using montmorillonite and kaolin showed that mono- and divalent cations play an important role in the adsorption of nucleic acids. The concentrations necessary for cation adsorption depend on both the valency of the cations and the nucleic acid species involved. Thus, double strand nucleic acids require higher cation concentrations than those with single strands in order to be adsorbed to the same extent at the clay surface; divalent cations are, of course, more efficient than are monovalent ions. It is not possible to remove clay cations by washing once the complex between the clay cation and the nucleic acid has been formed. The cations fulfil a "bridging" function between the negative charges of the clay mineral surface and the negative charges of the phosphate groups of the nucleic acids (Franchi et al., 2003).

A majority of biogenesis researchers seem to accept the RNA world, in spite of the many unanswered questions; a world in which protein synthesis, coded by nucleic acids and catalysed by ribosomes, did not yet occur. Leslie Orgel, probably the most profound connoisseur of this difficult métier, calls attention to important consequences which follow from acceptance of the RNA world. In summary, Orgel discusses two possibilities:

> The RNA world was *the first biological world*. If this is the case, we can learn or predict but little about the prebiotic chemistry of the RNA world from the biochemistry of today; perhaps only the fact that the formation and polymerisation of nucleotides were once prebiotic processes. Thus Orgel is not of the

opinion that we can learn anything about processes of prebiotic chemistry by extrapolating backwards from contemporary biochemical processes, as, for example, in the case of the enzyme carbon monoxide dehydrogenase/acetyl CoA synthase. This enzyme, which has an Fe/Ni/S centre, reduces CO_2 to CO, which is passed on to a similarly active centre and incorporated into acetyl CoA.

Huber and Wächtershäuser (1997) were able to demonstrate the incorporation of CO at an Fe/Ni sulphide catalyst. Orgel, however, doubts that it is justified to draw further conclusions from this result, since this enzyme mechanism may have developed after protein synthesis began or evolved. Orgel considers the use of the mechanism referred to above of CO incorporation in biomolecules to be a result of chemical determinism.

The RNA world was *not the first biological world*. In this case, the conclusions drawn above are not justified. We can then speculate that the monomers of an early genetic polymer can still be recognized today as important biochemical substances. Orgel considers it unlikely that RNA could have been formed prebiotically: ribonucleotides are too complex to have evolved on the primeval Earth in sufficient amounts, and with sufficient purity.

The still hypothetical RNA world must fill the gap. Orgel suspects that, in this still dark phase of chemical and molecular evolution, a class of molecules was present which could have been relatively easily formed under the conditions present on the young Earth, and which was found in rocks from meteorites: perhaps amino acids. These need not to have been the 20 proteinogenic amino acids which we know today, and which are in part quite complex, such as phenylalanine, histidine and tryptophane: simple examples would have sufficed.

The main problem, that of finding a polymeric compound built up from amino acids and capable of pairing (as, for example, in DNA), is clearly a structural problem. There are only two possibilities for its solution:

– The occurrence of β-linked peptides or
– The incorporation of L- and D-amino acids in the peptide chain.

Orgel considers that, among the many known polymers built up from amino acids and nucleobases, one group of polymers is particularly interesting: this was obtained by a group working at the Technical University of Munich in Germany (Diederichsen, 1996, 1997; Diederichsen and Schmitt, 1998). These are peptides with alternating D- and L-amino acids (in this case alanine), which are linked by substitution of one of the β-H atoms by guanine or cytosine (alanyl-PNA); they form a linear alanyl-PNA double strand.

Whether the polymers shown schematically in Fig. 6.21 (or similar forms) were ever of importance on the Earth as information carriers is an open question. Leslie Orgel considers the alanyl nucleotides synthesized by Diederichsen to be possible and suitable candidates for a pre-RNA world (Orgel, 2003).

Fig. 6.21 Section
(simplified) of a model of the
linear alanyl PNA double
strand. The regular peptide
bonds are alternately
configured, and thus take up a
β-sheet configuration
(Diederichsen, 1997)

The vast, open-ended area of nucleic acid chemistry, with its many variants and precursors, is likely to provide us with more surprising results in the future.

References

Bean HD, Anet FAL, Gold JR, Hud NV (2006) Orig Life Evol Biosphere 36:39
Cech TR (1986) Scientific American, November p 76
Cech TR (2000) Science 289:878
Crick FHC (1968) J Mol Biol 38:367
Culotta E (1992) Science 257:613
de Graaf RM, Schwartz AW (2005) Orig Life Evol Biosphere 35:1
Diederichsen U (1996) Angew Chem 108:458, Int Ed 35:445
Diederichsen U (1997) Angew Chem 109:1966, Int Ed 36:1886
Diederichsen U, Schmitt HW (1998) Angew Chem 110:312, Int Ed 37:302
Dworkin JP, Lazcano A, Miller S (2003) J Theor Biol 222:127
Egholm M, Buchhardt O, Nielsen PE, Berg RH (1992) J Amer Chem Soc 114:1895
Eigen M (1971) Nature 58:465
Eschenmoser A (1994) Orig Life Evol Biosphere 24:389
Eschenmoser A (1999) Science 284:2118
Eschenmoser A (2004) Orig Life Evol Biosphere 34:277
Ferris JP, Wos JD, Lobo AP (1974) J Mol Evol 3:311
Ferris JP (1994) Nature 369:184
Ferris JP, Hill AR, Liu R, Orgel LE (1996) Nature 381:58
Ferris JP (1998) Catalyzed RNA synthesis for the RNA world. In: Brack A (Ed.) The Molecular
    Origins of Life. Cambridge University Press, pp 255–268
Ferris JP (2006) Phil Trans Roy Soc B 361:1777
Franchi M, Ferris J, Gallori E (2003) Orig Life Evol Biosphere 33:1
Fuller WD, Sanchez RA, Orgel LE (1972a) J Mol Biol 67:25
Fuller WD, Sanchez RA, Orgel LE (1972b) J Mol Evol 1:249
Gilbert W (1986) Nature 319:618
Herdewijn P (2001) Angew Chem 113:2309
Huber C, Wächtershäuser G (1997) Science 276:245
Illangasekare M, Sanchez G, Nickles T, Yaros M (1995) Science 267:643
Inoue T, Orgel LE (1981) J Amer Chem Soc 103:7666
Inoue T, Orgel LE (1983) Science 219:859
Inoue T, Joyce GF, Grzeskowiak K, Orgel LE, Brown JM, Reese CB (1984) J Mol Biol 178:669
Joyce GF (1989) Nature 338:217

Joyce GF, Orgel LE (1999) Prospects for Understanding the Origin of the RNA World. In: Gesteland RF, Cech TR, Atkins JF (Eds.) The RNA World, 2nd Ed. Cold Spring Harbour Laboratory Press, Cold Spring Habor, New York

Koppitz M, Nielsen PE, Orgel LE (1998) J Amer Chem Soc 120:4563

Kuhn H (1972) Angew Chem 84:838, Int Ed 11:798

Levy M, Miller SL, Oró J (1999) J Mol Evol 49:165

Li T, Nicolaou (1994) Nature 369:218

Lohrmann P, Bridson PK, Orgel LE (1980) Science 208:1464

Merle L, Spach G, Merle Y, Sági J, Szemzö A (1993) Orig Life Evol Biosphere 23:91

Merle Y und L (1996) Orig Life Evol Biosphere 26:400

Miyakawa S, Yushi PC, Gaffey MJ, Gonzales-Torril E, Hyland C, Ross T, Rubij K, Ferris JP (2006) Orig Life Evol Biosphere 36:343

Müller D, Pitsh S, Kittaka A, Wagner E, Wintner CE, Eschenmoser A (1990) Helv Chim Acta 73:1410

Naylor R, Gilham PT (1966) Biochemistry 5:2722

Nelson KE, Miller SL (1996) Orig Life Evol Biosphere 26:345

Nelson KE, Levy M, Miller SL (2000a) Orig Life Evol Biosphere 30:259

Nelson KE, Levy M, Miller SL (2000b) Proc Natl Acad Sci USA 97:3868

Nielsen PE, Egholm M, Berg RH, Buchhardt O (1991) Science 254:1497

Nissen P, Hansen J, Ban N, Morre PB, Steitz TA (2000) Science 289:920

Nowick JS, Feng Q, Tjivikua T, Ballester P, Rebek J Jr (1991) J Am Chem Soc 113:88

Orgel LE (1968) J Mol Biol 38:381

Orgel LE (2000) Science 290:1306

Orgel LE (2003) Orig Life Evol Biosphere 33:211

Oró J, Basile B, Cortes S, Shen C, Yamrom T (1984) Origin of Life 14:237

Piccirilli JA, McConnell TS, Zaug AJ, Noller HF, Cech TR (1992) Science 256:1420

Ponamperuma C, Sagan C, Mariner R (1963) Nature 199:222

Pross A (2004) Orig Life Evol Biosphere 34:307

Rebek J Jr (1990) Angew Chem 102:261, Int Ed 29:245

Rebek J Jr (1994) Scientific American, July

Reimann R, Zubay G (1999) Orig Life Evol Biosphere 29:229

Rhodes MM, Rébelová K, Sponer J, Walter NG (2006) Proc Natl Acad Sci USA 103:13380

Robertson MP, Miller SL (1995) Nature 375:772

Rogers J, Joyce GF (1999) Nature 402:323

Saygin Ö, Ellmauerer E (1984) Origins of Life 14:139

Scheffler M, Dorenbeck A, Jordan S, Wüstefeld M, von Kiedrowski G (1999b) Angew Chem Int Ed 38:3312

Schöning K-U, Scholz P, Guntha S, Wu X, Krishnamurthy R, Eschenmoser A (2000) Science 290:1347

Schramm G, Groetsch H, Pollmann W (1962) Angew Chem 74:53, Int Ed 1:1

Schütz R, Cantin M, Roberts C, Greiner B, Uhlmann E, Leumann C (2000) Angew Chem 112:1305, Int Ed 39:1250

Schwartz AW, Orgel LE (1985) Science 228:585

Schwartz AW (1998) Origins of the RNA world. In: Brack A. (Ed.) The Molecular Origins of Life. Cambridge University Press, p 241

Shapiro R (1996) Orig Life Evol Biosphere 26:238

Shapiro R (1999) Planetary Dreams. Wiley & Sons, New York

Shapiro R (2000) Orig Life Evol Biosphere 30:243

Shapiro R (2002) Book of Program and Abstracts, ISSOL-02 Oaxaca, p 60

Shapiro R (2007) Scientific American 296:624

Sievers D, von Kiedrowski G (1994) Nature 369:221

Sleeper HL, Lohrmann R, Orgel LE (1979) J Mol Evol 13:203

Szathmáry E, Gladkih I (1989) J Theor Biol 138:55

Tjivikua T, Ballester P, Rebek J Jr (1990) J Am Chem Soc 112:1249

Unrau PJ, Bartel DP (1998) Nature 395:260
von Kiedrowski G (1986) Angew Chem 98:932, Int Ed 25:932
von Kiedrowski G (1993) Bioorganic chemistry frontiers, Vol 3 / 113
Wang K-J, Ferris JP (2001) Orig Life Evol Biosphere 31:381
Wilson DS, Stoszak JW (1999) Ann Rev Biochem 68:611
Wittung P, Nielsen PE, Buchardt O, Egholm M, Nordin B (1994) Nature 368:561
Woese C (1967) The genetic code. Harper & Row, New York
Yamagata Y (1999) Orig Life Evol Biosphere 29:511
Yamagata Y (2000) Orig Life Evol Biosphere 30:198
Zielinski WS, Orgel LE (1987) Nature 327:346
Zubay G (2000) Origins of Life on the Earth and in the Cosmos 2nd Ed. Academic Press, San
    Diego, New York
Zubay G, Mui T (2001) Orig Life Evol Biosphere 31:87

Chapter 7
Other Theories and Hypotheses

7.1 Inorganic Systems

More than 50 years ago, the English physical chemist J. D. Bernal (1901–1971) suggested that clay minerals may have played a key role in synthetic processes taking place on the primordial Earth; he was referring to the adsorption and concentration of organic substances at the surface of such minerals.

Four billion years ago, the Earth's thin crust consisted of geochemicals (i.e., compounds containing the elements Si, O, Al, Fe, Mg, Ca, K and Na, as well as traces of other elements). Thus, some biogenesis researchers believed that the first replicating material consisted of geochemical material rather than substances containing carbon and other bioelements. Clay minerals in particular were included in experimental and theoretical studies. The most important are kaolinite and montmorillonite; the latter was, and still is, used in many experiments carried out to simulate prebiotic reactions.

In 1963, Armin Weiss (then at the University of Heidelberg, Germany) reported the intercalation of amino acids and proteins in mica sheet silicates (Weiss, 1963). Some years later, U. Hoffmann, also from Heidelberg, published an article titled "Die Chemie der Tonmineralien" (The Chemistry of Clay Minerals), in which he mentioned possible catalytic activity of clays in processes which could have led to the emergence of life (Hoffmann, 1968).

The most radical proponent of a hypothesis of the origin of life based on processes occurring on clay minerals was Graham Cairns-Smith from the University of Glasgow (Cairns-Smith, 1966, 1985a, b). When his critics asked him why there was no evidence which supported his ideas, he answered with the following analogy: the construction of a vault requires stable scaffolding; when the vault is finished, the scaffolding is pulled down, and no trace of the support is left! Cairns-Smith considers it unwise to look for "smaller versions" of today's genetic materials. The first material capable of reproduction must have been much simpler than today's genetic material and must have consisted of chemical substances which were present everywhere on the young Earth. This first (as yet unknown) material with the ability to transfer information must have been replaced later by an also unknown, but more complex type of substance, in which the necessary properties were retained or enhanced. Cairns-Smith calls this move from one class of substance, or system, to another the "genetic takeover".

According to Cairns-Smith, the first primitive "gene" materials could have been clay minerals; these crystallize out everywhere on Earth from dilute silica solutions and hydrated solutions of metal ions. Both groups of substances are continually being formed by weathering processes. Two cycles keep this dynamic process going:

> The geological cycle: sediments from deep regions of the Earth, which are subject to high pressure and high temperature, are brought back to the Earth's surface by geological dynamics. The metamorphic rock is more labile than the starting material and can be leached more easily, so that a cycle is set up in which new clay species are formed.
> The water cycle.

Cairns-Smith is careful enough to concede that the first hypothetical information-carrying material was not of necessity a clay mineral; however, the basic features of the model can best be demonstrated using different clay species. Thus, for example, clays could have crystallized out in sandstone pores from solutions containing products derived from weathering. The result would have been clay layers, which could have been separated and transported further by external influences; replication under similar conditions would have followed. Such crystallization processes would have also involved errors, such as defects, vacancies, and the incorporation of other ions or atoms; these "inorganic mutations" would have been passed on, i.e., they would have been incorporated into the next sheet to be formed.

Fig. 7.1 Greatly simplified model of the formation of crystal layers by the alternation of growth and cleavage

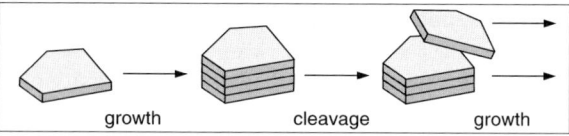

growth   cleavage   growth

Clays consist of parallel silicate layers: in kaolinite, two unsymmetrical types of layers are linked by hydrogen bonds. One consists of aluminium ions and hydroxyl groups, the other of silicon and oxygen ions. Cairns-Smith does not postulate a detailed mechanism, but only describes the main thrust of his argument. Critics complain that clear experimental results are not available (however, other proponents of new hypotheses often provide no evidence to back up their suggestions!).

Armin Weiss (1981) presented some results of experimental work: using montmorillonite, he was able to show that the complete information present in a matrix is passed on to the daughter layers. In principle, the intercalating synthesis of a new layer of montmorillonite from the nutrient solution can be compared to the replication of a DNA chain. The distance between the layers is of great importance in these experiments and acts as a performance-limiting factor.

Why can layer silicates serve as models for replication processes? The answer is simple: they have properties which are observed in replicating systems. Montmorillonite crystals contain similar parallel layers, the distance between which

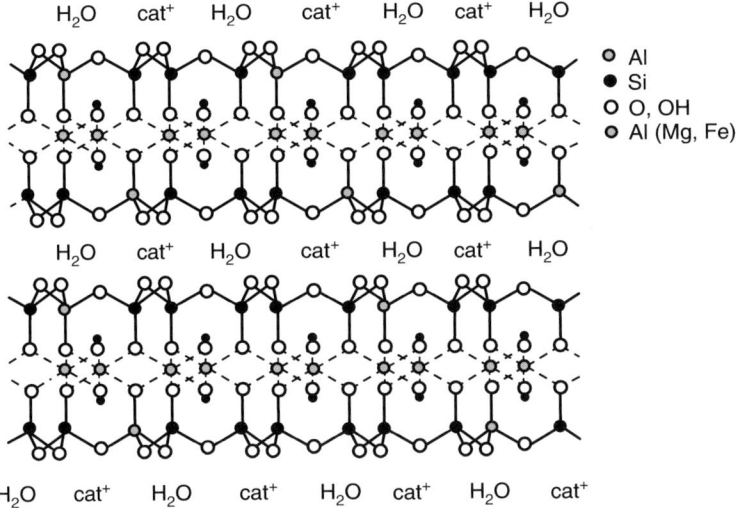

Fig. 7.2 Idealized schematic representation of the structure of a single montmorillonite layer

increases in an aqueous medium. A stepwise change in the distance between the layers is observed at relatively high electrolyte concentrations. If the concentration of a 1:1 electrolyte falls below 10^{-3} M, the interaction between the layers decreases, and the crystal aggregate separates into single layers. If highly diluted montmorillonite is introduced into a nutrient solution, the isolated layers serve as crystallisation or growth nuclei for the formation of new layers. If such matrices are present, the activation energy is a quarter to a third lower than for layer formation without a matrix (A. Weiss, 1981). This process would have needed to be reinforced by periodically occurring natural phenomena such as alternation between drought and rainy periods, or melted snow, with differing electrolyte content in the solvent water.

Another possibility is the method of intercalating synthesis of silicate layers: here the distance between the individual silicate layers is increased by using an electrolyte solution of suitable composition and concentration, so that the synthesis of a new layer can occur. It is unfortunately very difficult to obtain clear experimental evidence for the intercalating synthesis of silicate layers: if a new layer is identical to the matrix, the two cannot be distinguished! In spite of the many problems, A. Weiss (1981) was able to obtain evidence for the intercalating synthesis of new silicate layers; the starting material was, however, very carefully chosen and purified. The optimal composition of the nutrient solution (Na^+, K^+, Al^{3+} and $Si(OH)_4$) was first determined, and then a certain fraction of the Al^{3+} ions and the $Si(OH)_4$ was added as a catechol complex. Here again, the conditions are far different from prebiotic; however, we are dealing with a model system which can be used to derive possible principles and mechanisms. The analytical evidence for the reaction products (the "F_1 generation") was carried out using chemical and X-ray spectroscopic methods. The layer formation process could be followed for several individual steps; up to step 10 (the 10th generation), there were only small changes in charge density. From the 16th generation onwards, however, the number of transcription errors increased rapidly. An evaluation of these results is difficult: however, the layer silicates do represent an important model system, and further experiments on them might well be useful.

Let us conclude this section by returning to the theoretical work of Cairns-Smith, who presented his ideas in detail in the book *Seven Clues to the Origin of Life* (Cairns-Smith, 1985a). His theory can be summarized in the form of seven "clues", with the help of which he hopes to be able to unlock the secret of biogenesis. Numbers 1 and 2 involve biology and biochemistry, for example, the fact that DNA and RNA are substances whose synthesis would be too difficult under primitive conditions. The third contains references to building, while in 4 and 5, comparisons with the structures of ropes (gene fibres) lead to the history of technology: this shows how a series of developments led from simple machines to the highly complex processes we know today. In the last but one, inorganic crystals are characterized as "low-end" products, while in the last clue the ubiquitous clays are shown to be able to form both primitive genes and simple control structures such as primitive catalysts or membranes.

Only some biogenesis researchers will agree wholeheartedly with these "seven clues"; however, two of the concepts presented by Cairns-Smith will certainly be widely supported:

> The idea of "takeover": many biogeneticists are of the opinion that the RNA world must have had a simpler precursor "pre-RNA world".
> Some clay minerals are excellent catalysts in important chemical reactions, even though their ability to transfer information is less than convincing.

Cairns-Smith, as the leading proponent of the mineral theory, has also shown interest in both the "hydrothermal biogenesis" theory (Cairns-Smith, 1992) and the "iron-sulphur hypothesis" proposed by G. Wächtershäuser (see Sect. 7.3).

It appears that various possibilities of "mineral biogenesis" chemistry for prebiotically meaningful syntheses have not been investigated up to now. A recent report provided a surprise in the form of a successful experiment in the area of inorganic protocells. Cellular structures are formed spontaneously in a simple inorganic system when beads of $CaCl_2$ and $CuCl_2$ are introduced into an alkaline solution containing Na_2CO_3, NaI and H_2O_2. The "cells" formed by the system have a semipermeable membrane. Reactive substances can diffuse into the cell, while products can diffuse out. The system was far from thermodynamic equilibrium; copper ions acted as catalysts in the phase transfer oxidation of iodide by H_2O_2. The future will show whether this experiment will be of importance for further developments in prebiotic chemistry (Maselko and Strizhak, 2004).

An information science research group devised a new model which could explain information storage in the prebiotic phase of the biogenesis process. They assume that layered double hydroxide (LDH) minerals acted as proto-RNA molecules on the young Earth about 4 billion years ago. This hypothesis relates to Cairns-Smith's "genetic takeover" thesis, which thus again became the subject of discussion.

LDH minerals are lamellar host–guest compounds consisting of inorganic layers with positive charges, intercalated with anions in the interlayer between the sheets. Examples cited by the authors are brucite $[Mg(OH)_2]$ and hydrotalcite $\{[Mg_6Al_2(OH)_{16}][(CO_3) \cdot 4H_2O]\}$. The hypothetical LDH-like system forms crystal structures with a high degree of exactness when intercalated with anions. The replication fidelity is maintained.

The authors realize the many problems inherent in their hypothesis and hope that "someone" will be inspired to solve the problem of how a transition from inorganic-based life to life as we know it could have taken place (Greenwell and Coveney, 2006).

7.2 Hydrothermal Systems

7.2.1 Introduction

The discovery of the first hydrothermal vents between 1977 and 1979 was a sensation, not only for geologists and experts on geological deposits, but also for biologists and researchers who were trying to solve the problem of the origin of life. Hydrothermal vents occur along the mid-ocean ridges, where sea floor spreading allows magma to rise from the Earth's mantle and form a new sea floor. During this dynamic process, faults form in the rock; cold sea water flows into such faults, comes in contact with magma chambers and is forced back to the surface. The temperature of the water thus heated can be as high as 400°C; the ocean depth is between 2,000 and 3,000 m, so that the water is at a pressure of several hundred atmospheres. This superheated water dissolves minerals and gases from the basalt rock, which are set free or precipitated at the vent. The result is chimney structures between 1 and 3 m in height, which often spew out hot water saturated with metal sulphides ("black smokers"). The hot deep-sea vents sustain rich ecosystems; large colonies of clams, 25–30 cm in diameter, live alongside long tube worms (vestimentiferan pogonophorans) and white crabs on the basalt sea floor. The food chain in these remarkable ecosystems is based on bacteria which metabolize sulphides, hydrogen sulphide and CO_2 in the hydrothermal solutions (Macdonald and Luyendyk, 1981; Edmond and von Damm, 1983; Hékinian, 1984).

The variety of life forms to be found near hydrothermal vents does not, of course, mean that life itself originated there: these geological systems are much too unstable for that. The dynamics of tectonic plates cause the vents to disappear after some decades, or at most after a few hundred years. According to Nils Holm from the Department of Geology and Geochemistry at the University of Stockholm, the discovery of the hydrothermal vents led to intense, and in some cases controversial, discussions of the question as to whether hydrothermal systems were the birthplaces of life around four billion years ago. Many geologists believe that hydrothermal activity on the primeval Earth was probably stronger than it is today, as the thick continental crust had not yet formed, and the Earth's interior was even hotter than it is now (Holm, 1992).

There are other arguments which support the assumption that biogenesis occurred at thermal sources:

> The simplest organisms known to us are thermophilic microorganisms
>
> The earlier assumption of a reducing atmosphere has been modified in favour of a neutral one. It is hypothetically possible that iron vapour and reduced forms of carbon from meteorite impacts on the ocean could have led to limited regions with reducing properties.
>
> The hydrothermal systems, hundreds of meters under the surface of the ocean, would have protected evolving systems from the high-energy cosmic radiation as well as from meteorite impacts. Even partial evaporation of sea water, due to gigantic impacts, could have been resisted by molecular systems present at great depths (Holm and Andersson, 1995).
>
> The occurrence of supercritical liquids in hydrothermal systems cannot be excluded.

These four arguments have led scientists from various disciplines to look more closely at the theory of a possible biogenesis in hydrothermal systems in the deep sea.

Holm and Andersson have provided an up-to-date survey of simulation experiments on the synthesis under hydrothermal conditions of molecules important for biogenesis (Holm and Andersson, 2005). It is clear that several research groups have been able to show in the meantime, using simulation experiments, that the conditions present at deep sea vents appear suitable for the synthesis of very different groups of substances. However, it remains unclear how these compounds could have been stabilized and protected against rapid decomposition. At present, metal ions (as complexing agents) and mineral surfaces are the subject of discussion and experiment.

7.2.2 Geological Background

As the Earth's tectonic plates drift apart, a new oceanic crust is formed from the basalt rising from the depths; cold sea water acts as a cooling agent. Geologists distinguish between two types of hydrothermal systems (Holm, 1992):

> Sediment-free, on-axis systems at tectonic spreading centres, which have high temperature gradients (as they lie directly above magma chambers). The mean temperature of the ejected water is 620–640 K, the throughput 24 km^3/year.
>
> Off-axis systems, on the flanks of spreading centres, driven by free convection due to cooling of the ocean crust, where the mean water temperature is around 420 K.

The redox properties of these systems are vitally important for possible abiotic chemical syntheses; the main minerals present in young basalt ocean basins, at depths between 300 and 1,300 m, are pyrite (FeS_2), pyrrhotite (FeS) and magnetite (Fe_3O_4) (the PPM system).

At greater depths, the redox characteristics are mainly determined by another mineral system: fayalite (Fe_2SiO_4), magnetite and quartz (SiO_2) (the FMQ system). The buffer properties of the two systems can be expressed in terms of the following equations:

$$6\,FeS + 4\,H_2O \quad \rightarrow \quad 3\,FeS_2 + Fe_3O_4 + 4\,H_2 \qquad (7.1)$$

$$3\,Fe_2SiO_4 + 2\,H_2O \rightarrow 2\,Fe_3O_4 + 3\,SiO_2 + 2\,H_2 \qquad (7.2)$$

The latter reaction can also be formulated using oxygen:

$$3\,Fe_2SiO_4 + O_2 \rightarrow 2\,Fe_3O_4 + 3\,SiO_2 \qquad (7.3)$$

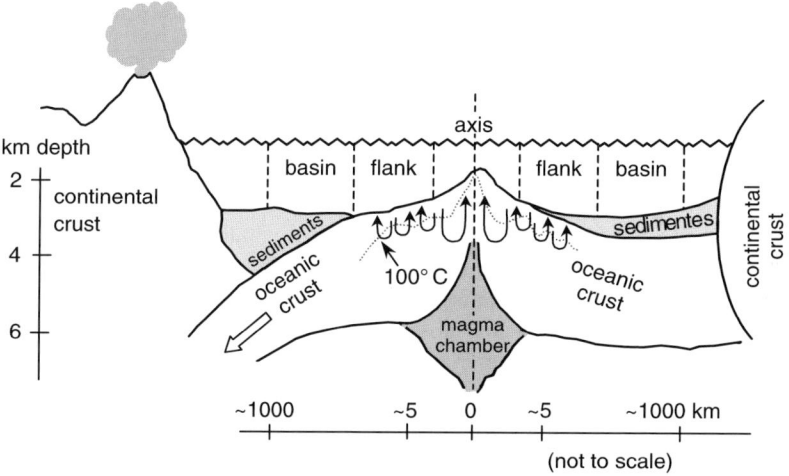

Fig. 7.3 Schematic representation of the geological conditions in hydrothermal systems. In the middle is the axis of the spreading centre, with hot springs ($\sim$300°C) in sediment-free regions. The springs with lower temperatures are on the flanks. After Holm (1992)

The two mineral systems in the ocean crust are compatible, as has been shown by calculations for the following reactions (Helgeson et al., 1978):

$$CO_2 + 4\,H_2 \rightarrow CH_4 + 2\,H_2O \qquad (7.4)$$

At temperatures of 620–670 K, with a ratio $CO_2/CH_4 > 1$ and
At temperatures below 520 K, with a ratio $CO_2/CH_4 < 1$.

This means that at high temperatures, CO_2 is formed (CO_2-degassing), while at lower temperatures, CH_4 is formed primarily.

According to Hennet et al. (1992), the conditions in the hotter zones are oxidising in nature, while the colder regions have reducing properties. This means that for the CO_2/CH_4 system:

At high temperatures (570–670 K), oxidising conditions are present, and CO_2 is stable.

At lower temperatures, reducing conditions are present (CH_4 is stable); this is typical for the oceanic crust. Most of the hydrothermal water circulates in the oceanic crust at a temperature of around 420 K, and the reducing conditions present there are mainly controlled by the PPM mineral mixture (Alt et al., 1989).

Hydrothermal systems are also important for the Earth's matter and energy turnover. A large proportion of the heat energy generated by the Earth is directed into the cold oceans by the hydrothermal vents. Calculations show that the total amount of sea water present on Earth passes through its geological system in 8–10 million years. This process is accompanied by huge turnovers of chemicals: thus, the metals copper and zinc are enriched by a factor of more than 10,000 in the sulphides precipitated at the hot springs.

7.2.3 Syntheses at Hydrothermal Vents

The first indication of a possible connection between geological processes occurring at the boundaries between tectonic plates of the mid-oceanic ridges and the biogenesis problem was provided by J. B. Corliss (1981). He considered the hydrothermal conditions to be "ideal reactors for abiotic synthesis": these ideal conditions were the water temperature gradients, the pH, and the concentrations of solutes in the hot springs. The presence of certain minerals which could act as catalysts, such as montmorillonite, clay minerals, iron oxide, sulphides etc., was also very important. The initial model presented for the hydrothermal synthesis of biomolecules (Corliss, 1981) was modified, particularly by Russell (1989) and Wächtershäuser (see Sect. 7.3).

After the discovery of hydrothermal systems, interest was initially concentrated on the extremely hot springs; only later was it realized that the off-axis regions (temperature $\sim$420 K) were much more interesting for prebiotic research. An important geochemical process, leading to the formation of hydrocarbons (in analogy to the Fischer-Tropsch principle) is the formation of serpentine from peridotite (the latter consists mainly of olivine and pyroxene). The oxidation of the iron(II) in the olivine fraction leads to the reduction of water and thus to the formation of hydrogen. This process was observed experimentally by Berndt et al. (1996), who studied the serpentinisation of olivine at 570 K and a pressure of 500 atm. The conversion of Fe(II) in olivine to Fe(III) in magnetite led to the formation of hydrogen, and thus to the synthesis of reduced carbon compounds from the dissolved CO_2. After 69 days, the result was as follows: the concentration of hydrogen rose from zero to 158 mM/kg and was accompanied by magnetite formation. The concentration of CO_2 fell from 8 mM/kg to 2.3 mM/kg, while those of methane, ethane and propane rose (to 84, 26 and 12 µM/kg, respectively). Carbon-containing material, mainly aliphatics, was also formed.

The controversial problem of the synthesis of biomolecules under simulated hydrothermal conditions was investigated by Hennet et al. (1992), who used mixtures of HCN, HCHO and NH_3 at 420 K and under 10 atm of pressure in the presence of the PPM redox buffer. The yield of the amino acids formed in this experiment was an order of magnitude higher than that obtained in a corresponding synthesis using spark discharges in a reducing atmosphere. Contamination can be excluded, since the amino acids were obtained only as racemates. Table 7.1 shows a comparison of the yields of carbon-containing products obtained using spark discharges and under hydrothermal conditions.

Table 7.1 Comparison of the yields of carbon-containing compounds obtained from an atmosphere of CH_4, NH_3, H_2O and H_2 using spark discharges with those obtained under hydrothermal conditions from a mixture of HCN, HCHO and NH_3 at 423 K and 10 atm in the presence of pyrite-pyrrhotite-magnetite redox buffer (Holm and Andersson, 1995)

Amino acids	Spark Discharge Yields in µM	Hydrothermal Reactions Yields in µM
Glycine	630	9275
Alanine	340	847
Serine	–	26
Aspartic acid	4	281
Glutamic acid	6	81
Isoleucine	–	52

Critics of these systems have often suggested that amino acids are highly labile under high temperatures and pressures. This point was studied under conditions which the critics had not thought of (Andersson and Holm, 2000): the stability of various amino acids was checked using the PPM buffer system referred to above. A mixture of the minerals potash feldspar, muscovite and quartz (KMQ) regulated the hydrogen ion activity. The system was heated in a Teflon-coated autoclave to 473 K at 50 atm. The reaction time of 48 h was extremely short in comparison to natural cosmic processes. The result was the formation, probably by decomposition of serine, of glycine, which was not present in the starting mixture. The alanine concentration increased, possibly due to the decomposition of other amino acids. What is important is the fact that the rates of decomposition of leucine, alanine, and aspartic acid in the redox-PPM-buffered system (which corresponds to natural hydrothermal conditions) were much lower than in unbuffered experiments; this result can thus be taken as proof of the redox buffer action of the KMQ system used.

One of the most important questions in prebiotic chemistry is as follows: Which reactions of simple building block molecules are robust enough that they can survive great changes in pressure, temperature, pH and irradiation and still lead to more complex systems?

Hazen and Deamer looked at the chemical and physical properties of the end products of hypothetical prebiotic reactions carried out under extreme conditions of pressure and temperature, for example in CO_2-rich regions of hydrothermal vents. The results of laboratory experiments indicate that prebiotic syntheses leading to a variety of products could have occurred in hydrothermal systems; some of these have amphiphilic properties, and would have been capable of self-organisation processes.

The authors chose pyruvic acid as their model compound; this C_3 molecule plays a central role in the metabolism of living cells. It was recently synthesized for the first time under hydrothermal conditions (Cody et al., 2000). Hazen and Deamer carried out their experiments at pressures and temperatures similar to those in hydrothermal systems (but not chosen to simulate such systems). The non-enzymatic reactions, which took place in relatively concentrated aqueous solutions, were intended to identify the subsequent self-selection and self-organisation potential of prebiotic molecular species. A considerable series of complex organic molecules was tentatively identified, such as methoxy- or methyl-substituted methyl benzoates or 2, 3, 4-trimethyl-2-cyclopenten-1-one, to name only a few. In particular, polymerisation products of pyruvic acid, and products of consecutive reactions such as decarboxylation and cycloaddition, were observed; the expected "tar fraction" was not found, but water-soluble components were found as well as a chloroform-soluble fraction. The latter showed similarities to chloroform-soluble compounds from the Murchison carbonaceous chondrite (Hazen and Deamer, 2007).

7.2.4 Other Opinions

In order to check the hypotheses discussed in the previous sections, Bernhardt et al. (1984) studied the stability of amino acids and proteins at 573 K and 265 atm: the amino acids were decomposed to a large extent. The authors based their arguments on information on the growth of bacteria at 520 K provided by Baross and Deming (1983).

Miller and Bada (1988) and Miller et al. (1989) attempted to answer the question as to whether biogenesis processes can occur at hydrothermal vents in regions of the deep sea. Miller states clearly that biomolecules could not have been formed under such conditions. Holm (1992) sums up Miller's arguments in the following four points:

> High temperatures (around 620 K) destroy organic compounds more quickly than they can be formed.
> The synthesis of polymers is not the same as biogenesis.
> There are no gradients close to the hot springs, but merely intermixture with the cold sea water (at 275 K).
> The vital role of minerals and clay minerals must first be proven.

In their book *The Spark of Life*, Wills and Bada (2000) expressed their particular affection for colleagues working on hydrothermal systems; in the book, this group of scientists is referred to as "ventrists". As a proof of the four points listed above, Miller and Bada (both working in San Diego) offered some data of their own on amino acid mixtures (523 K, 265 atm, neutral pH); the result is similar to that of Bernhardt et al. (1984), in that the molecules under study were almost completely destroyed.

E. L. Shock (1990) provides a different interpretation of these results: he criticizes that the redox state of the reaction mixture was not checked in the Miller/Bada experiments. Shock also states that simple thermodynamic calculations show that the Miller/Bada theory does not stand up. To use terms like instability and decomposition is not correct when chemical compounds (here amino acids) are present in aqueous solution under extreme conditions and are aiming at a metastable equilibrium. Shock considers that oxidized and metastable carbon and nitrogen compounds are of greater importance in hydrothermal systems than are reduced compounds. In the interior of the Earth, CO_2 and N_2 are in stable redox equilibrium with substances such as amino acids and carboxylic acids, while reduced compounds such as CH_4 and NH_3 are not. The explanation lies in the oxidation state of the lithosphere. Shock considers the two mineral systems FMQ and PPM discussed above as particularly important for the system seawater/basalt rock. The FMQ system acts as a buffer in the oceanic crust. At depths of around 1.3 km, the PPM system probably becomes active, i.e., N_2 and CO_2 are the dominant species in stable equilibrium conditions at temperatures above 548 K. When the temperature of hydrothermal solutions falls (below about 548 K), they probably pass through a stability field in which CH_4 and NH_3 predominate. If kinetic factors block the achievement of equilibrium, metastable compounds such as alkanes, carboxylic acids, alkyl benzenes and amino acids are formed between 423 and 293 K.

Shock's postulates led Bada et al. (1995) and others to react critically. Thus Bada, Miller and Zhav are of the opinion that quasi-equilibrium calculations provide flawed descriptions of the experimental observations. They believe that thermodynamic calculations cannot be used for organic compounds in high temperature vents.

In the same year, Miller and the biologist Antonio Lazcano (National Autonomous University of Mexico) spoke out against hypotheses that life could have originated at hydrothermal vents. They believe that the presence of thermophilic bacteria (the oldest life forms) does not prove that biogenesis occurred in the depths of the oceans. Stanley Miller sees a greater chance for successful prebiotic chemistry under the conditions of a cold primeval Earth rather than at high temperatures in hydrothermal regions (Miller and Lazcano, 1995).

To put it in a nutshell: hot or cold? That is the question!

7.2.5 Reactions under Supercritical Conditions

One problem which had previously been ignored in the dispute on the stability of biomolecules under extreme conditions was the influence of the properties of supercritical water. Water becomes supercritical at temperatures above the critical temperature of 647 K and pressures above the critical pressure (22.1 MPa). The physical properties of water change dramatically when the critical parameters are approached, and in the supercritical state, the densities of the liquid and gas phases are identical, so that they are indistinguishable. It is still unclear whether supercritical water was involved in syntheses in hydrothermal systems.

One example will show the manifold types of reactions studied by Mok et al. (1989). Lactic acid decomposes in supercritical water to give acetaldehyde, which then reacts further; it can also undergo dehydration to give acrylic acid, which is either hydrogenated to give propionic acid or decarboxylated to give ethene (Fig. 7.4).

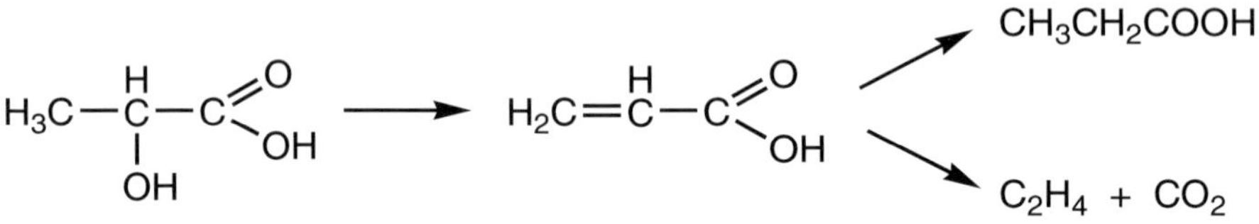

Fig. 7.4 Reaction scheme for the conversion of lactic acid to acrylic acid, propionic acid and ethene in supercritical water

Bröll et al. (1999) have provided detailed surveys of the variety of reaction mechanisms which can occur in supercritical water. It is possible that supercritical conditions were present in the vicinity of hydrothermal systems, where as yet unknown reactions took place.

The question of the elimination of water in polycondensation reactions still provides an unsolved problem. Solutions are being searched for in many laboratories, for example in Italy: Palyé and Zucchi from the University of Modena consider it possible that limited regions where liquid or supercritical CO_2 phases were present could have existed on the young Earth. Such regions, with non-aqueous media, could have been particularly favourable for some prebiotic reactions, such as those involving the elimination of water. Experiments to study this hypothesis are planned (Palyé and Zucchi, 2002; Holm and Andersson, 1998).

In order to study the behaviour of the amino acid glycine under extreme conditions, Japanese researchers injected glycine solutions into an apparatus which simulated hydrothermal conditions. Under the reaction conditions chosen, either sub- or supercritical conditions were present in the aqueous phase. Analysis of the reaction products showed that glycine was converted to diglycine and diketopiperazine (as well as traces of triglycine). The dipeptide yields reached a maximum at 623 or 648 K at pressures of 22.2 or 40.0 MPa. However, the highest glycine decomposition rates were also observed under these conditions (Alargov et al., 2002).

7.2.6 Fischer-Tropsch Type Reactions

Chemical reactions similar to the Fischer-Tropsch synthesis have been discussed for some years in connection with prebiotic chemistry; they are described as "Fischer-Tropsch type" reactions (FTT). In its technically optimized form, the FTT synthesis is carried out at about 453 K at atmospheric pressure, using activated Fe, Co and Ni catalysts; the general equation is:

$$n\,CO + 2n\,H_2 \rightarrow C_nH_{2n} + n\,H_2O \tag{7.5}$$

The reaction affords a broad range of hydrocarbons and is normally carried out in a water-free medium. Since other conditions are valid for abiotic syntheses, the term FTT is used.

The geological process of the formation of serpentine from peridotite probably involves the synthesis of carbon compounds under FTT conditions (see Sect. 7.2.3). The hydrogen set free in the serpentinisation process can react with CO_2 or CO in various ways. The process must be quite complex, as CO_2 and CO flow through the system of clefts and chasms in the oceanic crust and must thus pass by various mineral surfaces, at which catalytic processes as well as adsorption and desorption could occur.

Horita and Berndt (1999) studied the abiogenic formation of methane under conditions present at hydrothermal vents. Solutions of bicarbonate (HCO_3^-) were subjected to temperatures of 470–670 K and a pressure of 40 MPa. Under these conditions, CO_2 was reduced only very slowly to methane. Addition of a nickel–iron alloy, which corresponds closely to the minerals in the Earth's crust, led to a clear increase in the reaction rate of methane synthesis. The following reaction is assumed to occur:

$$HCO_3^- + 4H_2 \rightarrow CH_4 + OH^- + 2H_2O \tag{7.6}$$

The yields of methane are directly proportional to the amount of alloy added. Abiotic methane formation is probably more general than had previously been thought, in particular since nickel–iron alloys have been found in the oceanic crust.

Prebiotic chemistry must cope with many problems; a particularly difficult one is contamination. Prebiotic experiments often lead to the formation of important molecular species in extremely low concentrations. The successes of the synthesis may sometimes appear sensational, but there is always the danger that artefacts may be involved. Control experiments carried out with ultrapure deionised water showed that, at higher temperatures (>373 K), synthetic polymers in components of the apparatus could provide a source of organic contaminants such as formate, acetate or propionate ions. Stainless steel had a catalytic effect on the decomposition of formate, so that the use of titanium alloys in the apparatus is recommended.

These laboratory results show clearly that control experiments should always be carried out in order to determine the extent of contamination (Smirnov and Schoonen, 2003).

7.3 The Chemoautotrophic Origin of Life

None of the biogenesis hypotheses previously discussed differs so clearly from all the other models as the "chemoautotrophic theory" proposed by the Munich patent attorney Dr. Günter Wächtershäuser.

Wächtershäuser gained attention with a fundamental article published at the end of the 1980s (Wächtershäuser, 1988a). Active support for his entry into the (almost) closed society of biogeneticists was provided by Karl Popper, whose authority in the philosophy of science was recognized worldwide. Wächtershäuser formulated his ideas on the basis of Popper's scientific theory. The new theory (or hypothesis) *negates*:

> The prebiotic primeval soup, i.e., the oft-cited mixture of organic molecules in the primordial ocean, or in ponds which could have arisen in many ways, e.g., in the atmosphere or the hydrosphere; the substances concerned could also have been "delivered" from outer space.
> Connected with this: the development of heterotrophic systems, i.e., of systems which use up organic substances for the formation of new compounds.
> The presence of other "worlds", such as the RNA world or the clay crystal world.
> That the primeval Earth must have seen the emergence of a functioning, if primitive, genetic apparatus prior to the development of metabolism.

It is understandable that these ideas were found interesting by biogenesis researchers across the world (though interest does not necessarily signify agreement).

In the same year as his publication on the "theory of surface metabolism", Wächtershäuser introduced his theory on possible nucleic acid precursors (Wächtershäuser, 1988b). He suggested that pre-DNA was composed only of purine bases, while the pyrimidines were later synthesized enzymatically. In this model, adenine and xanthine link via two hydrogen bonds, while guanine and isoguanine use three hydrogen bonds (xanthine and isoguanine being bonded to the sugar via the N_3 nitrogen of the base). Although this suggestion was interesting, the so-called "pyrite theory" led to more scientific controversy. A series of important insights in biochemistry, and now also in biogenesis research, point to the great importance of the two elements iron and sulphur; they are able to form Fe-S complexes of defined structure, the so-called Fe-S clusters. "The Interface Between the Biological and Inorganic Worlds: Iron-Sulfur Metalloclusters" is the title of an article in *Science* written by Rees and Howard (2003).

Fe-S complexes have important functions in today's living systems, in enzymes such as the ferredoxins and oxidoreductases, as well as in electron transport proteins. It is striking that these redox reactions mainly involve elements and compounds such as CO, H_2 and N_2, which were probably also components of the primeval Earth's atmosphere. Thus, the assumption of an active involvement of Fe-S clusters in a (hypothetical) "Fe-S world" in processes which finally led to biogenesis appears completely reasonable! We now have a background to the "theory of the chemoautotrophic origin of life".

The RNA world requires a system capable of self-replication as a precondition for the beginnings of life. In contrast, the surface metabolism theory proposed by Wächtershäuser postulates that the initial step is metabolism, from which complex replication systems can evolve later. This metabolism would have occurred at the surface of minerals which many geologists believe to have been present, in considerable amounts, at the primeval Earth's surface.

The oxidative formation of pyrite from iron and hydrogen sulphide supplied the reduction equivalents vital for the fixation of the carbon to be used in the synthesis of key molecules from simple building blocks (such as CO or CO_2).

$$FeS + H_2S \rightarrow FeS_2 + 2H^+ + 2e^- \tag{7.7}$$

The energy source postulated here must fulfil several conditions:

  It must be geochemically realistic.
  It must be selective and mild (thus, Wächtershäuser excludes the influence of solar UV irradiation).
  Electron flow must occur directly from the reducing agent to the CO_2.
  The reduction potential must suffice for all the reduction reactions involved in the metabolic system (Wächtershäuser, 1992).

The pyrite formed in the above reaction is well known as "fool's gold", and was certainly present on the young Earth.

A fundamental question for all reactions which could have been involved in the early phase of chemical evolution is that of the origin of the reduction equivalents necessary for the autotrophic synthesis. For example, the synthesis of one molecule of glucose from carbon dioxide requires 24 electrons, while the synthesis of the amino acid cysteine requires as many as 26 electrons per molecule of amino acid:

$$6CO_2 + 24H^+ + 24e^- \rightarrow C_6H_{12}O_6 + 6H_2O \tag{7.8}$$

$$3CO_2 + NO_3^- + SO_4^{2-} + 26e^- + 29H^+ \rightarrow$$
$$H_2C(SH) - CH(NH_2) - COOH + 11H_2O \tag{7.9}$$

(de Duve, 1994)

Fig. 7.5 Pyrite (FeS_2) crystals, with quartz

Wächtershäuser assumes that the newly formed pyrite has a positively charged surface, which readily binds anions such as $-COO^-$, $-OPO_3^{2-}$ or $-S^-$. Fixation of CO_2 at the pyrite surface affords anionic groups, causing bonding of products in their growing state. Retention times at the pyrite depend on the strength of bonding to the surface. Strongly bonded substances have time to go through reactions, forming a two-dimensional reaction system, i.e., that of surface metabolism. Weakly bonded molecules, however, exit the system.

According to Wächtershäuser, the difference between the primeval soup theory and the pyrite surface theory involves different roles for the surfaces: some proponents of the primeval soup theory do, however, use clay minerals in their experiments. Clay has mostly negative surface charges, which weakly attract positively charged organic molecules. Pyrite, however, is positively charged, and there is strong interaction with anionic organic molecules; these were not just adsorbed, but were formed at the pyrite surface. In addition, they are less susceptible to hydrolysis than in the "primeval soup solution".

Wächtershäuser provides a detailed discussion of whether particular crystal modifications of pyrite could have led to homochirality in the biomolecules produced, but his theory appears to be extremely speculative.

The surface metabolism hypothesis can also be used to deal with questions on the formation of the first cell structures: for example, molecules required for the construction of cell membranes could have been formed via the so-called reductive Krebs cycle (citric acid cycle).

$$HOOC-(CH_2)_n-COOH \longrightarrow HOOC-(CH_2)_n-\underset{\underset{O}{\|}}{C}-COOH \tag{7.10}$$

$$HOOC-(CH_2)_n-\underset{\underset{O}{\|}}{C}-COOH \longrightarrow HOOC-(CH_2)_{n+1}-COOH \tag{7.11}$$

(These are not balanced equations!)

Wächtershäuser's prime candidate for a carbon-fixing process driven by pyrite formation is the reductive citrate cycle (RCC) mentioned above. Expressed simply, the RCC is the reversal of the normal Krebs cycle (tricarboxylic acid cycle: TCA cycle), which is referred to as the "turntable of metabolism" because of its vital importance for metabolism in living cells. The Krebs cycle, in simplified form, can be summarized as follows:

A carrier molecule containing four carbon atoms (the C_4 unit) takes up a C_2 unit (the "activated acetic acid"), which is introduced into the cycle. The product is a six-carbon molecule (the C_6 unit), citric acid, or its salt, citrate. CO_2 is cleaved off in a cyclic process, so that a C_5 unit is left; this loses a further molecule of CO_2 to give the C_4 unit, oxalacetate. In the living cell, this process involves ten steps, which are catalysed by eight enzymes. However, the purpose of the TCA cycle is not the elimination of CO_2, but the provision of reduction equivalents, i.e., of electrons, and thus H^+ ions, for processes occurring in the respiratory chain. The net equation of the TCA cycle can be summarized as follows:

$$Acetyl\text{-}CoA + 3\,NAD^+ + FAD + GDP + P_i + 2\,H_2O \rightarrow$$
$$2\,CO_2 + 3\,NADH + FADH_2 + GTP + 2\,H^+ + CoA \qquad (7.12)$$

Thus, in one cycle, eight hydrogen atoms ($H^+ + e^-$) are transferred to hydrogen-transmitting coenzymes and later oxidized to water in the respiratory chain. This process is linked to oxidative phosphorylation, i.e., the synthesis of ATP from ADP and inorganic phosphate.

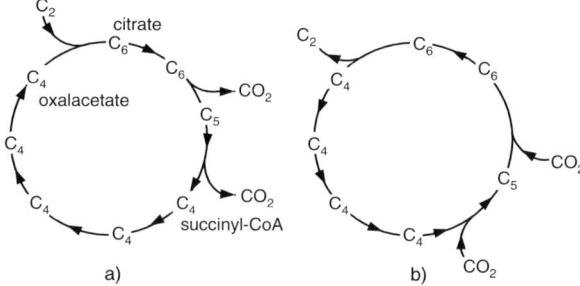

Fig. 7.6 A greatly simplified representation of the "normal" Krebs cycle (**a**) which occurs in to-day's living cells and the corresponding "reductive" process (**b**), which was possibly important in the metabolism of evolving systems

The RCC proceeds formally in exactly the opposite direction, from the C_4 molecule to the C_6 molecule via uptake of CO_2, but without the help of highly specialized enzymes. However, it is not a hypothetical archaic phenomenon, but has been observed in cells which exist today, for example in the photosynthetic bacterium *Chlorobium thiosulfatophilum* (Evans et al., 1966).

The RCC can only exist if a correspondingly high reduction potential is present; this must be delivered by the FeS/H_2S system. If this condition is fulfilled, CO_2 can be taken up at the pyrite surface and incorporated into the growing carbon-containing molecules. However, the archaic RCC was not necessarily exactly the same as that in existence today, but probably underwent changes during long evolutionary phases. In any case, the RCC was an autocatalytic process which depended on the presence of an intact FeS_2 surface.

According to Lahav (1999), the complete process can be summarized as follows:

$$Oxalacetate + 4\,CO_2 \rightarrow 2\,Oxalacetate$$

This archaic cycle contains either many, or only a few, realistic steps, depending on one's attitude to Wächtershäuser's hypotheses.

The RCC requires small amounts of primer to get started (for example, succinate or acetate), but when it is "running" it should continue autocatalytically. The processes of CO_2 fixation at the pyrite surface can be summarized in one single equation:

$$4\,CO_2 + 7\,FeS + 7\,H_2S \rightarrow (CH_2COOH)_2 + 7\,FeS_2 + 4\,H_2O \qquad (7.13)$$

$$\Delta G^\circ = -420\,kJ \cdot mol^{-1}$$

(Wächtershäuser, 1990a)

G. Wächtershäuser formulated his suggestions on the initial metabolic processes on the primeval Earth (as described above) in the form of six postulates and eleven theses (Wächtershäuser, 1990b). Laboratory experiments planned to check, and perhaps confirm, these theories have already been attempted. The reductive Krebs cycle (rTCA cycle) has recently been the subject of much discussion. Smith and Morowitz consider that the rTCA cycle is a universal, possibly primordial, core process which could have provided substances for the synthesis of biomolecules on the young Earth (Smith and Morowitz, 2004).

There are new ideas and experiments on the rTCA cycle. A group from Harvard University studied some reaction steps in the rTCA cycle which were kept going by mineral photochemistry. The authors assumed that solar UV radiation can excite electrons in minerals, and that this energy is sufficient to initiate the corresponding reaction steps. In this photocatalytic process, semiconductor particles were suspended in water in the presence of a zinc sulphide colloid (sphalerite); the experiments were carried out in a 500 mL reaction vessel at 288 K. Irradiation involved a UV immersion lamp (200–410 nm) in the photoreactor. Five reactions out of a total of 11 in the rTCA cycle were chosen to check the hypothesis:

1. Oxalacetate → malate
2. Fumarate → succinate
3. Succinate → oxoglutarate
4. Oxoglutarate → oxalosuccinate
5. Oxalosuccinate → isocitrate

Of these reactions, 1 and 2 gave high stoichiometric yields of 75% and 95%, reaction 4 of only 2.5%. The other two reaction yields were in the region of the detection limit.

If the rTCA cycle were to have functioned on the primeval Earth, all the reaction steps (both redox and non-redox) must have proceeded in high yield. One single metal surface can certainly not drive the whole rTCA cycle; more complex mixtures of active mineral surfaces are required (Zang and Martin, 2006).

There are also voices critical of the rTCA cycle: Davis S. Ross has studied kinetic and thermodynamic data and concludes that the reductive, enzyme-free Krebs cycle (in this case the sequence acetate–pyruvate–oxalacetate–malate) was not suitable as an important, basic reaction in the life evolution process. Data on the Pt-catalysed reduction of carbonyl groups by phosphinate show that the rate of the reaction from pyruvate to malate is much too low to be of importance for the rTCA cycle. In addition, the energy barrier for the formation of pyruvate from acetate is much too high (Ross, 2007).

It is possible that "colloidal photochemistry" will provide a new approach to prebiotic syntheses. The work described previously on redox reactions at colloidal ZnS semiconductor particles has been carried on successfully by S. T. Martin and co-workers, who studied reduction of CO_2 to formate under UV irradiation in the aqueous phase. ZnS acts as a photocatalyst in the presence of a sulphur hole scavenger; oxidation of formate to CO_2 occurs in the absence of a hole scavenger. The quantum efficiency for the formate synthesis is 10% at pH 6.3: acetate and propionate were also formed. The authors assume that the primeval ocean contained semiconducting particles, at the surface of which photochemical syntheses could take place (Zhang et al., 2007).

The first experiments on chemoautotrophic theory were carried out by Stetter at the University of Regensburg. It was found that synergy in the FeS/H_2S system determined the reductive effect, for example, in the conversion of nitrate to ammonia or of alkynes to alkenes. The conditions used corresponded to those present in hydrothermal systems: aqueous phase, 373 K, almost neutral pH and anaerobic conditions (Blöchl et al., 1992). Two years later, the formation of an amide bond without the use of a condensation agent was successfully demonstrated in the same laboratory (Keller et al., 1994).

Experiments in which CO and CH_3SH were converted to the activated thioester methyl thioacetate, which hydrolysed to acetic acid, were the subject of great interest (Huber and Wächtershäuser, 1997): R. H. Crabtree (1997) from Yale University wrote an interesting introduction to this publication. The reactions were naturally carried out under conditions corresponding to those at deep sea vents: 373 K, various pH values and autogeneous pressure. FeS and NiS were precipitated in situ. If the FeS/NiS is modified by adding catalytic amounts of selenium, only acetic acid and CH_3SH are formed from CO and H_2S. The authors suggest a (hypothetical) reaction mechanism for the formation of acetic acid from CO and CH_3SH in the FeS/NiS system (Fig. 7.7).

Using experimental conditions similar to those described above, Huber and Wächtershäuser were able to detect the formation of peptide bonds in reactions involving three amino acids; the main product, however, was only a dipeptide.

A research group from the Geophysical Institute of the Carnegie Institution in Washington, DC, carried out experiments under unusual conditions (Cody et al., 2000a, b); G. Wächtershäuser commented on the results in *Science* (Wächtershäuser, 2000). The catalytic properties of iron sulphide were studied, using a mixture of FeS, 1-nonanethiol and formic acid (as the source of a reactive carbon atom) at 523 ± 0.2 K, in equilibrium with CO_2 and H_2 via the water gas conversion reaction; the pressures used were 50, 100 and 200 MPa. The reaction vessel was coated with gold, which had no effect on the experimental results (as shown by control experiments). Higher pressures led to an increase in the amount of organometallic compounds present; spectroscopic studies (UV, visible and Raman) permitted the characterisation of a number of organometallic species. The successful synthesis of 1-decanoic acid indicates that carbonylation occurs under the conditions used, while that of methylated compounds shows that a fraction of the CO was reduced

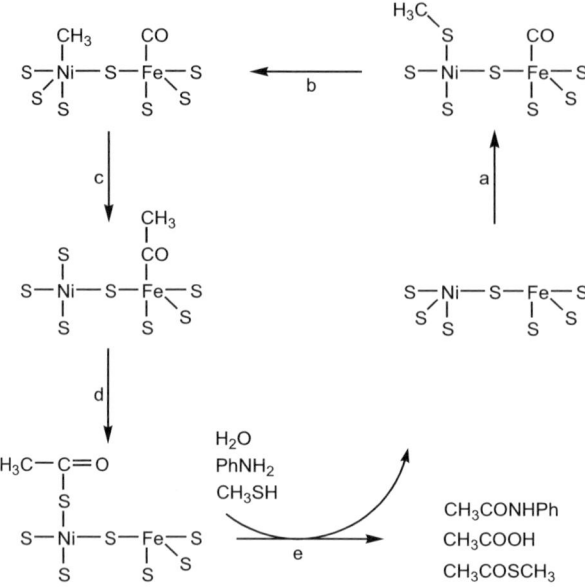

Fig. 7.7 A possible reaction mechanism for the formation of acetic acid from CO and CH$_3$SH at NiS/FeS (Huber and Wächtershäuser, 1997). The reaction pathways are as follows:

(**a**) Uptake of CO by the iron centre and of CH$_3$SH by the nickel centre.
(**b**) Formation of the methyl-Ni moiety.
(**c**) Migration of the methyl group to the carbonyl group with formation of an acetyl group bonded to Fe (or Ni).
(**d**) Migration of the acetyl group to a sulphido (or sulphydryl) ligand, with formation of a thioacetate ligand at Ni (or Fe).
(**e**) Formation of acetic acid by hydrolysis. The free valences of the sulphur ligand are either bonded to another metal centre or to H (or to CH$_3$). As an alternative to step **d**, the acyl residue at the CH$_3$–S-ligand can migrate with formation of the methylthioester, which then separates off.

to methyl residues. FeS is consumed, leading to the formation of a carbonylated iron-sulphur cluster and elemental sulphur, for example, according to the equation:

$$2\,FeS + 6\,CO + 2\,RSH \rightarrow Fe_2(RS)_2(CO)_6 + 2\,S^0 + H_2 \qquad (7.14)$$

The results obtained appeared quite promising, but the real sensation was the detection of pyruvate, the salt of 2-oxopropanoic acid (pyruvic acid), which is one of the most important substances in contemporary metabolism. Pyruvic acid was first obtained in 1835 by Berzelius from dry distillation of tartaric acid. The labile pyruvate was detected in a reaction mixture containing pure FeS, 1-nonanethiol and formic acid, using simulated hydrothermal conditions (523 K, 200 MPa). The pyruvate yield, 0.7%, was certainly not overwhelming, but still remarkable under the extreme conditions used, and its formation supports Wächtershäuser's theory. Cody concludes from these results that life first evolved in a metabolic system prior to the development of replication processes.

About three years after Wächtershäuser's first publication appeared, an article by Christian de Duve and Stanley Miller was published in the *Proceedings of the National Academy of Sciences* under the title "Two-Dimensional Life?"; the title alluded to the theory of reactions at positively charged pyrite surfaces (de Duve and Miller, 1991). Their criticisms of the chemoautotrophic theory were directed particularly towards certain kinetic and thermodynamic aspects, but also to theoretical statements for which no experimental support was available.

Experiments were carried out in Miller's laboratory on the Wächtershäuser system FeS/H_2S and the reduction of CO_2 to give biomolecules. The group was unable to obtain any positive results under their conditions, i.e., the synthesis of amino acids or nucleobases was *not successful*! The authors believed that the system used was not sufficiently robust in the geological-chemical sense and thus could not have played a role in the evolution of primitive metabolism on the young Earth (Keefe et al., 1995). Work in this area continued in San Diego, however: it was confirmed that, in the presence of H_2S, FeS is a strong reducing agent which can convert alkenes, alkynes and thiols to saturated hydrocarbons. However, this system is not able to synthesize amino acids or purines using CO_2 as the carbon source. The group was also unable to obtain amino acids via reductive amination of carboxylic acids (Keefe et al., 1996).

As early as 1974, the importance of iron sulphides, in particular pyrite (FeS_2) and pyrrhotite ($Fe_{0.86}S$), in reactions which could lead to the synthesis of biomolecules was identified. It was also suspected that they were involved in the formation of primitive ferredoxins and in electron transfer in photochemical processes (Österberg, 1974). However, Österberg is no longer so sure that metabolic sulphur cycles were involved in the prebiotic biogenesis process (Österberg, 1997). The sulphur concentration under reducing equilibrium conditions was probably very low ($< 10^{-8}$ M). According to Österberg, metabolic sulphur cycles first became important as oxygen evolved, i.e., when life was already present. The estimated sulphur concentrations referred to above probably represent values with respect to a homogeneous distribution in the total volume of water; the possibility of locally higher concentrations in limited regions was not taken into account.

The importance, and the performance, of the FeS/FeS_2 system were studied by Kaschke et al. (1994) in Glasgow. They were able to show that the system is capable of reducing carbonyl groups; their model substance was cyclohexanone. Although the reactions cannot be regarded as prebiotic, the results agree with thermodynamic calculations on the reductive power of the Fe/S system.

A detailed theoretical study of the properties of the redox system FeS/FeS_2 was carried out in the Department of Geosciences of SUNY Stony Brook (Schoonen et al., 1999). The authors conclude that the hypothetical reduction of CO_2 (by the FeS/FeS_2 redox pair) formulated in Wächtershäuser's early work, and the carbon fixation cycle on the primeval Earth associated with it, probably could not have occurred. This judgement is made on the basis of a theoretical analysis of thermodynamic data; other conditions would naturally have been involved if CO had reacted rather than CO_2. It is not known whether free CO existed in the hydrosphere, or if so, at what concentrations.

Positive news arrived from Holland, however; the FeS/H$_2$S/CO$_2$ problem was studied at the University of Nijmegen. The following products were obtained from reactions of FeS with H$_2$S in the presence of CO$_2$ under anaerobic conditions: hydrogen, thiols, small amounts of CS$_2$ and dimethyldisulphide. The same compounds are formed when H$_2$S is replaced by HCl in the H$_2$S-generating system FeS/HCl/CO$_2$; product analysis was carried out using GC-MS. The synthesis of hydrogen and thiol depends on the FeS/HCl or FeS/H$_2$S ratios and the temperature; the formation of H$_2$ is considerably accelerated at temperatures above 323 K, the thiol synthesis above 348 K. The reduction of CO$_2$ was detected by using DCl and D$_2$O (D is the hydrogen isotope deuterium). The authors believe that these reactions could also have occurred under the conditions present at hydrothermal vents on the primeval Earth (Heinen and Lauwers, 1996).

Photoelectrochemical experiments on the pyrite/H$_2$S system, as well as theoretical considerations, led Tributsch et al. (2003) to the conclusion that CO$_2$ fixation at pyrite probably could not have led to the syntheses proposed by Wächtershäuser. The reaction mechanism involved in such reactions is likely to be much more complex than had previously been assumed. The Berlin group supports the objection of Schoonen et al. (1999) that, apart from other points, the electron transfer from pyrrhotine to CO$_2$ is hindered by an activation energy which is too high. There is, thus, no lack of different opinions on the model of chemoautotrophic biogenesis; hopefully future studies will shed more light on the situation!

If CO$_2$ is replaced by CO in the pyrite/H$_2$S system, the conditions become much more favourable, as the positive results obtained by Huber and Wächtershäuser (1998) and Cody (2000a) show.

New evidence for the postulate "metabolism first" has been provided by a group from the Technical University of Munich; using CO as an activation agent under hot conditions in an aqueous phase (to simulate a hydrothermal system), they were able to show that α-amino acids (in this case phenylalanine) undergo polycondensation to give peptides. The necessary preconditions are: CO (at various pressures) and freshly precipitated colloidal (Fe, Ni)S. The main product is the dipeptide, which can be broken down to give the starting material under conditions identical to those of the synthesis. This catabolic mechanism proceeds via N-terminal hydantoin or urea derivatives, both of which were identified. The authors postulate a cyclic process with anabolic and catabolic phases. The hydantoin derivative is a heterocyclic compound related to the purines. The group suggests that this could be a pointer to the simultaneous occurrence of peptides and nucleic acids in the prebiotic biogenesis process. The results are considered by the authors to be a further confirmation of the thesis of the chemoautotrophic origin of life (Huber et al., 2003).

Further experiments by Huber and Wächtershäuser on chemoautotrophic biogenesis under hydrothermal conditions have shown that a number of α-amino acids and α-hydroxyacids could have been formed, subsequent to the binding of carbon (in the form of CO and CN$^-$) to catalytically active transition metal precipitates. The general structure of such compounds is R–CHA–COOH, with R = H, CH$_3$, C$_2$H$_5$ or HOCH$_2$ and A = OH or NH$_2$.

The reactions were carried out at temperatures between 353 and 393 K and CO pressures up to 75 atm; reaction times were between 20 h and 10 days. Of vital importance are the catalytically active precipitates of Ni or Ni/Fe with carbonyl, cyano and methylthio ligands as carbon sources. Calcium or magnesium hydroxide were used as buffers to prevent the system from becoming too acidic (Huber and Wächtershäuser, 2006).

These experiments were, however, not completely accepted by some chemists working in the area of prebiotic chemistry; these critics were unhappy with some synthetic conditions used, such as the KCN concentrations (0.1–0.2 M at 373 K). They felt that cyanide in such concentrations would have been relatively rapidly hydrolysed at 373 K in addition, CO pressures of 75 atm were considered incommensurate with real conditions in hydrothermal systems.

The answer to these criticisms clears up some misunderstandings, for example, that Huber and Wächtershäuser had used highly stable cyanide complexes rather than free cyanide ions in solution. Experiments were carried out and described not only at high pressures but also at atmospheric pressure. The authors remark that "the criticism . . . is made from the perspective of the prebiotic soup theory"; this has nothing to do with the chemoautotrophic theory (Bada et al., 2007; Wächtershäuser and Huber, 2007). Wächtershäuser's article "The Origin of Life and its Methodological Challenge" is recommended to those readers who wish to move from earthbound experiments and complicated calculations to the higher spheres of scientific philosophy (Wächtershäuser, 1997).

Finally, an observation on Wächtershäuser made by R. Shapiro in his book *Planetary Dreams*: he recounts Leslie Orgel's judgement on Wächtershäuser's ideas, ". . .that he[1] considered the work to be the most important finding in the last century" (Shapiro, 1999). The *Science* journalist M. Hagmann (Zürich, Switzerland) has provided a comprehensive profile of Wächtershäuser (Hagmann, 2002).

Pyrite is not only one of the key compounds in Wächtershäuser's theory, but could also have fulfilled an important function for phosphate chemistry in prebiotic syntheses. A group in Rio de Janeiro studied the conditions for phosphate sorption and desorption under conditions which may have been present in the primeval ocean. In particular, the question arises as to the enrichment of free, soluble inorganic phosphate (P_i), which was probably present in low concentrations similar to those of today (10^{-7}–10^{-8} M) (Miller and Keffe, 1995). Experiments show that acid conditions favour sorption at FeS_2, while a weakly alkaline milieu works in an opposite manner. Sorption of P_i can be favoured by various factors, such as hydrophobic coating of pyrite with molecules such as acetate, which could have been formed in the vicinity of hydrothermal systems, or the neutralisation of mineral surface charges by Na^+ and K^+.

An increase of the pH in the aqueous medium, and capture of SO_4^{2-} by the mineral surface, could lead to the liberation of P_i. In addition, there seem to be self-regulation mechanisms for the P_i sorption–desorption process, depending on the SO_4^{2-} concentration at the interface. Processes such as enrichment and liberation

---

[1] Orgel (author's note)

of P_i could have provided important preconditions for successful phosphorylation, such as that of nucleosides (de Souza-Barros et al., 2007).

7.4 De Duve's "Thioester World"

The discovery of the deep sea hydrothermal systems, and the sulphur-metabolising bacteria which live in them, caused some researchers to look more closely at the element sulphur. It seemed obvious to consider a link between sulphur bacteria—primitive life forms—and the emergence of the simplest forms of life. de Duve, 1974 Nobel Prize winner for medicine, joined the ranks of the biogenesis researchers in the 1980s.

The physiologist de Duve concentrated his efforts on a material link between the prebiotic phase of the primeval Earth and the state of development at which RNA (or a similar type of molecule) determined the further progress of the evolution process. In particular, this connecting link needed to have been able to transfer chemical energy, since without such a procedure, the RNA synthesis appears impossible. The molecular species which Christian de Duve favours for this important function is that of the thioesters. The exact reasoning as to why this is the case is discussed in detail in his book *Vital Dust: Life As a Cosmic Imperative* (de Duve, 1996).

The importance of the thioesters was realized at the beginning of the 1950s by Theodor Wieland from the University of Frankfurt am Main (Wieland and Pfleiderer, 1957), who used aminoacyl mercaptans as activated amino acids in peptide syntheses (see Sect. 5.3). Thus, 30 years later, this area of basic research came to be useful for prebiotic chemistry.

L. Weber, then at the Salk Institute in San Diego, was able to form "high-energy" thioesters from glycerinealdehyde and N-acetylcysteine. The reaction occurred under anaerobic conditions, at pH 7, in an aqueous solution of sodium phosphate. Of the aldehyde, 0.3% was converted to the lactoyl thioester per day of reaction (Weber, 1984).

$$H_3C-\overset{\overset{\displaystyle H}{|}}{\underset{\underset{\displaystyle OH}{|}}{C}}-C\overset{\displaystyle \nearrow O}{\underset{\displaystyle \searrow S-R}{}}$$

Three years previously, Weber had been able to obtain the thioester N,S-diacetylcysteine from UV irradiation of an aqueous solution of acetaldehyde and N,N'-diacetylcysteine (Weber, 1981).

The thioester world postulated by de Duve should in fact be called the "sulphuriron world", since iron ions are essential for the redox processes occurring in such a thioester world. de Duve (1991, 1996) asks a question which is vital for the whole of prebiotic chemistry: where did the redox equivalents necessary for the construction of biomolecules on the primeval Earth come from? This question becomes largely irrelevant if the strongly reducing atmosphere postulated by Miller/Urey and

Oparin is accepted. However, since most experts do not now accept this hypothesis, the question remains valid. The term "redox equivalent" can be equated with the presence or availability of electrons. So where do the required electrons come from?

De Duve (and other authors) consider the source to be iron, which as Fe^{2+} provides one electron per Fe atom when subject to solar UV irradiation. The Fe^{3+} generated can be precipitated from solution with Fe^{2+} as a mixed oxide ($FeO \cdot Fe_2O_3$ = Fe_3O_4). This is found today in the form of black-band ironstone, an ore which is between 1.5 and 3.5 billion years old and was produced by interaction of Fe^{2+} with oxygen produced by light-converting bacteria. (The figure of 3.5 billion years is, however, not accepted by some scientists; see Sect. 10.1).

The process outlined above led to the provision of the necessary quantities of hydrogen for the reduction of CO_2, CO, NO_3^- and other oxidized starting materials, which were in turn converted to biomolecules in further reaction steps. The question as to whether contemporary living cells contain relics, in the form of thioesters or thio compounds, which indicate the great importance of this class of substances, can clearly be answered positively.

Many substances which are necessary (and even essential) for life functions contain sulphur: for example, the amino acids cysteine and methionine, the tripeptide glutathione or coenzyme A (CoA), with the latter containing the SH group of cysteamine as the terminal functional group. CoA acts as a coenzyme in all important biochemical acylations. The cysteamine SH group bonds to carboxylic acids to give thioesters:

$$R-\underset{\underset{O}{\|}}{C}-S-CoA$$

Hydrolysis of the high-energy S-acetyl compound formed sets free about 34 kJ/mol; this value corresponds approximately to the amount of energy set free in the hydrolysis of ATP.

Keefe et al. (1995) from Stanley Miller's laboratory reported a possible prebiotic synthesis of pantetheine, the part of the CoA molecule without its ADP moiety. They were able to synthesize the CoA precursor from β-alanine, pantoyllactone and cysteamine. This condensation requires concentration of the reaction mixture; the "warm lagoon" theory is required here in order to achieve prebiotic conditions!

Could thioesters have been formed on the young Earth? This is quite probable, as there are clear indications of the formation of thiols in H_2S sources. Choughuley and Lemmon (1966) were able to detect the S-containing amino acid cysteinic acid and its decarboxylation product taurine under β-irradiation in a reducing H_2S atmosphere. Hydrogen sulphide has a broad absorption spectrum, ranging from 270 nm to the vacuum UV. Sagan and Khare (1971) and Khare and Sagan (1971) consider H_2S to be the initial photon acceptor. The presence of thiols on the primeval Earth can be taken as certain. In order for them to be available for prebiotic reactions, an energy barrier—an "uphill" reaction—must be surmounted.

$$R-SH + R'-C\diagdown^{O}_{OH} \rightleftharpoons R'-C\diagdown^{O}_{S-R} + H_2O \quad (7.15)$$

How could this problem be solved? Only traces of thioesters are formed from free carboxylic acid and thiols in aqueous solution, i.e., the equilibrium reaction 7.15 is shifted to the left. According to de Duve (1991), there are two possibilities for spontaneous thioester synthesis under conditions present on the primeval Earth:

> According to equation 7.15, at higher temperatures and low pH values (around pH 2). As already stated, Weber (1984) was able to carry out a thioester synthesis starting from glycerinealdehyde and N-acetylcysteine.
> From the oxidative synthesis of thioesters from aldehyde and thiol:

$$R'-SH + R-C\diagdown^{O}_{H} \rightleftharpoons R-C\diagdown^{O}_{S-R'} + 2\,e^- + 2\,H^+ \quad (7.16)$$

or from α-ketoacid and thiol:

$$R'-SH + R-C(=O)-C\diagdown^{O}_{OH} \rightleftharpoons R-C\diagdown^{O}_{S-R'} + 2\,e^- + 2\,H^+ + CO_2 \quad (7.17)$$

In the latter case, electron uptake occurs after decarboxylation of the ketoacid via Fe^{3+}, which is able to take up electrons in strongly oxidized regions of the black-band iron sediments. Fe-ions act catalytically on the process of thioester formation, which can then occur without the help of enzymes. Thus, it was solar UV irradiation which carried the prebiotic thioesters across the energy threshold.

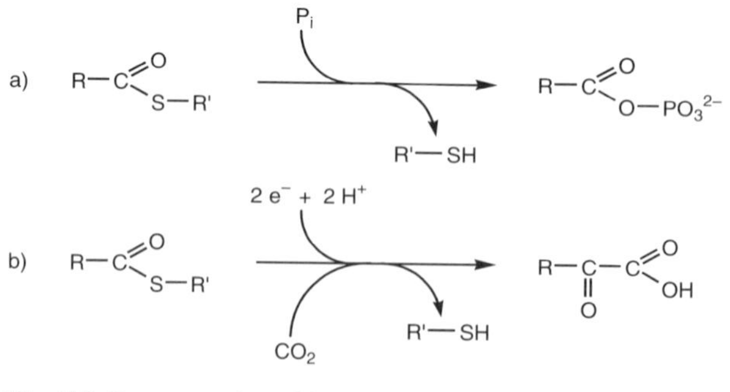

Fig. 7.8 Two examples of important reactions in the "thioester world": **a** the phosphorolysis of thioesters leads to acylphosphates and **b** α-ketoacids are formed in the reductive cleavage of thioesters with accompanying carbonylation (de Duve, 1991)

According to de Duve, thioesters on the young Earth were capable of a broad spectrum of chemical reactions: for example, of phosphorolysis leading to acylphosphates (Fig. 7.8a) or of reductive thioester cleavage, which (after carbonylation) made possible the synthesis of ketoacids (Fig. 7.8b).

Compounds of crucial importance for the RNA world or a precursor phase could have been formed in this or similar ways. Thus, a thioester world seems to support the RNA world hypothesis, although de Duve can be considered as a careful critic of the latter hypothesis. His opinions were expressed in a short article in *Nature* under the provoking title "Did God make RNA?" (de Duve, 1988).

When the Fe^{2+}/Fe^{3+} system is included, the thioester hypothesis, with its roots in sulphur chemistry, shows clear links with the "iron-sulphur world" of Wächtershäuser's chemoautotrophic biogenesis model (see Sect. 7.3).

The thioester hypothesis can be summed up as follows: the formation of thiols was possible, for example, in volcanic environments (either above ground or submarine). Carboxylic acids and their derivatives were either formed in abiotic syntheses or arrived on Earth from outer space. The carboxylic acids reacted under favourable conditions with thiols (i.e., Fe redox processes due to the sun's influence, at optimal temperatures and pH values) to give energy-rich thioesters, from which polymers were formed; these in turn (in part) formed membranes. Some of the thioesters then reacted with inorganic phosphate (P_i) to give diphosphate (PP_i). Transphosphorylations led to various phosphate esters. AMP and other nucleoside monophosphates reacted with diphosphate to give the nucleoside triphosphates, and thus the RNA world (de Duve, 1998). In contrast to Gilbert's RNA world, the de Duve model represents an RNA world which was either supported by the thioester world, or even only made possible by it.

Important prebiotic sequences, for example, the linking up of amino acids to form proteins, involve acylation reactions (see Sects. 5.1 and 5.2). Condensation agents are often not very efficient in aqueous phases; condensation reactions may involve drastic conditions, such as high temperatures or an acidic environment. Activated amino acids, for example, thioester derivatives, can be considered as starting

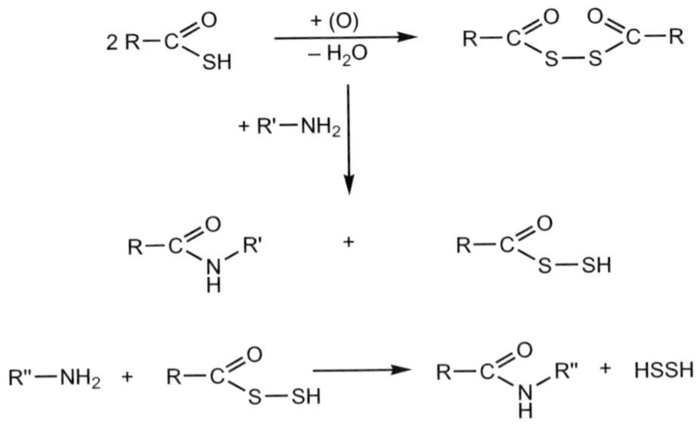

Fig. 7.9 Oxidative acylation of thioacids. The second reaction step is fast, the third slow (Liu and Orgel, 1997)

materials for the construction of peptides. A closely related reaction pathway for the formation of peptide bonds could have taken place via oxidative acylation with the help of thioacids (Liu and Orgel, 1997). These authors acylated the amino acids phenylalanine and leucine in aqueous solution with thioacetic acid and the oxidising agent $Fe(CN)_3$. The reaction was carried out under mild conditions, and yields were very good. The authors consider the oxidative acylation to be a generally applicable method for the activation of carboxylic acids, including amino acids, and suggest that it may well have been feasible under prebiotic conditions. The assumed reaction mechanism is shown in Fig. 7.9.

Arthur L. Weber (1998), now working at the Seti Institute of the Ames Research Center at Moffett Field, reports the successful synthesis of amino acid thioesters from formose substrates (formaldehyde and glycolaldehyde) and ammonia; synthesis of alanine and homoserine was possible when thiol catalysts were added to the reaction mixture. On the basis of his experimental results, Weber (1998) suggests the process shown in Fig. 7.10 to be a general prebiotic route to amino acid thioesters.

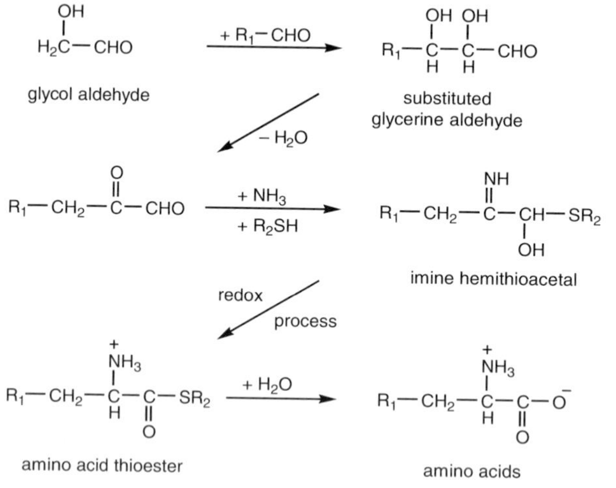

Fig. 7.10 A general prebiotic synthetic route to amino acid thioesters (Weber, 1998)

These can either undergo hydrolysis to give the free amino acids or react with a further molecule of amino acid to give a dipeptide.

7.5 Prebiotic Reactions at Low Temperatures

We do not yet know what the optimal temperatures for the evolution of life were. It is generally accepted that the prebiotic chemistry on the primeval Earth must have taken place at moderate temperatures. It is, however, also possible that various temperature ranges were present during the process, which may have taken a total of between 200 and 300 million years.

It has been pointed out that some important biomolecules have short half-lives at higher temperatures, as is clearly shown by laboratory experiments. The synthesis of adenine in the HCN oligomerisation demonstrated that chemical processes can also take place at lower temperatures. After hydrolysis of the oligomerisation products, adenine was isolated after 60–100 days from 0.01 M solutions at a pH of 9.2. Addition of glycol nitrile caused the yield to increase by a factor of four, i.e., to 48 µg/L (Schwartz et al., 1982).

A rationale for low-temperature prebiotic reactions on the young Earth came from a completely different quarter: a research group from New Zealand found that a viable mechanism for RNA folding clearly requires temperatures which can be regarded as cool or even cold. Both practical and theoretical studies were carried out: the results on RNA folding agree very well (Moulton et al., 2000). S. Miyakawa and co-workers from Stanley Miller's laboratory in San Diego also studied the importance of low temperatures for prebiotic reactions: their calculations showed that eutectic freezing out of HCN and formamide solutions is necessary in order to reach concentrations at which reactions can occur. This means that the two compounds could never have undergone polymerisation reactions in a warm primordial ocean (Miyakawa et al., 2002a). The same group carried out a remarkable low-temperature experiment on the synthesis of purines and pyrimidines: a dilute solution of ammonium chloride was allowed to react at 195 K for 25 years! After hydrolysis of the polymerisation products (both acidic and at pH 8) they could detect the presence of the purines adenine and guanine, as well as some pyrimidines (uracil, 5-aminouracil and orotic acid) in a yield of $2–30 \times 10^{-3}\%$ (with respect to carbon). Orotic acid was formed in highest yield, but no cytosine was present. Although the amounts of product were low, the eutectic freezing out effect showed itself to be a powerful concentration procedure for prebiotic syntheses on the primeval Earth (Miyakawa et al., 2002b).

P. A. Monnard et al. from the laboratory of D. W. Deamer also worked on ice/eutectic phases at 255 K. They studied the influence of solutions of inorganic ions (such as Na^+, Cl^-, Mg^{2+}, Ca^{2+} and Fe^{2+}) both on the formation of vesicles and on non-enzymatic polymerization of activated RNA monomers (Monnard et al., 2002).

These results show that there is an alternative to the theory of prebiotic chemistry in hydrothermal systems, not just theoretically but also experimentally. H. Trinks carried out experiments in Spitzbergen ice floes and developed a theory of the possible origin of life, or of precursors of life, in ice structures. Back in the laboratory, he and Christof Biebricher studied the oligomerisation of nucleotides in the cold to give RNA oligomers. The reaction system is well known; the oligomerisation of activated adenylic acid at a poly-U matrix. Reactions were carried out in artificial ice with temperature variations (249–266 K) in order to simulate natural periodic processes. After one year, polyadenylic acid was obtained in good yield with chain lengths up to 400 nucleotides and predominant $3' \rightarrow 5$ bonding; the condensation agent 2-methylimidazole was used to activate the process.

The physical properties and the structure of the water ice are certainly also important for prebiotic processes; these have been studied by one of the authors (H. T.) both in natural surroundings and in the laboratory. For example, one cubic metre of ice forms around 10^{14}–10^{15} tiny compartments, as well as a network of channels with a total surface of 10^5–10^6 square metres! Temperature changes cause temperature gradients in the ice, which in turn cause density gradients to occur. These lead to concentration gradients of the seawater electrolytes which influence chemical reactions. Although the experimental results are very revealing, what are still missing are prebiotically relevant activation conditions for the nucleotides (Trinks et al., 2005; Schiermeier, 2006).

7.6 Atomic Carbon in Minerals

IR studies of highly pure magnesium oxide carried out about 20 years ago indicated the presence of atomic carbon, which is known to be highly mobile. It is found not only in MgO crystals, but also in minerals from plutonic rock. The one chosen for study was olivine, a magnesium iron silicate; using techniques from nuclear physics, the mobility of the carbon atoms was demonstrated at temperatures as low as 77 K. Hydrocarbons (not only methane, but also unsaturated hydrocarbons and aromatics) were liberated from basalts on heating: thus, 5×10^{-5} g of HC per gram of olivine were detected. Methanol is formed if the surface of the crystals is treated with water just prior to heating, while exposure of olivine to ammonia leads to the formation of amines such as methylamine or dimethylamine. This means that under optimal conditions, each cubic kilometre of degassing basaltic magma could set free 3–6×10^9 kg CO_2 and 1–3×10^8 kg of abiotically synthesized HCl.

In the context of the discovery of amines and oxygen-containing organic compounds, the question arises as to whether the presence of atomic carbon in olivine or MgO crystals could have led to the formation of amino acids. Knobel et al. (1984) reported the detection of amino acids in liquid extracts of the reaction mixtures at the 1983 ISSOL conference; yields were, however, extremely low, the total yield being 1.5–3.0×10^{-7} g per gram of MgO. These results were the subject of considerable attention in the media.

This area of research is not being followed up at present; however, the processes occurring at hydrothermal vents pose some new questions. Carboxylic and dicarboxylic acids (such as glycolic, malonic, oxalic and succinic acids) were detected in tetrahydrofuran and aqueous extracts from large synthetic MgO crystals (Freund et al., 1999). The authors suggest that magmatic and metamorphic rocks, which come to the surface of a tectonically active planet such as the Earth and undergo weathering there, could be an important source of abiotically formed organic compounds. The amounts of organic material set free per unit rock volume are unknown, but estimates have been made. On the assumption that one to ten cubic kilometres of weathered rock were formed, Freund (2000) has calculated a rate of formation of the order of 10^{10}–10^{12} g of organic material per year. It is possible that hydrothermal liquids increased the estimated values by leaching rocks and minerals. Using highly sensitive IR-spectroscopic methods, Freund et al. (2001) detected organic proto-molecules in both highly pure synthetic MgO and natural olivine from the Earth's upper mantle.

References

Andersson EM, Holm NG (2000) Orig Life Evol Biosphere 30:9
Alargov DK, Deguchi S, Tsuji K, Horikoshi K (2002) Orig Life Evol Biosphere 32:1
Alt JC, Anderson TF, Bonnell L (1989) Geochim Kosmochim Acta 53:1011
Bada JL, Miller SL, Zhao M (1995) Orig Life Evol Biosphere 25:111
Bada JL, Feglex Jr B, Miller SL, Lazcano A, Cleaves HJ, Hazen RM, Chalmers J (2007) Science 315:937
Baross JA, Deming JW (1983) Nature 303:423
Berndt ME, Allen DE, Seyfried Jr WE (1996) Geology 24:351
Bernhard G, Ludemann HD, Jaenicke R, König H, Stetter KO (1984) Naturwissenschaften 71:583
Blöchl E, Keller M, Wächtershäuser G, Stetter KO (1992) Proc Natl Acad Sci USA 89:8117
Bröll D, Kaul C, Krämer A, Krammer P, Ricter T, Jung M, Vogel H, Zehner P (1999) Angew Chem 111:3180, Int Ed 38:2998
Cairns-Smith AG (1966) J Theor Biol 10:53
Cairns-Smith AG (1985a) Seven Clues to the Origin of Life. Cambridge University Press
Cairns-Smith AG (1985b) Scientific American, June p 74
Cairns-Smith AG (1992) Orig Life Evol Biosphere 22:161
Choughuley ASU, Lemmon RM (1966) Nature 210:628
Cody GD, Boctor NZ, Blank J, Brandes J, Filley TR, Hazen RM, Yoder H Jr (2000a) Orig Life Evol Biosphere 30:187
Cody GD, Boctor NZ, Filley TR, Hazen RM, Scott JH, Sharma A, Hatten S, Joder Jr (2000b) Science 289:1337
Corliss JB, Baross JA, Hoffman SE (1981) Ocean. Acta No. SP 59-69
Crabtree RH (1997) Science 276:222
de Duve C (1988) Nature 336:209
de Duve C, Miller SL (1991) Proc Natl Acad Sci USA 88:10014
de Duve C (1991) Blueprint for a Cell: The Nature and Origin of Life. Patterson, New York
de Duve C (1996) Vital Dust: Life as a Cosmic Imperative. HarperCollins Canada / Basic Books
de Duve C (1998) Clues from present-day biology: the thioester world. In: Brack A (Ed.) The Molecular Origins of Life. Cambridge University Press, p 219
de Souza-Barros F, Braz-Levignard R, Ching-Sang Jr Y, Monte MMB, Bonapace JAP, Montezano V, Vieyra A (2007) Orig Life Evol Biosphere 37:27
Edmond JM, von Damm K (1983) Scientific American, April p 70
Evans MCW, Buchanan BB, Aron DI (1966) Proc Natl Acad Sci USA 55:928
Freund F, Gupta AD, Kumar D (1999) Orig Life Evol Biosphere 29:489
Freund F (2000) Orig Life Evol Biosphere 30:145
Freund F, Staple A, Scoville J (2001) Proc Natl Acad Sci USA 98:2142
Hagmann M (2002) Science 295:2006
Greenwell HC, Coveney P (2006) Orig Life Evol Biosphere 36:13
Hazen RM, Deamer DW (2007) Orig Life Evol Biosphere 37:143
Heinen W, Lauwers A-M (1996) Orig Life Evol Biosphere 26:131
Hékinian R (1984) Scientific American, July p 34
Helgeson HC, Delaney JM, Nesbitt HW, Bird DK (1978) Amer J Sci 271:1
Hennet J-C, Holm NG, Engel NH (1992) Naturwissenschaften 79:361

Holm NG (1992) Why are hydrothermal systems proposed as a plausible environment for the origin of life? In: Holm NG (Ed.) Marine Hydrothermal Systems and the Origin of Life. Kluwer, Dordrecht Boston London, p 5
Holm NG, Andersson EM (2005) Astrobiologie 5:444
Holm NG, Andersson EM (1995) Planet Space Sci 43:153
Holm NG, Andersson EM (1998) Hydrothermal Systems. In: Brack A (Ed.) The Molecular Origins of Life. Cambridge University Press, p 86
Hoffmann U (1968) Angew Chem 80:736, Int Ed 7:681
Horita J, Berndt ME (1999) Science 285:1055
Huber C, Wächtershäuser G (1997) Science 276:245
Huber C, Wächtershäuser G (1998) Science 281:670
Huber C, Wächtershäuser G (2006) Science 314:630
Huber C, Eisenreich W, Hecht S, Wächtershäuser G (2003) Science 301:938
Kaschke M, Russell MJ, Coll WJ (1994) Orig Life Evol Biosphere 24:43
Keefe AD, Miller SL, Mc Donald G, Bada J (1995) Proc Natl Acad Sci USA 92:11904
Keefe AD, Newton GL, Miller SL (1995) Nature 373:683
Keefe AD, Miller SL, Mc Donald G, Bada J (1996) Orig Life Evol Biosphere 26:230
Keller M, Blöchl E, Wächtershäuser G, Stetter KO (1994) Nature 368:336
Khare BN, Sagan C (1971) Nature 232:577
Knobel R, Breuer H, Freund F (1984) Origins of Life 14:197
Lahav N (1999) Biogenesis. Oxford University Press, New York Oxford
Liu R, Orgel LE (1997) Nature 389:52
Macdonald KC, Luyendyk BP (1981) Scientific American, May p 86
Maselko J, Strizhak P (2004) J Phys Chem B 108:4937
Miller SL, Orgel LE (1974) The Origin of Life on the Earth. Prentice-Hall, Englewood Cliffs NY
Miller SL, Bada JL (1988) Nature 334:609
Miller SL, Bada JL, Friedmann N (1989) Orig Life Evol Biosphere 19:536 (abstr)
Miller SL, Lazcano A (1995) J Mol Evol 41:689
Miyakawa S, Cleaves AJ, Miller SL (2002a) Orig Life Evol Biosphere 32:195
Miyakawa S, Cleaves AJ, Miller SL (2002b) Orig Life Evol Biosphere 32:209
Mok DS-L, Antal MJ Jr, Jones M Jr (1989) J Org Chem 54:4596
Monnard P-A, Apel CL, Kanavarioti A, Deamer DW (2002) Astrobiology 2:139
Moulton V, Gardner P, Pointon RF, Creamer LK, Jameson GB, Penny DJ (2000) J Mol Biol 51:416
Österberg R (1974) Nature 249:382
Österberg R (1997) Orig Life Evol Biosphere 27:481
Pályi G, Zucchi C (2002) Book of Program and Abstracts, ISSOL-02 p 91
Rees DC, Howard JB (2003) Science 300:929
Ross DS (2007) Orig Life Evol Biosphere 37:61
Russell MJ, Hall AJ, Turner D (1989) Terra Nova 1:238
Sagan C, Khare BN (1971) Science 173:417
Schiermeier Q (2006) Nature 440:20
Schoonen M, Xu Y, Bebie J (1999) Orig Life Evol Biosphere 29:5
Schwartz AW, Joosten H, Voet AB (1982) Biosystems 15:191
Shapiro R (1999) Planetary Dreams. Wiley, New York
Shock EL (1990) Orig Life Evol Biosphere 20:331
Smith E, Morowitz HJ (2004) Proc Natl Acad Scie USA 101:13168
Smirnov A, Schoonen MAA (2003) Orig Life Evol Biosphere 33:117
Tributsch H. Fiechter S, Jokisch D, Rojas-Chapana J, Ellmer K (2003) Orig Life Evol Biosphere 33:129
Trinks H, Schröder W, Biebricher CK (2005) Orig Life Evol Biosphere 35:429
Valley JW (2005) Scientific American October p 42
Wächtershäuser G (1988a) Microbiol Rev 52:452
Wächtershäuser G (1988b) Proc Natl Acad Sci USA 85:1134
Wächtershäuser G (1990a) Orig Life Evol Biosphere 20:173

Wächtershäuser G (1990b) Proc Natl Acad Sci USA 87:200
Wächtershäuser G (1992) Order out of Order: Heritage of the Iron-sulfur World. In: Trân Thanh Vân J und K, Mounolou JC, Schneider J, Mc Kay C (Eds.) Frontiers of Life. Editions Frontiers, Gif-sur-Yvette, pp 21–39
Wächtershäuser G (1997) J Theor Biol 187:483
Wächtershäuser G (2000) Science 289:1307
Weber AL (1981) J Mol Evol 17:103
Weber AL (1984) Origins of Life 15:17
Weber A (1998) Orig Life Evol Biosphere 28:259
Weiss A (1963) Angew Chem 75:113, Int Ed 2:134
Weiss A (1981) Angew Chem 93:843, Int Ed 20:850
Wieland T, Pfleiderer G (1957) Advances in Enzymology 19:235
Wills C, Bada J (2000) The Spark of Life. Perseus, Cambridge Mass.
Zhang XV, Martin ST (2006) J Amer Chem Soc 128:16032
Zhang XV, Ellery SP, Friend CM, Holland HD, Michel FM, Schoonen MAA, Martin ST (2007) J Photochem Photobiol A – Chem 185:301

Chapter 8
The Genetic Code and Other Theories

8.1 The Term "Information"

The term "information" often comes up in discussions of biogenesis, so a brief survey of its meaning is called for. Just as with the problem of the definition of the term "life", there seems to be no generally accepted definition of the term "information".

The state of the physical world can be described in terms of three basic quantities:

Mass
Energy
Information

Expressed in a simplified manner, but to the point, we can say that mass and energy describe the state of a system, while information describes the distribution of this state in time and space.

According to Shannon, the well-known relationship between the information content of an item of news and its expectation probability can be calculated using the formula:

$$I_i = \operatorname{ld} p_i^{-1}$$

where I_i is the information content, p_i the probability and ld is the logarithm to the base 2. According to this definition, the information content of an item of news is determined only by its probability. Information can be divided into various species, which Küpers calls "dimensions":

The *syntactic* dimension, which encompasses the relationship between the symbols

The *semantic* dimension, which encompasses the relationship between the symbols and what they stand for

The *pragmatic* dimension, which encompasses the relationship between the symbols, what they stand for, and what this means for the emitter and the acceptor in terms of a demand for action (Küpers, 1990)

Pragmatic means oriented towards action: pragmatic information causes change in the acceptor (Jantsch, 1980).

The book *Information and the Origin of Life* by Küpers (1990) can certainly be recommended; details of new developments in the field of quantum information can be found in the book edited by Bruß and Leusch (2006).

8.2 The Genetic Code

"The genetic code remains an enigma, even though the full codon catalog was deciphered over 30 years ago". This is the first sentence in an article by Knight et al. (1999) from the Department of Ecology and Evolutionary Biology at Princeton University. Scientists have attempted to solve this riddle with the help of a variety of hypotheses and models.

The genetic code controls the assignment of the nucleotide triplets to the amino acids, and thus forms the interface between nucleic acids and proteins. A short look back at its history reveals that about fifty years ago, just after the publication of the double helix model by Watson and Crick, George Gamow suggested that there must be a relationship between the various amino acids and certain regions in the DNA structure which are formed by nucleotides (Gamow, 1954; see Sect. 5.2). A few years later, it was realized that the amino acid sequence of the proteins in the cells of all living things are determined (with few exceptions) by the same reaction mechanisms. The survival of the individual organisms requires that this complex process operates exactly. In summary, the main steps are as follows:

The information contained in the DNA (i.e., the order of the nucleotides) is first transcribed into RNA. The messenger RNA thus formed interacts with the amino-acid-charged tRNA molecules at specific cell organelles, the ribosomes. The loading of the tRNA with the necessary amino acids is carried out with the help of aminoacyl-tRNA synthetases (see Sect. 5.3.2). Each separate amino acid has its own tRNA species, i.e., there must be at least 20 different tRNA molecules in the cells. The tRNAs contain a nucleotide triplet (the anticodon), which interacts with the codon of the mRNA in a Watson–Crick manner. It is clear from the genetic code that the different amino acids have different numbers of codons: thus, serine, leucine and arginine each have 6 codewords, while methionine and tryptophan are defined by only one single nucleotide triplet.

How could such a complex information transfer system have evolved? No witnesses from the archaic period, three to four billion years ago, have survived; even the analysis of meteoritic rock does not help here.

Knight discusses the thesis that the genetic code could possibly have three different "faces":

Selection
History
Chemistry

The genetic code, modified by *selection*, represents an adaptation of optimized functions, for example, for the minimisation of coding errors (arising from mutations

Table 8.1 The genetic code as a three-letter code, with the normal abbreviations for the 20 proteinogenic amino acids. The first base corresponds to the 5′ end, the third to the 3′ end of the nucleotide chain

		Second base				
		U	**C**	**A**	**G**	
First base	**U**	Phe	Ser	Tyr	Cys	U C
		Leu		Stop	Stop	A
					Trp	G
	C	Leu	Pro	His	Arg	U C
				Gln		A G
	A	Ile	Thr	Asn	Ser	U C A
		Met		Lys	Arg	G
	G	Val	Ala	Asp	Gly	U C
				Glu		A G

Phe	phenylalanine	Thr	threonine	Asp	aspartic acid
Leu	leucine	Ala	alanine	Glu	glutamic acid
Ile	isoleucine	Tyr	tyrosine	Cys	cysteine
Met	methionine	His	histidine	Arg	arginine
Val	valine	Gln	glutamine	Gly	glycine
Ser	serine	Asn	asparagine	Trp	tryptophane
Pro	proline	Lys	lysine		

or incorrect translations). The *historic* component reflects the interplay between amino acids and the code across a long period of time. The *chemical* component may be due to the fact that certain codon characteristics could have been formed by favourable chemical interactions between certain amino acids and short nucleic acid sequences. If these interactions had not been present, the amino acids in question would have been excluded from incorporation into peptides and proteins.

If the genetic code in its present form still poses so many questions, the elucidation of its development three to four billion years ago will be even more difficult! Some researchers feel that an exact reconstruction of the process of its construction may never be possible, while others see the genetic code as being purely fortuitous, a system which was "frozen" at some time in history. It appears plausible that the code, just like other organism properties, is the product of natural selection (Vogel, 1998).

Some of the many hypotheses and models will be presented briefly. The physicochemical hypothesis refers to a minimalisation of the liability of the genetic code to cause errors in information transmission. The error rate can fall when amino acids with similar codons have similar properties, such as the presence of hydrophilic residues. This would mean that the exchange of a base in the base triplet would cause only a small change in the character of the protein.

Model calculations indicate that the genetic code cannot be the product of chance, but that it was optimized by selection processes. Computer simulations show the insensitivity of the contemporary genetic code, since (in model calculations) it withstood errors better than a million other codons (Vogel, 1998).

Another hypothesis was put forward by Tze-Fei Wong, who suggested that there is a connection between the development of the genetic code and the formation of new amino acids. Here he agrees with several other biogenesis researchers that only a few amino acid species were present at the beginning of the prebiotic development process: these would have made use of all the then existing codons. Thus, codons would have to have been "given up" by the previous "owners" to accommodate the newly formed amino acids. Wong, now at the University of Hong Kong, assumed that the syntheses which evolved, in which new amino acids were obtained from those already present (or their precursors), involved small numbers of enzymatically catalysed reaction steps (Wong, 1975).

Table 8.2 shows the amino acid conversions which led to contemporary amino acids, as suggested by Wong. The evolutionary relationships between the code words for the various amino acids are shown in Table 8.1.

Table 8.2 According to the co-evolution theory proposed by Wong, the biosynthetic routes to amino acids from their precursors could perhaps provide information on the evolution of the genetic code (Wong, 1975)

Glu → Gln	Asp → Lys	Ser → Trp
Glu → Pro	Glu → His	Ser → Cys
Glu → Arg	Thr → Ile	Val → Leu
Asp → Asn	Thr → Met	Phe → Tyr
Asp → Thr		

It is worth looking at the various base sequences with the help of the genetic code. All the pairs of codon units which are linked by arrows in Table 8.2 differ by only one single base exchange.

Another hypothesis was provided by Mikio Shimitso (1982) on the basis of studies of steric effects in molecular models. It had been noted years previously that the fourth nucleotide at the $3'$ end of the tRNA molecules (referred to as the discrimination base) might have a recognition function. In the case of certain amino acids (i.e., their tRNA–amino acid complexes) this base pair, in combination with the anticodon of the tRNA molecule, can select the amino acid corresponding to the tRNA species in question; this is done on the basis of the stereochemical properties of the molecule. Since the anticodon of a tRNA molecule and the fourth nucleotide of the acceptor stem are far apart in space, two tRNA molecules must complex in a head-to-tail manner. The "pocket" thus formed can then fit specifically to the corresponding amino acid.

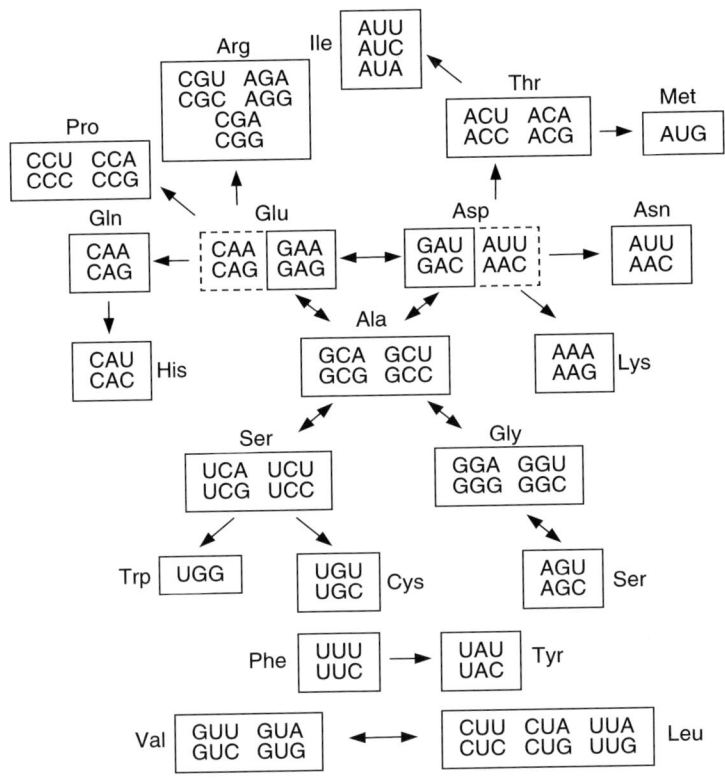

Fig. 8.1 Wong's "evolutionary map" shows possible relationships between code words. The codons in boxes correspond to today's code words (compare Table 8.1). The codons for Asp and Glu in dotted boxes refer to these amino acids in a very early evolutionary stage of the genetic code. Single-headed arrows refer to the biosynthetic relationships between precursor and product, while double-headed arrows refer to reciprocal formation possibilities. All pairs of codon units (irrelevant of whether they are linked by single- or double-headed arrows) differ only in one single base change (Wong, 1975)

M. Di Giulio from the Institute of Genetics and Biophysics at the University of Naples has been working on this problem for several years. He studied tRNA sequences with the help of statistical analyses (Di Giulio, 1997; Di Giulio and Meduquo, 1998, 1999, 2000, 2001) and was in principle able to confirm Wong's co-evolution theory. Three out of 15 tRNA nucleotide sequences, those for Ile, Met and Val, contain the common element GAGCGC. This result is interpreted to mean that these three tRNA species have a common evolution history. This conclusion agrees with the fact that the three amino acids isoleucine, valine and methionine are related in terms of their biosynthesis (Di Giulio, 1994).

The reaction of R. Arminovin and S. Miller (1999) to one of Di Giulio's articles on the co-evolution theory (or hypothesis) illustrates the wide range of feelings on the code problem: although the critics find this theory "attractive", they feel that it cannot be proven by a comparison of biosynthetic relationships.

An important factor in the evolution of the genetic code is certainly provided by the aminoacyl-tRNA synthetases (see Sect. 5.3.2). It is clear that the two synthetase classes are not randomly distributed across the matrix of the amino acid assignment of the genetic code. For example, with one exception, all XUX codons code for class 1 synthetases, while all XCX codons code for class 2 aminoacyl-tRNA synthetases. A possible explanation could be that the synthetases and the genetic code evolved simultaneously. However, it is more likely that these enzymes evolved when the genetic code had already been established (Wetzel, 1995).

The question as to the age of the genetic code was answered by M. Eigen via a statistical evaluation of comparative rRNA sequences. According to Eigen, the genetic code must have originated about 3.6 billion years ago, in agreement with the time at which the first life forms are assumed to have developed. Although there are still some open questions about the evolution of the genetic code, there are several points where there is (more or less) consensus within the biogenetics community (de Duve, 1994):

> The genetic code evolved in a stepwise manner. The initial steps must have been primitive, and thus inexact. (L. Orgel once suggested that the primitive code may have consisted of only two codons: one for hydrophilic and one for hydrophobic amino acids).

> The first proteins contained only a few amino acids, and not the 20 which are now involved.

> Although only a few types of amino acids were available during the first few million years of the development of the code, it is likely that three-letter codes were used from the beginning. F. Crick also believes in a three-letter code, because of the "continuity principle". It is possible that the third base had no function in the early days but was only necessary for steric reasons; for example, because with a two-letter code (only two bases) in the mRNA strand, there would not have been enough room for two amino acid-bearing tRNA molecules.

M. Eigen suggested that the primeval forms of tRNA consisted of iterative PurXPyr, thus, for example, GXC triplets. This assumption is present in the title of the publication "Transfer-RNA, ein frühes Gen?" (Transfer RNA, an early gene?) (Eigen and Winkler-Oswatitsch, 1981). As shown above, a nucleotide chain with a repeating pattern naturally leads to the generation of a complementary antiparallel nucleotide sequence:

$$............PurXPyr\ PurXPyr.............$$
$$............PyrXPur\ PyrXPur.............$$

X can be any of the four nucleobases—G, A, C or U; thus, four anticodons can be formed: GGC, GAC, GCC and GUC. The antiparallel structure contains the codons GCC, GUC, GGC and GAC. Today, these codons code for the four amino acids alanine, glycine, valine and aspartic acid. These are, astonishingly, the four protein building blocks produced in the best yields in the Miller-Urey experiment, and they have also been detected in the Murchison meteorite. It is difficult to believe that these are purely accidental results.

Another model is based on the fact that the genetic code shows a number of regularities, some of which have already been mentioned above. It is suspected that codons beginning with C, A or U code for amino acids which were formed from α-ketoacids (or α-ketoglutarate, 1-KG), oxalacetate (OAA) and pyruvate. This new model, which is quite different from the previous models, assumes that a covalent complex formed from two nucleotides acted as a catalyst for chemical reactions such as the reductive amination of α-ketoacids, pyruvate and OAA. More recent analyses suggest that the rTCA cycle (see Sect. 7.3) could have served as a source of simple α-ketoacids, including glyoxylate, pyruvate, OAA and α-KG. α-Ketoacids could, however, also have been formed via a reductive acetyl-CoA reaction pathway. The bases of the two nucleotides specify the amino acid synthesized and were retained until the modern three-letter codes were established (Copley et al., 2005).

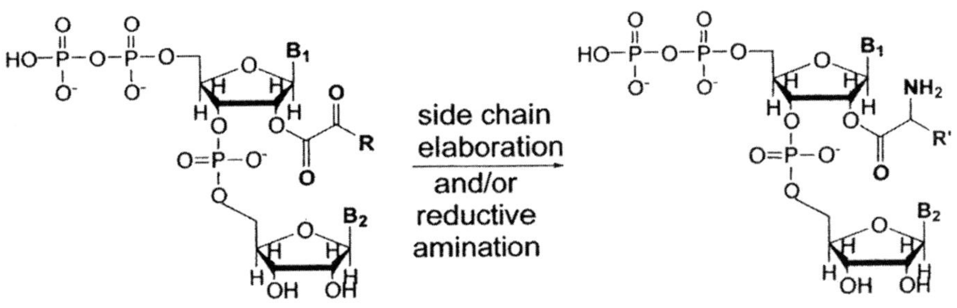

side chain elaboration and/or reductive amination

Fig. 8.2 Model for the synthesis of amino acids from alpha-keto acid precursors covalently attached to dinucleotides. The dinucleotide that is capable of catalyzing synthesis of particular amino acids is proposed to contain the first two bases of the codon specifying that amino acid (Copley et al., 2005)

One important question is that of the order in which the basic mechanisms of evolution processes, leading eventually to the emergence of life, occurred. As far as the development of the genetic code is concerned, it is not clear whether the code evolved prior to the aminoacylation process, i.e., whether aminoacyl-tRNA synthetases evolved before or after the code. A tRNA species which is aminoacylated by two different synthetases was studied: if this tRNA had important identity elements such as the discriminator base and the three anticodon bases for the two synthetases, this would be evidence that the aminoacyl-tRNA synthetases had developed *after* the genetic code. Dieter Söll's group, which is experienced in working with this family of enzymes, came to the conclusion that the universal genetic code must have developed before the evolution of the aminoacylation system (Hohn et al, 2006).

A further (mathematical) model for the evolution of the genetic code was devised by Carl Woese and co-workers. This dynamic theory provides information on the evolvability and universality of the genetic code. One conceptual difficulty was due to the fact that it had been overlooked that the genetic code was highly communal from the very beginning of its development. The Woese model distinguishes three different phases; these are (highly simplified):

Weak communal evolution
Strong communal evolution
Individual evolution

A detailed treatment of the evolution of the genetic code requires modelling physical components of the translational process; this includes the dynamic processes of the tRNAs and the aminoacyl-tRNA synthetases (Vetsigian et al., 2006). Thus, in spite of considerable advances in the search for the roots of the genetic code, there is still much to do!

8.3 Eigen's Biogenesis Theory

Manfred Eigen presented his theory on self-organisation and evolution of living systems more than 35 years ago (Eigen, 1971a, b). The Nobel Prize winner Eigen, working at the Max Planck Institute for Biophysical Chemistry in Göttingen, extended Darwinian evolution by means of a mathematically based theory, which goes back to the fundamental molecular processes which could have led to biogenesis.

According to F. Dyson (1985) (see Sect. 8.5), Eigen has stood Oparin's hypothesis (first cells, then enzymes, and finally genes) on its head by reversing this sequence: first genes, then enzymes, and finally the cell. Dyson finds the Eigen theory popular for two reasons:

The experiments used RNA as the working material. RNA replication is considered as a fundamental process.
The elucidation of the DNA double helix showed that genes are structurally more simple than enzymes (i.e., proteins).

This is assuming that biogenesis went through the following phases:

– The formation of small molecules,
– Self-organisation of macromolecules to give functioning, self-reproducing units,
– The formation of cell structures and
– Biological evolution from the unicellular organism to multicellular entities.

Eigen's theory concerns itself with phase two, the focal point of the genesis of the replication process. Apart from the first, each phase requires the previous one as its precondition.

The term "evolution" is often used in several senses in the biogenesis discussion. Thus, phase 1 is mostly referred to as "chemical evolution", while phase 2 is called "molecular evolution". Evolution is now often also referred to in physics, for example in the evolution of the cosmos, and also in the sense of further development or advancement.

Eigen's theory describes the self-organisation of biological macromolecules on the basis of kinetic considerations and mathematical formulations, which are in turn based on the thermodynamics of irreversible systems. Evolutionary processes are irreversibly linked to the flow of time. Classical thermodynamics alone cannot describe them but must be extended to include irreversible processes, which take account of the "arrow of time" (see Sect. 9.2). Eigen's theory is based on two vital concepts:

> The quasi-species and
> The hypercycle model.

The term "quasi-species" refers to a population of genetically related RNA molecules which are not identical. They act as matrices for the next generation of RNA molecules, which then also belong to the quasi-species, i.e., to the larger "clan". The molecules belonging to a quasi-species have certain morphological properties in common. Eigen assumes that there are natural selection processes (similar to the Darwinian process) which favour a replication of those molecules which are similar to the quasi-species. Eigen represents this situation in terms of a series of equations which represent the state which is formed between the (Darwinian) selection process and the replication errors due to chance.

A "hypercycle" is a more complex organisation form. Its precondition is the presence of several RNA quasi-species which are able to amalgamate chemically with certain proteins (enzymes or their precursors). If such a protein is linked to a quasi-species, the resulting duo favours the replication of a second quasi-species. According to Dyson, the linked populations get stuck in a stable equilibrium. Problems occur at this level! Any theory on the origin of replication has the central problem that the replication process must occur perfectly in order to ensure "survival". If there are replication errors, these will increase from generation to generation, until the system collapses: the "error catastrophe" has then occurred!

M. Eigen describes this process roughly as follows: let us consider a self-replicating system which is characterized by a quantity of information equal to N bits. The probability that a bit is incorrectly copied is w, and the selection reacts to errors by means of a selection factor S. In other words: an error-free system has a selection advantage S over a system with an error. The survival criterion is then:

$$N \cdot w < \log S$$

1. If the above condition is fulfilled, the selection advantage of the error-free system is so large that a number of errors in the total population can be tolerated.
2. If it is not fulfilled, the error catastrophe occurs. On the left-hand side of the equation is the number of bits of information which the system loses if the errors are copied in each new generation. The right-hand side corresponds to the number of bits which are caused by selection.

The consequence is as follows: if more information is lost than can be additionally delivered, the system is threatened. If the above condition is to be fulfilled, the error rate cannot be larger than $1/N$. This condition can hardly be fulfilled by today's organisms, as they have values of N of around 10^8 and of w of around 10^{-8} (Dyson, 1985).

However, complex error corrections (by repair enzymes) permit a great reduction in the error rate. Such repair systems were not available to primitive replicators, so they needed to survive with error rates of more than 1:100; this reduced the size of the genome to around 100 bases (nucleotides). This became obvious in work done by Saul Spiegelmann (1967) and the Eigen group (Biebricher et al., 1981).

The fundamental experiments carried out by Saul Spiegelmann (Columbia University) entered the annals of science: he was the first person to provide evidence for Darwinian evolution "in a test tube". The object under study was the bacteriophage Q beta, an RNA phage which can infect E. coli cells. For his in vitro incubation experiment, Spiegelmann used: purified Q beta-RNA, Q beta-replicase and the four ribonucleoside triphosphates (ATP, GTP, UTP and CTP). After 20 min, he transferred part of the reaction mixture to a fresh reaction medium (with replicase and nucleotides): 75 such transfer reactions were carried out. The incubation time was successively reduced, in order to select the RNA species initially formed; the rate of RNA synthesis increased from reaction to reaction. The length of the virus RNA decreased considerably and was finally only 17% of the initial length. Those genes which the phage no longer required to survive under the in vitro conditions were eliminated, since the replicase was continually added to the system from outside. It was only important that the 3' end of the RNA contained the initiation sequence essential for the activity of the Q beta-replicase. An even shorter mutant with only 220 of the 4220 nucleotides originally present in the wild type was obtained. These short Q beta-RNA strings are referred to rather disrespectfully as "Spiegelmann monsters".

M. Eigen carried out similar experiments with the Q beta phage (Biebricher et al., 1981). He decreased the concentration of added viral RNA stepwise but still obtained a good yield of mutant RNA, the chain length of which was, however, decreased. New RNA was formed even without the addition of RNA. This astounding result was initially considered to be due to tiny amounts of RNA having been allowed to enter the system; however, further experiments showed these qualms to be unfounded (Eigen et al., 1982). An exact analysis of the short-chain RNA showed that it was similar to those obtained previously by Spiegelmann.

However, if a triplet genetic code system really did exist around 3.5 billion years ago, an RNA strand containing about 100 nucleotides would only have been able to code for a maximum of 33 amino acids. With 33 amino acids, the polypeptide formed would have been only two thirds as long as the insulin molecule, and it is doubtful whether such a chain length would have sufficed for an active replication system.

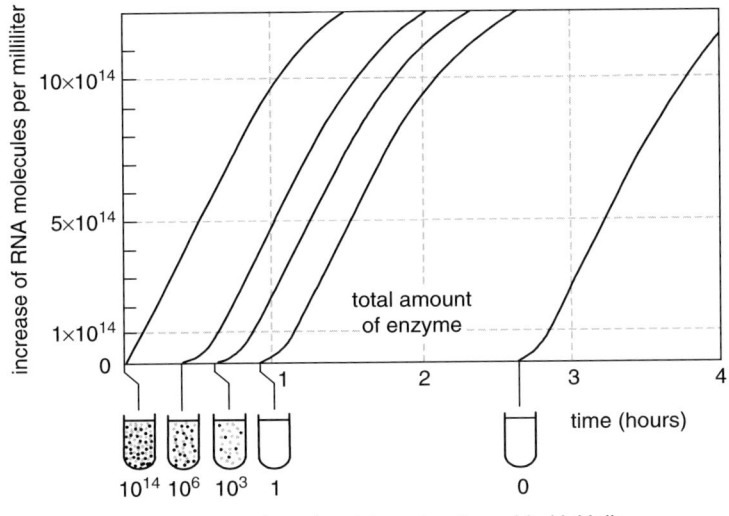

Fig. 8.3 The proliferation curves of RNA strands (the Q beta system) for decreasing concentrations of added matrix molecules. If the number of matrix molecules is larger than that of the enzymes, a linear proliferation is observed (first curve). This slows down at high concentrations, due to product inhibition. RNA proliferation is exponential if the amount of enzyme is larger than that of the matrix. If no matrix is added, the system goes through an incubation phase and then forms an RNA sequence which is related to certain Q beta fragments (Eigen et al., 1982)

The "information crisis", i.e., the fact that, because of the error frequency, longer RNA chains have so many errors after only a few reproduction steps that they can no longer be replicated, cries out for catalysts which can guarantee more exact replication. While only protein catalysts (enzymes) had been discussed until recently, ribozymes are now possible candidates. More complex catalysts would have required more complex matrices; but where did the matrix molecules come from? This serious problem, referred to by Eigen himself as an information crisis, is sometimes referred to as "Eigen's dilemma" (Blomberg, 1997).

This dilemma could be overcome by the hypercycle model: hypercycles are in fact not theoretical concepts, but can be observed (in a simple form) in today's organisms, where an RNA virus transfers the information for an enzyme in the host cell, which is able to carry out the preferred synthesis of new virus RNA. This RNA synthesis is supported by host factors, and an RNA minus-strand is formed. The following RNA replication affords a plus-strand. The process corresponds to a double feedback loop and involves the enzyme coded by the RNA matrix and the information present in the matrix in the form of a nucleotide sequence. Both factors contribute to the replication of the matrix, so that there is second-order autocatalysis (Eigen et al., 1982).

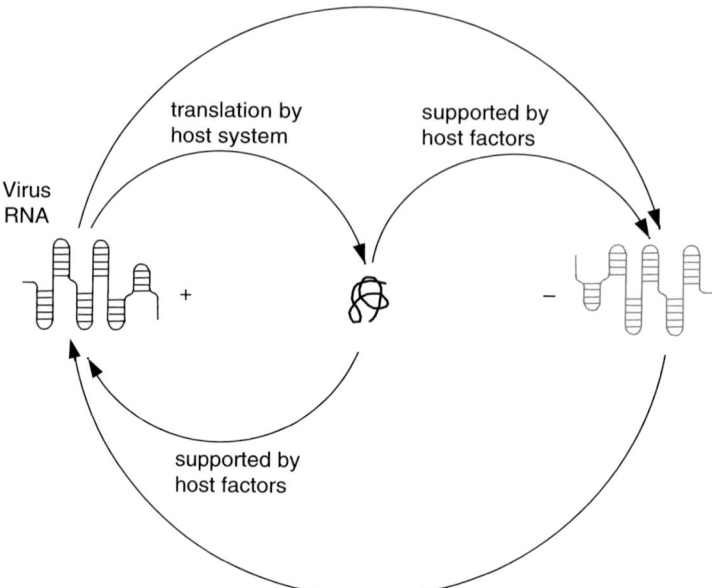

Fig. 8.4 Hypercycle phenomena can be observed when a cell is infected by an RNA virus. The virus provides the host cell with information for an enzyme favouring only the reproduction of viral information, i.e., of an RNA strand. This RNA is converted by the host cell into a protein (a replicase) which forms a new RNA minus-strand. The latter is then replicated to give a plus-strand (Eigen et al., 1982)

The hypercycle models developed later by Eigen were much more complex. Since both protein enzymes and nucleic acids contribute to hypercycles, the latter could only have come into operation at a later stage of the (hypothetical) RNA world. It seems possible that the protein enzymes on the primeval Earth could have been replaced by ribozymes.

In a self-reproducing, catalytic hypercycle (second order, because of its double function of protein and RNA synthesis) the polynucleotides N_1 contained not only the information necessary for their own autocatalytic self-replication but also that required for the synthesis of the proteins E_1. The hypercycle is closed only when the last enzyme in the cycle catalyses the formation of the first polynucleotide. Hypercycles can be described mathematically by a system of non-linear differential equations. In spite of all its scientific elegance and general acceptance (with certain limitations), the hypercycle does not seem to be relevant for the question of the origin of life, since there is no answer to the question "how did the first hypercycle emerge in the first place?" (Lahav, 1999).

Doubts about the problem of the error catastrophe were expressed in Freiburg im Breisgau. Computer simulations showed that, apart from the error catastrophe, other fatal events threaten evolving populations (Bresch et al., 1980; Niesert et al., 1982;

Niesert, 1987). The Freiburg group found three possible catastrophes, which can be summed up as follows:

The egoistic RNA: this catastrophe occurs when, after a mutation, one RNA molecule learns to replicate faster than the others, "forgetting" to act as a catalyst.

The short circuit: this happens when an RNA molecule in the hypercycle is changed so much by mutation that it does not catalyse the next reaction in the chain, but a later one. The hypercycle is then short-circuited to become a simple cycle.

Collapse: when statistical variations cause one of the important components of the cycle to die off, the complete cycle collapses.

Computer simulations showed that the first two of these catastrophes become more probable as the size of the molecular population increases. In order to avoid them, the population of a hypercycle would need to be kept as small as possible. The probability of collapse, however, decreases with increasing population. Because of these contradictions, Ursula Niesert gave one of her articles the title "The Origin of Life between Scylla and Charybdis", because computer simulations indicate that there is only a small interval of hypercycle populations in which all three of the above catastrophes can be excluded.

As expected, a response to the hypercycle criticisms appeared, in fact in the same issue of the *Journal of Theoretical Biology* (Eigen et al., 1980). According to this, the Freiburg investigations refer to one particular evolution model, in which the occurrence of mutants with different, selective values is ignored. In such realistic models, the error threshold loses its importance for the stability of the wild type. If the latter reaches a finite fitness value, it can always be the subject of selection, as no rivals are present.

Freeman Dyson considers that any theory on the origin of life which begins with cooperative organisation in a large population of molecules, and makes no provision for short circuits in the metabolic pathways, will be met by the criticism just described (Dyson, 1985).

The critical discussion of some aspects of Eigen's theories and hypotheses by no means diminishes the great merits which Manfred Eigen has gained in more than 30 years of work in this area of science. His experimental and theoretical contributions to the question of the evolution of genetic information are of decisive importance, even if corrections of his theories may be necessary from time to time.

8.4 Kuhn's Biogenesis Models

Hans Kuhn, who described his own models in an article on the "Self-organisation of Molecular Systems and the Evolution of the Genetic Apparatus" (Kuhn, 1972), also worked in the Max Planck Institute for Biophysical Chemistry in Göttingen. Eigen and Kuhn presented two differently conceived approaches which attempt to provide an explanation for the problem of the evolution of replication.

H. Kuhn developed a model which shows how it is possible to proceed in small, clear, calculable steps from one development phase to the next. Starting from certain situations or states of the system, possible conditions for moving to the next steps are estimated. In the development of his model, Kuhn proceeds in a manner similar to that involved in quantum mechanics: here, suitable test functions were generated which provided approximate solutions for wavefunctions in order to be able to explain chemical bonding phenomena better.

The biogenesis process can be reduced to a few subprocesses; Fig. 8.5 shows this in a greatly simplified schematic form (Kuhn and Waser, 1982).

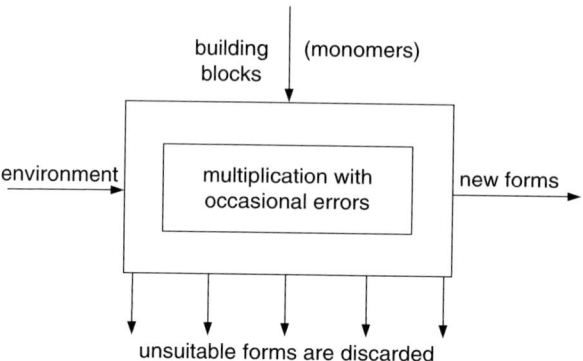

Fig. 8.5 Schematic representation of the basic features of the evolution process (Kuhn and Waser, 1982)

As early as 1972, Kuhn described a model in which he assumed that RNA replication with a certain error rate could have occurred without the participation of enzymes. Natural phenomena with cyclic behaviour are an important factor in Kuhn's thinking; these drive duplication processes. Examples are summer and winter, day and night, or high and low tide (whereby the latter were probably subject to greater variations on the primeval Earth than they are today). These rhythms were often linked with considerable temperature variations, which, for example, made possible the transition from double to single strand RNA (and vice versa). It can be assumed that the cyclic variations involved reactions in which monomers were linked to form polymers.

Other environmental conditions, such as changes of the pH value of sea water, or in reactant concentration, could have led to decomposition reactions. Kuhn's initial model distinguishes between divergent and convergent evolutionary phases:

Divergent phases: many molecular species survived with similar probabilities, resulting in a diverse population.

Convergent phases: these are highly selective. The only mutants which survived were those which best served the new purpose.

Fig. 8.6 Knowledge (K) as a function of the length of the evolution process (K as the ability to survive in a certain environment) (Kuhn, 1976)

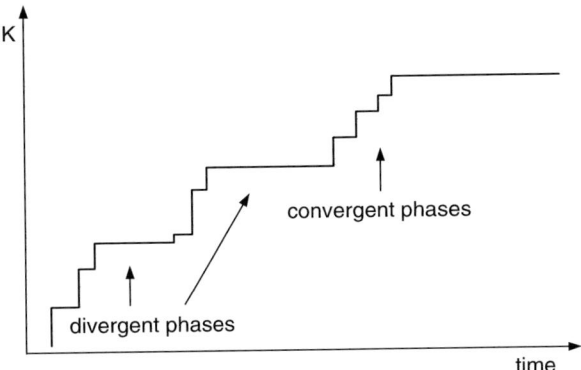

Phase reverse occurs at a turning point, although it is generally very unlikely that such a point will be reached. However, phase reverse may become necessary if the population is sufficiently large. The formation of aggregates could occur in tiny spaces, such as rock pores; the aggregates are RNA polymers, the outer surfaces of which could by chance be catalytically active (was this a premonition that ribozymes could exist?). Kuhn refers to RNA molecules with particular tertiary structures as "nucleation molecules", which could be capable of increased aggregate formation if the nucleation molecules have a "collector strand", i.e., a free strand end with no tertiary structure. Certain regions of the nucleation molecules are suitable for the occurrence of important reactions such as the attachment of external molecules, perhaps activated amino acids, which could lead to the formation of proteins or protoproteins.

Kuhn and Waser presented an extended model at the beginning of the 1980s (Kuhn and Waser, 1981, 1983). Here again, the evolution of replicating systems is described as a series of single reaction steps, important preconditions for which are periodic temperature changes and a structured environment. A new mechanism for the way in which RNA strands can take up particular structures was developed; these could have the form of hairpin strands involving Watson–Crick pairing. The formation of a "collection apparatus" (collector strand) would be a great advantage for selection, as it could facilitate the construction of suitable hairpin structures (Fig. 8.7).

The assumption was that, at a later phase of development, amino acids would attach themselves to the RNA structures, which would reach the approximate length of today's tRNA molecules. The structures could be stabilized, for example, by Ca^{2+} ions in the space between the strands (with their negatively charged phosphate groups).

The Kuhn model is presented in detail (and in parts mathematically justified) in the articles referred to above. It is unclear why Kuhn's ideas have not always received due attention in the literature in comparison with other theories. Without going into detail, some features will be mentioned briefly.

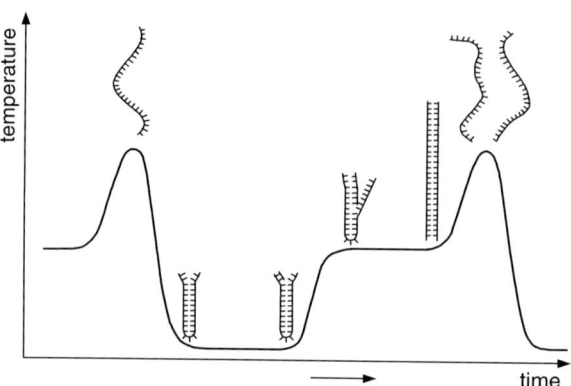

temperature

time

Fig. 8.7 Schematic representation of an RNA strand replication. Periodically varying temperatures lead to deconvolution (high temperatures) or double strand formation (low temperatures), for example, in hairpin conformations. In this form, replication can only begin at the end of a strand, from whence it continues as the temperature rises, i.e., base pair bonds are weakened. A further temperature programme is required in order to ensure both the assembly of the folded strands to give higher aggregates and their later separation (Kuhn and Waser, 1981)

Kuhn's model treats the question of the origin of life primarily as a logistical problem, and only secondarily as a physicochemical one. The question then arises as to how the logistical requirements of biogenesis can be dealt with by physico-chemical models (Kuhn and Waser, 1983). The starting point is not the thermodynamic conditions necessary for the occurrence of dissipative structures in a homogeneous medium. Instead, the search is for conditions necessary for aggregates to be formed at a certain point from a few macromolecules; the aggregates then replicate themselves. A spatially and temporally structured environment is an important precondition for all the processes developed in the model; otherwise the important building blocks cannot be prevented from diffusing out of the regions in question. This provides an impetus both for replication and a certain system dynamic. It is understandable that the Kuhn model provides no exact chemical mechanisms, e.g., for the RNA replication processes, but other models can also not do this.

The translation model was refined and developed further in a later article (Lehmann and Kuhn, 1984). It is assumed that in the early stages of evolution, guanosine and cytidine were the most important building blocks for RNA syntheses, as they allow strong base pairing (via three hydrogen bonds). Quantitative considerations lead to an upper limit of about 50 nucleotides for a reasonable chain length.

Later articles dealt with further elaboration of ideas on the driving forces which would have led to the formation of higher aggregates from RNA and amino acids. As had been suggested 20 years earlier, these processes could have taken place in rock pores and could have been driven by hydration and dehydration phases (Kuhn and Waser, 1994). The tiny pores in rocks act as minute test tubes, so optimal compositions could have been determined and replicated using many millions of systems. According to this model, none of the synthetic processes taking place would have required the presence of protein enzymes (see also Lahav, 1999). Just as other theories on the possible first steps of biogenesis do, Kuhn assumed that certain questions and problems were already solved, for example, the problem of the synthesis of the nucleotides, or amino acid activation. This means that such evolution models only deal with the main thrusts of the argument, while the details still need to be cleared up.

A computer simulation carried out by C. Kuhn (2001) was able to confirm certain critical phases in the first steps of Kuhn's theory, such as the formation of aggregates (collector strand and hairpin strand). In this simulation, the process of the development of a simple genetic apparatus took place in three stages:

> Construction, i.e., simulation of the formation of aggregates, and diffusion and fitting together of the molecular strands.
>
> Selection.
>
> The multiplication phase: the aggregates fall apart to give single strands, which now act as matrices. Complementary copying occurs, accompanied by a number of copying errors. Matrix and replication strand separate.

Kuhn's biogenesis models were developed further (Kuhn and Kuhn, 2003). The basic principle remains unchanged: replication first! As before, no exact single steps are elaborated, but only the main aspects of the biogenesis process are dealt with.

The authors suggest that it was the structural diversity of the environment which made biogenesis possible; in other words, there was an enormous selection of regions with different properties and states on the young Earth which acted as stimuli for the increasing complexity of the evolving systems. As complexity increased, those regions of the primeval Earth which were not available for earlier, more primitive systems could be "colonized".

This new approach embodies the idea that it was environmental factors which were decisive in making possible the complex steps of the development of life, or even forced them to occur. To be more specific, the spontaneous formation of short strands occurred initially at many places in tiny niches such as the rock pores already mentioned. The formation of longer strands capable of replication is unlikely under these conditions. However, the fusion of short strands to give longer ones, which could now replicate, could have occurred in a different environment or under more favourable conditions.

The theory is there: what are now needed are inventive, adventurous experimentalists who are prepared to try to validate its concepts.

8.5 Dyson's "Origins" of Life

Freeman Dyson, the American physicist from the Institute of Advanced Studies at Princeton, advanced an interesting biogenesis theory (Dyson, 1985). He expressly used the term "origins", as did several other biogeneticists, since even the question "one origin or several?" is still the subject of contention. According to Dyson, there are two logical possibilities for the origins of life:

Life had only *one* origin: in this case, the two properties which characterize life, replication and metabolism, must both have been present together (at least in a rudimentary form).

Life began twice in two different systems: one was capable of metabolism but not replication, the other of replication but not metabolism.

If life began twice—and that is the central premise in Dyson's thinking—the process must have had its origin in proteins and not in nucleic acids or their precursors. Thus, Dyson considers the systems studied by Eigen and Orgel (and Kuhn, who Dyson does not mention, but who must be included) to be of the type which fit in with the second case above. The detailed theory presented by Manfred Eigen, starting in 1971, is thus (according to Dyson), not in fact a theory of the origin of life, but in fact of the origin of replication.

Dyson, with his clear credo "metabolism first", is in basic agreement with the biogenesis process described by G. Wächtershäuser, though they disagree on some points. In order to provide the metabolism version of the origin of life with a well-based (mathematical) theory, Dyson, a theoretical physicist, developed a theory which he refers to as the "toy model". He tries, using a mathematical model, to quantify the processes of development from metabolism to life, while realising that he is in a difficult initial situation compared with that of Eigen, since metabolism is not a well-defined concept. Replication, however, can be described and defined much more easily: it involves copying, a process which can either be exact or can involve a certain error rate. Eigen describes the variation of molecule populations with time in terms of equations, and Dyson tries to describe Oparin's theory, with metabolism first, in a similar manner.

The difficulties in defining the term metabolism, or the inability to do so, are solved by Dyson in two stages:

He describes molecular populations mathematically in the way physicists calculate classical dynamic systems. Very exact dynamic equations are devised, while the laws of interaction are left very general. This leads to a general theory of molecular systems, which makes it possible to define what is understood by the origin of metabolism (Dyson, 1999).

The general theory is now reduced to a "toy model", using the following assumption: there are simple, arbitrary rules for the probability of a molecular interaction. The complex network of biochemical reaction chains is expressed by one single formula.

We shall not go into the details of Dyson's model here but only present the important conclusions. The model is first defined using ten assumptions, for example:

Assumption 1:   First there were cells, then enzymes, and genes came much later.

Assumption 2: (shortened)   Cells are sluggish drops with a population of polymeric molecules which do not leave the cell. The polymers are built up from monomers and contain an exact number (N) of monomers. These can diffuse into and out of the cell, and energy can be supplied externally.

Dyson introduces a series of mathematical and physical simplifications and then reaches his important conclusions. These involve the reciprocal relationship of three parameters; when these are chosen, the model is completely defined. The parameters are:

a: defines the variety of the monomers (e.g., different amino acids)
b: a measure of the number of different types of chemical reactions which could be catalysed by primeval life forms (a type of quality factor)
N: the size of the molecular population

It is now necessary to find out whether the model shows "interesting behaviour" for certain values of a, b and N which are consistent with the facts of organic chemistry. Interesting behaviour means a jump from disorder to an ordered state. This can occur only in a narrow range of a, b and N and is independent of chemical and physical constants. Thus, the model makes statements about the matter which could have composed the first living system. The favoured ranges for interesting behaviour are:

a, from 8 to 10: the number of monomer types should lie between 8 and 10. Thus, with respect to today's number of protein monomers, the 20 amino acids, primeval proteins could have needed only 10 amino acids to give functional polymers. Dyson's model would not have been viable with $a = 3$: this would have meant a beginning of life involving four types of nucleotides, which would not have had the biochemical variety necessary for the transition from disorder to order. Thus Dyson assumes that the proteins came first.

b, from 60 to 100: This range appears reasonable for the discrimination factor b. Modern polymerases have a discrimination factor of 5,000–10,000, while a simple peptide catalyst, with an active centre involving 4 to 5 amino acids, has a b-factor of 60–100.

N, from 2,000 to 20,000: a value of 10,000 can suffice to demonstrate a system which shows typical features of a life form. It is, however, also small enough to permit the statistical transition from disorder to order.

Why could this model be successful? It is able to tolerate high error rates, so that it avoids the error catastrophe by not replicating. A replicating system can only occur spontaneously if N is not larger than 100. But in the Dyson system (with a and b in the ranges given above), the error rate is around 25–30%. In spite of this, a polymer with 10,000 or more monomers can make the jump from disorder to order with a relatively high probability. An arrangement in which only 3 of 4 bonds in the chain were correctly placed would have been fatal for a replicating system, while a non-replicating system can tolerate such a state.

Each point in the phase diagram in Fig. 8.8 corresponds to a certain value of a and b, i.e., it represents the possible chemical composition of a molecular population. Variable a forms the horizontal axis, $(1 + a)$ being the number of monomer types. The b axis represents the quality factor of the polymer catalysis. The transition region contains those populations which can have both ordered and disordered

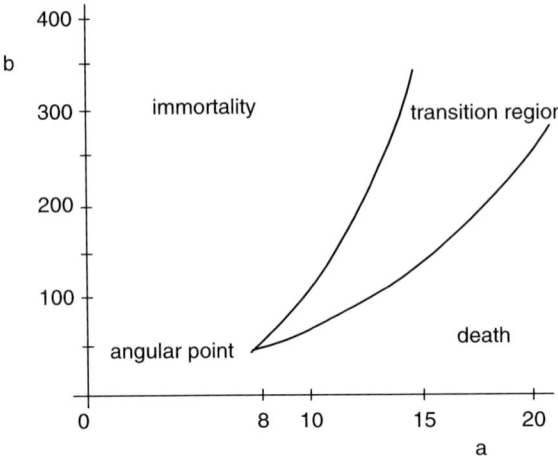

Fig. 8.8 The principle of the Dyson model. Each point in the phase diagram represents a possible composition of a molecular population. The horizontal axis is a, where $(a + 1)$ is the number of monomer types. On the vertical axis, b represents the quality factor of the polymeric catalysis. The "transition" region consists of populations which can have both an ordered and a disordered equilibrium state. In the "death region" there are only disordered states, while in the "immortal" region (in the "Garden of Eden"), there is no disordered state (Dyson, 1988)

equilibrium states. In the "death region" there is no ordered state: here a is too large, so that the chemical variation is too large, and b too small, i.e., the catalytic activity is too low.

In the "immortal" region, there is no disordered state, i.e., here a is too small (very low chemical variation) and b too large (catalytic activity too high). The biologically important molecules are to be found at the angular point of the curve, since they are subject to high error rates and have large populations which can take up an ordered state.

Some of the mathematical basis for the equations is due to the geneticist Moto Kimura, who is considered a resolute proponent of the "neutral theory of evolution", which states that statistical, chance fluctuations are more important in the formation of new species than is Darwinian natural selection. Evolution via such chance fluctuations is referred to as "genetic drift". Dyson considers that both forms of evolution are important (Dyson, 1999).

Dyson's model has been the subject of careful criticism as well as well-meaning agreement. Shneior Lifson (1997) found fault in particular with Dyson's assumption that metabolism (and other properties) could have developed without natural selection. In his third assumption, Dyson postulates that: "There is no Darwinian selection. Evolution of a molecule population occurs via genetic drift" (Dyson, 1999). Lifson (1997) points out that, while Dyson stresses the role of primitive metabolism, its adaptability, error tolerance etc., he himself considers that such properties can only evolve via natural selection.

Irrelevant of one's personal opinion, Freeman Dyson's "toy model" has enriched and enlarged the potpourri of (sometimes) bold theories by one further bouncy, glitzy hypothesis.

8.6 The Chemoton Model

In the same year that Manfred Eigen published his ideas on the "Self-organisation of Matter and the Evolution of Biological Macromolecules" in *Naturwissenschaften* and the *Quarterly Reviews of Biophysics* (1971), a book called *The Principles of Life* written by Tibor Gánti, a chemical engineer who was interested in biological problems, was published; an English translation (with commentary) appeared many years later (Gánti, 2003). He developed, and later refined, a model which included all the properties of primitive life. The chemoton model differs from Eigen's hypercycle model in that three different autocatalytic cycles are coupled together to give a chemical, stoichiometric supermodel; the individual cycles can only function in combination.

The author considers his chemoton model not as an intermediate station of biogenesis, but as a basic model for living systems. Greatly simplified, the three cycles are:

Cycle 1: an autocatalytic chemical cycle which provides material for the other two cycles.

Cycle 2: the matrix replication cycle. The matrix molecule can multiply by polymerisation in the presence of a component from cycle 1; it contains all the genetic information required by the whole system.

Cycle 3: this is the membrane growth cycle. The starting material for the outer envelope of the system comes from cycle 1 and is capable of self-organisation processes which lead to the membrane (T. Gánti, 1997).

According to Popa (2004) "the chemoton model is a virtual chemical network and does not promote any specific chemistry". Without doubt, Gánti's model represents an important, well-thought-through model for a primitive living system.

References

Amirnovin R, Miller SL (1999) J Mol Evol 48:254
Biebricher C, Eigen M (1981) J Mol Biol 148:369
Blomberg C (1997) J Theor Biol 187:541
Bresch C, Niesert U, Harnasch D (1980) J Theor Biol 85:399
Bruß D and Leusch G (Eds.) (2006) Lectures on Quantum Information. John Wiley & Sons Ltd.
Copley SD, Smith E, Morowitz HJ (2005) Proc Natl Acad Sci USA 102:4442
de Duve C (1991) Blueprint for a cell: The Nature and the Origin of Life, Carolina Biological Supply Company
Di Giulio M (1994) Orig Life Evol Biosphere 24:425

Di Giulio M (1997) J Theor Biol 187:573
Di Giulio M, Meduquo M (1998) J Mol Evol 46:615
Di Giulio M, Meduquo M (1999) J Mol Evol 49:1
Di Giulio M, Meduquo M (2000) J Mol Evol 50:258
Di Guilio M, Meduquo M (2001) J Mol Evol 52:372
Dyson F (1985) Origins of Life. Cambridge University Press
Eigen M (1971a) Naturwissenschaften 58:465
Eigen M (1971b) Quart Rev Biophys 4:149
Eigen M, Gardiner WG, Schuster P (1980) J Theor Biol 85:407
Eigen M, Winkler-Oswatitsch R (1981) Naturwissenschaften 68:282
Eigen M, Gardiner WG, Winkler-Oswatitsch R (1982) Ursprung der genetischen Information. In: Evolution. Spektrum, Heidelberg
Eigen M (1992) Steps Towards Life. Oxford University Press
Gamow G (1954) Nature 173:318
Gánti T (1997) J Theor Biol 187:583
Gánti T (2003) The Principles of Life. Oxford University Press
Hohn MJ, Park H-S, o'Donoghue P, Schnitzbauer M, Söll D, Proc Natl Acad Sci US 103:18095
Jantsch E (1980) The Self-organizing Universe: Scientific and Human Implications of the Emerging Paradigm of Evolution. Pergamon Press, Oxford.
Knight RD, Freeland SJ, Landweber LF (1999) Trends Biochem Sci 24:241
Küpers B-O (1990) Information and the Origin of Life. MIT Press, Cambridge, Mass.
Kuhn H (1972) Angew Chem 84:838, Int Ed 11:798
Kuhn H (1976) Naturwissenschaften 63:68
Kuhn H, Waser J (1981) Angew Chem 93:495, Int Ed 20:500
Kuhn H, Waser J (1982) Selbstorganisation der Materie und Evolution früher Formen des Lebens. In: Hoppe W, Lohmann W, Markl H, Ziegler H (eds.) Biophysics. Springer, Berlin Heidelberg New York
Kuhn H, Waser J (1994) FEBS Letters 352:259
Kuhn C (2001) Proc Natl Acad Sci USA 98:8620
Kuhn H, Kuhn C (2003) Angew Chem 115:272, Int Ed 42:262
Lahav N (1999) Biogensis. Oxford University Press, New York Oxford, p 252
Lehmann U, Kuhn H (1984) Origins of Life 14:497
Lifson S (1997) J Mol Evol 44:1
Niesert U (1987) Origins of Life 17:155
Niesert U, Harnasch D, Bresch C (1981) J Mol Evol 17:348
Popa R (2004) Between Necessity and Probability: Searching for the Definition of Life, Springer, Berlin Heidelberg New York, p. 184
Shimizu M (1982) J Mol Evol 18:297
Spiegelman S (1967) American Scientist 55:221
Vetsigian K, Woese C, Goldenfeld N (2006) Proc Natl Acad Sci USA 103:10696
Vogel G (1998) Science 281:329
Wetzel R (1995) J Mol Evol 40:545
Wong J T-F (1975) Proc Natl Acad Sci USA 72:1909

Chapter 9
Basic Phenomena

9.1 Thermodynamics and Biogenesis

This section provides a short introductory survey of an area of science which is not only mathematically exacting, but also of fundamental importance for certain aspects of biogenesis. Thermodynamics, a sub-discipline of physics, deals not only with "heat" and "dynamics", but formulated more generally, thermodynamics is concerned with energy and entropy and deals with theorems which are valid across almost all areas of physics.

Thermodynamic processes play an important, or even dominant, role in all branches of science, from cosmology to biology and from the vastness of space to the microcosmos of living cells. Energy and entropy determine and direct all the processes which occur in the observable world. Thermodynamics only describes the properties of large populations of particles: it cannot make any statements about the behaviour of single atoms or molecules. The most important properties of a system are determined by:

Its temperature
Its pressure
Its volume
The composition of the system

An "equation of state" documents these quantities mathematically and identifies the system, which must be in equilibrium. There are three laws of thermodynamics, the first two of which are the best known (or should be). C. P. Snow (1993) became widely known because of his essay "The Two Cultures", i.e., the sciences and the humanities. Snow relates that, in discussions with highly educated people, the idea that a scientist could know almost nothing about Shakespeare caused them to cringe, while when asked if they know "the second law of thermodynamics," they reacted distantly, and even critically!

But back to our subject: the *first law of thermodynamics* deals with energy and is also known as the law of the conservation of energy. It can be formulated as follows: "The increase in the internal energy of a thermodynamic system is equal to the amount of heat energy added to the system minus the work done by the system on the surroundings." Energy can occur in various forms, for example, chemical,

H. Rauchfuss, *Chemical Evolution and the Origin of Life*,
© Springer-Verlag Berlin Heidelberg 2008

mechanical or electrical energy. These forms of energy can be interconverted, and they can all do work. The role of the German doctor Robert Mayer in the discovery of the principle of the conservation of energy in the mid-nineteenth century should not go unmentioned.

Work can be completely converted to heat, but—and this is important—a complete conversion of heat to work is not possible in an isothermal system. This problem is dealt with by the *second law of thermodynamics*, with its statement on entropy: "The entropy of an isolated system not in equilibrium will tend to increase over time, approaching a maximum value at equilibrium."

A short historical review will make the connections clearer. The two laws of thermodynamics are due to the work of Sadi Carnot and James Joule in the middle of the nineteenth century. But the physicist Robert Clausius (1822–1888), who taught at the universities of Zürich, Würzburg and Bonn, was responsible for the first clear formulations of the laws. Clausius recognised that the work of Carnot and Joule showed that the theory of heat could not refer only to heat applied, but that instead, two independent principles were required: the law of the equivalence of heat and energy, and the law stating the impossibility that heat energy could be transferred from a cooler to a warmer body. In other words, the transfer of a certain amount of heat energy occurs of its own accord in such a manner that the original temperature difference is reduced. If, however, a temperature difference is to be induced, work (for example in the form of electrical energy) must be done, as exemplified by the refrigerator. In such processes, a certain amount of energy "loses value" by falling from a higher to a lower temperature. Clausius introduced a new quantity, the entropy S, as a quantitative expression for this loss in value.

In other words, entropy is a quantitative measure of the "loss of value" which a certain amount of energy suffers on going from a higher to a lower temperature. Clausius carried out calculations involving a schematic, reversible heat engine, a "Carnot machine", and was able to show that each small amount of energy dQ_{rev} introduced reversibly into a system led to a corresponding change in entropy dS.

$$dS = \frac{dQ_{rev}}{T}$$

It is certainly not an exaggeration to refer to the second law of thermodynamics as an important, universally valid physical principle, as did the well-known English astronomer A. S. Eddington (1932).

There are various expressions of the two laws, the most succinct being that of Rudolf Clausius, who (in 1872) wrote: "The energy of the world is constant. Its entropy tends to a maximum" (some authors translate the original German word "Welt" as "universe"). In the 1930s, the second statement led to the idea of the "heat death of the universe" and caused much speculation.

The work of Ludwig Boltzmann (1844–1906) in Vienna led to a better understanding, and to an extension, of the concept of entropy. On the basis of statistical mechanics, which he developed, the term entropy experienced an atomic interpretation. Boltzmann was able to show the connections between thermodynamics and the phenomenon of order and chance events; he used the term entropy as a measure of disorder, and thus for the most probable state of a system. This successful physicist's life ended tragically in suicide; his headstone in the Vienna central cemetery bears the equation:

$$S = k_B \log W$$

which states that the entropy S is proportional to the logarithm of W. W is the number of possible microscopic states which a system can take up (also known as the thermodynamic probability). k_B is a constant named in honour of Boltzmann; the above equation is due to Max Planck, who used it in the discovery of the Planck constant h.

If a system, when left to itself, tends towards a state of greatest possible disorder, i.e., maximum entropy, then negative entropy (named negentropy by Brioullin) refers to a highly-ordered state of the system. With respect to the cosmos, this means that there are two opposing tendencies present in the universe:

A dispersive tendency
An order tendency

The dispersive tendency dominates in a high-temperature system containing only a few particles, while the order tendency is important in a system in which the particles are themselves ordered, as in a crystal or the DNA helix. The real states of systems of matter lie somewhere between these two extremes.

But what does all this have to do with the origin of life? The answer is simple: a great deal! The second law, with its consequences, has often been used by the opponents of evolution theory and modern biogenesis research as the basis for their own interpretation of the history of creation. According to them, application of the second law of thermodynamics means that order (i.e., life) could never have emerged from disorder without the help of an almighty creator.

This argument, however, ignores an essential aspect of the second law, which is valid only for *closed systems*, i.e., systems for which neither matter nor energy can be exchanged with the environment. The chemical and physical processes which led to biogenesis on the primeval Earth did not take placed in closed systems, but in open ones. Matter and energy exchange were in fact essential for the successful execution of reaction sequences, as for example in the formation and synthesis of biomonomers.

It is, however, possible—apparently in disagreement with the second law of thermodynamics—that order can be formed from disorder in a system, for example by means of self-organization processes; in many cases, these have been observed experimentally. This order, however, is formed at the expense of an increase in disorder outside the system under observation: order is "pumped out of" the system which is striving to reach an ordered state. Thus, disorder increases outside the system, and the second law is still valid.

The third law of thermodynamics, one version of which is "as a system approaches absolute zero of temperature, all processes cease and the entropy of the system approaches a minimum value", is of little importance for biogenesis. It means that it is impossible to cool a system down to absolute zero (even via an infinite number of steps).

This section will end by clearing up an important point of nomenclature. By a "system", we mean the part of our universe which we are dealing with. This can be the whole Earth, the solar system, or one tiny little cell. We distinguish between three types of systems:

An *open* system exchanges matter and energy with its environment.
A *closed* system does not exchange energy with its environment.
An *isolated* system exchanges neither matter nor energy with its environment.

Figure 9.1 illustrates the three types of systems schematically.

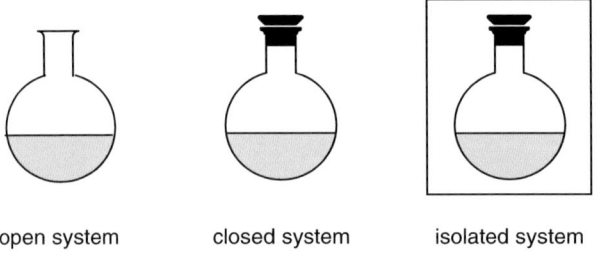

open system   closed system   isolated system

Fig. 9.1 The three possible states of a physical system

9.2 The Thermodynamics of Irreversible Systems

Some of the terms used in classical thermodynamics which refer to equilibrium states and closed systems have become important outside the boundaries of physics: one example is the term "adapted state" in Darwinian evolution theory, which represents a type of equilibrium state between the organism and its environment.

Most of the events occurring in Nature consist of irreversible processes: genuinely reversible processes are seldom encountered and are of mainly theoretical interest. A closer analysis of the so-called "flow equilibrium" (a term coined by the biologist Ludwig von Berthalanffy) brought together thermodynamics and the problems of living systems; flow equilibria are present in open systems. More than 50 years ago, von Berthalanffy postulated that the theory of flow equilibria would make possible the construction of exact mathematical equations for basic phenomena of life such as metabolism and growth.

Only in the last decades has the thermodynamics of open systems been treated intensively and successfully. The "thermodynamics of irreversible systems" was studied initially by Lars Onsager, and in particular by Ilya Progogine and his "Brussels school"; both studied systems at conditions far from equilibrium. Certain systems have the capacity to remain in a dynamic state far from equilibrium by taking up free energy; as a result, the entropy of the environment increases (see Sect. 9.1).

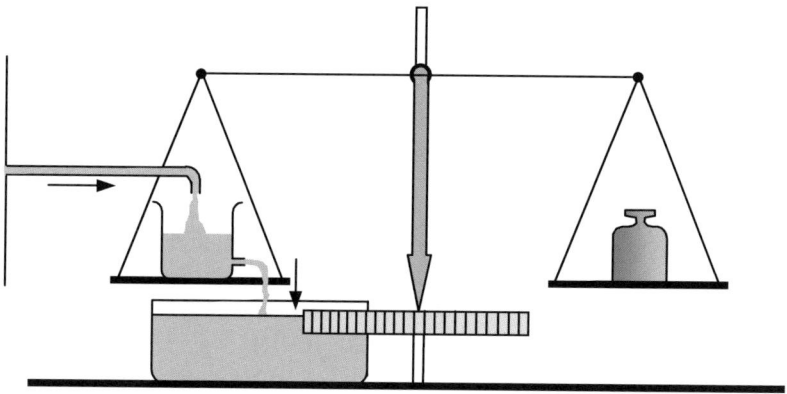

Fig. 9.2 Simple model for a flow equilibrium, also known as a steady state; these states are an important property of living systems

Equilibrium thermodynamics was developed about 150 years ago. It is concerned only with the achievement of an equilibrium state, without taking into account the time which a system requires for the transition from an initial to a final state. Thus, only the thermodynamics of irreversible processes can be used to describe processes which lead to the formation of self-organising systems. Here, the time factor, and thus also the rate at which material reactions occur, is taken into account. Evolutionary processes are irreversibly coupled with temporal sequences, so that classical thermodynamics no longer suffices to describe them (Schuster and Sigmund, 1982).

As already mentioned, a continual inflow of energy is necessary to maintain the stationary state of a living system. It is mostly chemical energy which is injected into the system, for example by activated amino acids in protein biosynthesis (see Sect. 5.3) or by nucleoside triphosphates in nucleic acid synthesis. Energy flow is always accompanied by entropy production (dS/dt), which is composed of two contributions:

The *flows*: these enter the system, and also leave it again (d_eS/dt).

The *internal entropy production*: this represents the time-related entropy growth generated within the system (d_iS/dt). The internal entropy production is the most important quantity in the thermodynamics of irreversible systems and reaches its maximum when the system is in a stationary state. The equation for the entropy production is then:

$$\frac{dS}{dt} = \frac{d_eS}{dt} + \frac{d_iS}{dt}$$

The theory of the thermodynamics of irreversible systems (Prigogine, 1979; Prigogine and Stengers, 1986) shows that the differential quotient of entropy with time (the change of entropy with time) can be expressed as the sum of products, the terms of which contain a force factor and a flow factor. In chemical systems, the force factor corresponds to the chemical affinity, the driving force of the reaction, and the flow quantity to the reaction rate.

Starting from the second law of thermodynamics, it is possible to derive a principle according to which the change of entropy production in the neighbourhood of a stationary state is always negative if the flows in the system are kept constant and only the forces varied. As already mentioned, the entropy production reaches a minimum value in the stationary state of the system. If it is at a minimum, and a positive fluctuation occurs, the system reverts to the minimum, and a stable state is again reached.

If the entropy production in the stationary state falls, a negative fluctuation occurs, and the system becomes unstable: the stable stationary state of the system is disturbed or destroyed. The system reacts by changing its composition until a new stable state is reached. In this new state, the system is characterised by a lower entropy content than that present prior to the fluctuation, since only a negative entropy change can occur. However, the lower entropy corresponds to a higher degree of order in the system.

How can negative fluctuations in entropy production occur or be triggered? As Manfred Eigen shows in his evolution theory, fluctuations in entropy production can be caused by the coming into being of a self-replicating molecular species which is capable of selection. Autocatalytically active mutants can also have the same effect. Looked at this way, the phenomenon of evolution consists of a continuous series of instabilities, i.e., collapses of stationary states.

The importance of the work of Prigogine and his Brussels school for a deeper understanding of evolution processes is mainly due to the fact that it is possible

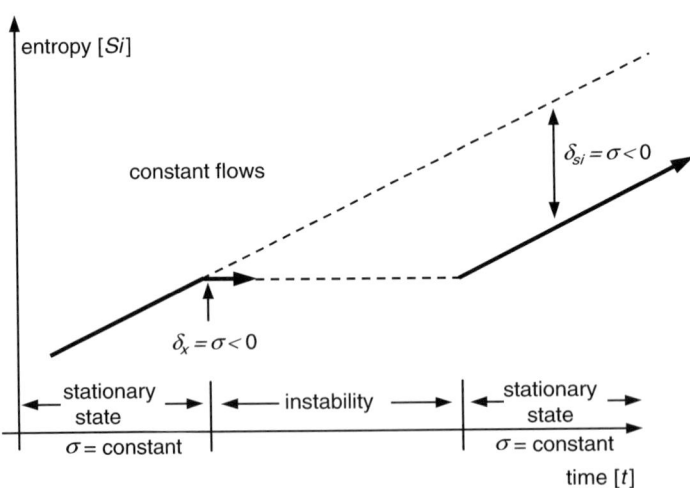

Fig. 9.3 Entropy–time diagram of an evolution process. If a negative fluctuation of the internal entropy production σ occurs in a system, the controlling stationary state is terminated. An instability occurs, starting from which a new stable state is taken up. The change in the internal entropy is always negative ($\delta S_i < 0$). The new stationary state has a lower entropy, i.e., the order of the system is increased (Eigen, 1971a, b)

to explain the formations of structures in chemical reactions and biological objects with the help of the thermodynamics of dissipative systems and the principle of "order through fluctuations".

9.3 Self-Organisation

The term "self-organisation" has very much become an "in" expression. We find it in articles throughout the literature of different scientific disciplines, including processes occurring in the submicroscopic range. For example, organic nanotubes are formed by self-organisation (Bong et al., 2001). Self-organisation is a broad term which can have many meanings. One possible definition is that "self-organisation is a process in which structures form spontaneously from the components of a system".

According to Erich Jantsch, self-organisation is the dynamic principle behind the emergence of the rich world of biological, ecological, societal and cultural structures (Jantsch, 1980). As can be seen from the definition, self-organisation processes are highly complex. They can include:

> The formation of patterns
> Growth
> Various types of cyclic processes
> Multistability
> Reproduction

It is often helpful to take a brief look back at aspects of the history of science in order to make complex topics easier to grasp. The history of physics can be divided into three phases:

> The first was dominated by the ideas of Galilei and Newton.
> The second includes the introduction of the theory of relativity and quantum mechanics in the first decades of the twentieth century.
> The history of science will see the third phase as the "physics of complexity".

According to Galilei, the observation of natural phenomena using suitable measuring instruments provides certain numerical values which must be related to one another; the solution of the equations derived from the numbers allows us to forecast future developments. This led to the misunderstanding that knowledge could only be obtained in such a manner. The result was deterministic belief, which was disproved for microscopic objects by Heisenberg's uncertainty principle. On the macroscopic scale, however, it appeared that the deterministic approach was still valid. Determinism was only finally buried when "deterministic chaos" was discovered.

Self-organisation is a property of "complex systems"; these are an important area of physics and have been studied intensively in the last few years. Since 1993, the Max Planck Society has had an institute for "Physics of Complex Systems" in Dresden. This is an interdisciplinary research area, dealing with problems which span the range from the cosmos to the living cell.

According to Stuart A. Kauffman (1991) there is no generally accepted definition for the term "complexity". However, there is consensus on certain properties of complex systems. One of these is deterministic chaos, which we have already mentioned. An ordered, non-linear dynamic system can undergo conversion to a chaotic state when slight, hardly noticeable perturbations act on it. Even very small differences in the initial conditions of complex systems can lead to great differences in the development of the system. Thus, the theory of complex systems no longer uses the well-known "cause and effect" principle.

Hermann Haken, the founder of "synergetics", describes how the American meteorologist E. N. Lorenz showed that the interaction of even a few degrees of freedom leads to the emergence of a new state in the system under observation, a deterministic chaos. Such systems are characterised by their sensitivity towards initial conditions. The oft-cited "beating of a butterfly's wings in Peking" (an expression coined by Lorenz), which can lead to a change in the weather on the west coast of the USA, has often been used as a drastic example of deterministic chaos. The "irregular motion" which characterises deterministic chaos has been explained very graphically by Haken using the simple example of a system consisting of a steel ball and a razor blade. If a small steel ball is allowed to fall from above onto a razor blade which is held in a vertical position, tiny variations in the position of the steel ball will cause it to be deflected in a wide arc either to the right or to the left.

As so often occurs in complex processes, the phenomenon of self-organisation can occur in various forms:

> Self-organisation via inherent properties: in this case, the self-organisation is determined by the properties of the particles forming it, such as base pairing in the DNA double helix, the formation of ice crystals, or RNase reconstruction (Fig. 9.4).
>
> Self-organisation in ontogenesis, i.e., the development of an individual organism from an ovum to the complete organism.
>
> Genuine self-organisation, i.e., self-organisation as a property of the system. Here, a system with a high degree of complexity organises itself under certain conditions. A typical example is Eigen's hypercycle model (see Sect. 8.3).
>
> Self-organisation as a physical principle: since the "big bang", self-organisation is considered to be a physical attribute of matter.

Jantsch (1980) provides a broad differentiation of the phenomenon of self-organisation, distinguishing between:

> *Conservative*, i.e., structure-conserving self-organisation and
> *Dissipative* self-organisation, for example in evolving systems.

Systems with *dissipative* self-organisation are important for processes which lead to biogenesis. These are open systems, the internal state of which is dominated by a disequilibrium far away from the equilibrium state.

Here are some examples of chemical and biochemical self-organisation processes:

The Belousov-Zhabotinskii reaction
Bénard cells
Ribonuclease (RNase) reconstruction
The tobacco mosaic virus

The Belousov-Zhabotinskii reaction is a typical oscillating chemical reaction. Spiral structures form periodically, disappear and reappear as the result of an autocatalytic reaction, the oxidation of Ce^{3+} and Mn^{2+} by bromate (Jessen, 1978).

The formation of Bénard convection cells takes place as follows: if water is heated from below in a vessel, macroscopic convection currents occur under certain conditions; seen from above, these have the structure of uniform, honeycomb-shaped cells.

An impressive biochemical example of self-organisation at the molecular level is the refolding of the enzyme RNase after complete denaturing. Christian Anfinsen was able to effect complete denaturing and reduction of the four disulfide bridges in the protein by treating it with urea and β-mercaptoethanol. The three-dimensionally folded protein takes up an adventitious bundle-like form, the four disulfide bridges undergoing conversion into eight SH groups: the enzymatic activity is completely lost. If the reduced protein is treated with atmospheric oxygen, the SH groups are reoxidised to the original disulfide bridges in the correct geometric form, so that the enzyme regains (almost) its full activity. In 1972, Christian Anfinsen was awarded the Nobel Prize for chemistry, in particular for this work (in a wider context).

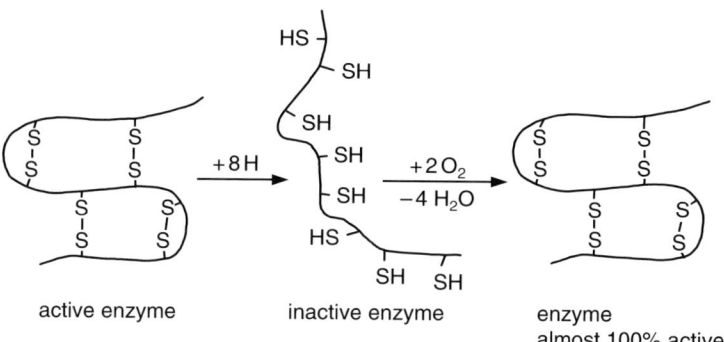

active enzyme   inactive enzyme   enzyme almost 100% active

Fig. 9.4 The enzyme RNase can be unfolded by reduction agents (urea and mercaptoethanol). Oxidative removal of the reduction agents causes the molecule to take up its three-dimensional structure again; it regains almost its full enzymatic activity

Completely different mechanisms are involved in the self-assembly of the tobacco mosaic virus (TMV). This virus consists of single-strand RNA, which is surrounded by 2,130 identical protein units, each of which consists of 158 amino acid residues. A virus particle, which requires the tobacco plant as a host, has a rod-like structure with helical symmetry ("Stanley needles"). It is 300 nm long, with a diameter of 18 nm. The protein and RNA fractions can be separated, and the viral components are able to return to their original form via self-assembly processes at a certain pH and under the necessary ionic conditions; the virus is then again completely active.

Although these examples involve completely different forces and mechanisms, they all lead to the same result: the formation of "structures" via the phenomenon of self-organisation. The term "structure" depends on the distribution of matter; we speak of structure when the distribution is not uniform, i.e., when it deviates from the most likely distribution.

The attractiveness of the term self-organisation has led to its widespread use, which does not, however, always stand up to critical scrutiny. The Australian physicist Paul Davies notes in his book *The Fifth Miracle* that, when looked at critically, some examples of self-organisation turn out to be only examples of a spontaneous self-order, such as the formation of the hexagonal convection cells, which resemble order in crystals more than they do the complete state of order in living systems (Davies, 2000). Thus, the phenomenon of self-organisation does not lead directly to living systems, but it is a precondition or an essential factor in the formation and emergence of a system which can be described as "living".

According to Stuart Kauffman, self-organisation processes initiate a trend which leads to more complex states of the system. In living systems, there are two forces which determine order (Kauffman, 1995):

    Self-organisation
    Selection

A living system can only survive because of the complex interaction between the two phenomena. Kauffman, theoretical biologist (and medical scientist) at the Santa Fé Institute, has been working for many years on questions regarding the emergence of life and the origins of molecular evolution. His models are mathematical in nature (Kauffman, 1991), and they lead to the postulate that complex systems can best adapt when they walk a thin line "at the edge of chaos". Put in general terms, this means that systems which are developing must on the one hand have the complexity which characterises the phenomenon of "life", but on the other must also have a certain amount of stability and mobility in order to be able to survive when the environmental conditions change. This corresponds to walking the fine line between structural chaos and life-threatening rigidity.

Kauffman also considers that autocatalytic reactions are a necessary precondition for biogenesis processes, as they can wind themselves up via self-amplifying feedback processes until a critical boundary has been reached. The next step would then be the transition from autocatalysis to self-organisation, similar to the transition from unstructured water to convection cells (the Bénard instability) (Davies, 2000).

Kauffman's conclusions arise only from computer simulations, and this is also the main reason for criticism, in particular for biologists. Even though extensive mathematical deliberations and sophisticated computer simulations could show that networks containing many elements can tend towards transitions which give systems of organized complexity, some important factors which lead to a living system are still missing.

9.4 The Chirality Problem

The term "chirality" was introduced into the vocabulary of science by Lord Kelvin in the 1880s and comes from the Greek word "χειρ = cheir", the hand, since chiral molecules behave like our hands, like mirror images which cannot be superimposed upon each other.

In nature, but also in chemical syntheses, many molecules occur in two spatially different structures which behave like mirror images. This can also be seen in two dimensions: one of the letters "A" (Fig. 9.5) can be superimposed upon the second "A" if it is moved to the right (or left). This is, however, not possible with the letters "G". Thus, "A" is described as not chiral, or achiral, and "G" as chiral.

The situation becomes more complex in three-dimensional structures: if we lay the palms of our two hands together, the mirror plane lies exactly between them, so that the hands are chiral. Whether or not a molecule is chiral depends upon symmetry relationships:

> Chiral molecules are characterised by symmetry elements of the first kind, for example, axes of rotation.
> Achiral molecules are characterised by the presence of symmetry elements of the second kind, for example, planes of symmetry, inversion centres or rotation–reflexion axes.

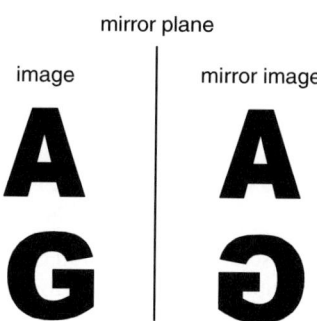

Fig. 9.5 Mirror images in two-dimensional space

But why is the chirality problem a subject for the question of biogenesis? Two of the most important biological molecules contain building blocks with only one of the two possible molecular species:

> With only a few exceptions, all proteins in today's life forms contain only L-amino acids.
> In the two nucleic acids, the sugar component consists solely of D-ribose or D-deoxyribose.

Fig. 9.6 Mirror image
behaviour of enantiomeric
molecules. The left-handed
L-α-amino acid is converted
to the right-handed
D-α-amino acid by reflection

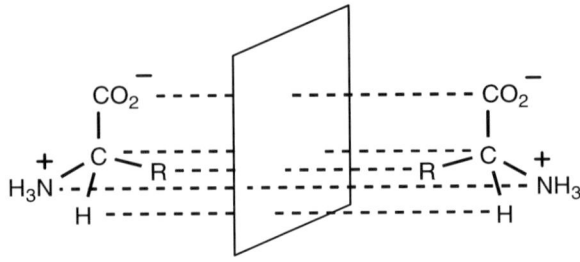

This clear preference for one of the two enantiomeric forms is called "homochirality of biomolecules" and is seen by many biogeneticists as a precondition for the evolution of life forms.

Although equal amounts of the two possible forms are generated in every synthesis of amino acids, (almost) only L-amino acids are incorporated into proteins and peptides. This phenomenon is valid for the amino acids in all life forms, from the bacterium to the elephant. But there are exceptions: some antibiotics contain D-amino acids in their proteins, and these also occur in a few components of cell walls. Here, the D-amino acids have a certain protecting function with respect to degradation enzymes, which are specialised to deal with L-amino acids.

As discussed in Chap. 6, nucleic acids contain D-ribose or D-deoxyribose. The corresponding L-form is banished from the information carriers RNA and DNA. So the question is: why L-amino acids and D-sugars and not the opposite? Two hypotheses are at the fore in this scientific discussion:

> The *conventional* hypothesis: in prebiotic times, both forms were present in equal amounts. A still unknown event caused only one of the two (enantiomeric) molecular species to be built into the corresponding macromolecules. Thus, a non-linear evolution process determined the direction (D- or L-form).

> The *deterministic* hypothesis: this assumes that fundamental natural laws are responsible for the preference for one of the two molecular species.

The scientific discussion of these hypotheses is not over yet; while the conventional hypothesis seems to be losing ground at present, there is still no conclusive clarification of the problem.

In the first case, it could be imagined that a protoenzyme on the primeval Earth, which catalysed the polycondensation of amino acids to (proto)proteins, "decided" for unknown reasons to favour the L-form. This decision would have needed to be passed on to subsequent sequences. The question comes up: chance or necessity? Could the protoenzyme have selected the D-amino acid with an equal probability? Work to find an answer is still in progress. There are many publications on the deterministic hypothesis, both theoretical and experimental.

The experimental result that the oligomerisation of monomers is inhibited when the monomers consist of a mixture of equal amounts of right- and left-handed units led to the assumption that life could only evolve in an environment in which one enantiomeric form was clearly in the majority. However, this assumption also means that models which describe the origin of life must be based on an abiogenic origin for the molecular violation of parity.

About 50 years ago, physicists were amazed to discover that the universe, which had previously been regarded as completely symmetrical, had a certain preference for left-handedness. It had been considered impossible that basic natural laws would distinguish between left and right. This assumption formed the basis for the physical law of the "conservation of parity"; according to this, the sum of the parities before and after each physical process must be equal. In other words: the mirror image of each physical phenomenon is also a real phenomenon (Ball, 1994).

The experts were amazed when, in 1956, two young Chinese scientists living and working in the USA, the physicists Chen Ning Yong and Tsung Dao Lee, put forward the hypothesis that parity does not always remain unchanged. However, they had no experimental evidence for their daring suggestion; this was soon provided by the physicist Chien-Shiung Wu from Columbia University. She studied the decomposition of ^{60}Co, an isotope which does not occur in nature but must be prepared in a nuclear reactor by neutron bombardment of natural ^{59}Co. The half-life of the unnatural isotope is five years, so there is enough time to carry out experiments with it. The β-decay of ^{60}Co leads to ^{60}Ni plus one electron plus one antineutrino. In the cobalt nucleus, a neutron is converted by the β-process to a proton, an electron and an antineutrino (or to be more exact, an electron-antineutrino). The Co nucleus is thus converted to a Ni nucleus. The β-decay is due to interactions between the nuclear particles which are caused by the "weak interaction". During the decomposition of the ^{60}Co nuclei, electrons are mainly emitted in the direction of the magnetic poles. If this process were to obey the law of conservation of parity, electrons would be expelled from both poles with equal probability. The sample had to be cooled down almost to absolute zero (to 0.01 K) in the experiments in order to freeze out the thermal motion of the particles which were to be aligned. A strong magnetic field aligned the particles, and the number of electrons emitted from each pole was determined. The experimental result was that more electrons were set free at one pole than at the other; this was the first example of a break in parity. This confirmation of the Lee/Yang hypothesis led to the award of the Nobel Prize for physics to these two scientists in 1957.

What was the importance of this research result for the chirality problem? One difficulty is provided by the fact that the interaction responsible for the violation of parity is in fact not so weak at all, although it only acts across a very short distance (smaller than an atomic radius). Thus, the weak interaction is not noticeable outside the atomic nucleus, except for β-decay. It would thus have either no influence on chemical reactions or only a very limited effect on chemical reactions, as these almost completely involve only interactions between the electron shells.

The influence of the weak interaction on chemical reactions can be calculated; since it favours left-handedness, it has an effect on the energy content of molecules and thus on their stability. In the case of the amino acids, the L-form would be more stable than the corresponding D-form to a very small extent. Theoretical calculations (using ab initio methods), in particular by Mason and Tranter (1983), indicated that the energy difference between two enantiomers due to the parity violation is close to 10^{-14} J/mol (Buschmann et al., 2000). More recent evidence suggests that the energy (or stability) difference lies about an order of magnitude higher (Basakov et al., 1995). However, some scientists working on the chirality problem consider that this difference is much too small to provide an explanation for the emergence of a homochiral animate world.

By taking into account the latest results on the behaviour of systems far away from equilibrium, Kondepudi and Nelson (1985) were able to show by calculation that L-amino acids are slightly favoured. There is a very tiny stabilisation effect due to the weak interaction; amplification mechanisms cause this effect to reach 98% of the probability that L-enantiomers of amino acids are favoured for incorporation into polymers. The amplification mechanisms are explained by the thermodynamics of irreversible systems.

Vitali Goldanski from Moscow has proposed an unusual model and is very critical of some of the assumptions made by Kondepudi. The Goldanski model assumes that chemical reactions can take place in the dark clouds of interstellar dust, even at temperatures close to absolute zero. He considers it possible that the processes involved favour one of the two enantiomers. The products of such reactions must then have been transported to Earth via meteorites or comets (Goldanski and Kuzmin, 1991). Reactions at very low temperatures can be explained by invoking quantum-mechanical tunnel effects. Such suggestions can be supported by experimental evidence: thus at the temperature of liquid helium (4.2 K), it was possible to polymerise formaldehyde to give chains consisting of several hundred monomer units. However, this process required either high-energy electron radiation in a particle accelerator or γ-radiation (^{60}Co) (Goldanski, 1986).

Another attempt to explain the homochirality of biomolecules is based on autocatalysis. The great advantage of asymmetric catalysis is that the catalyst and the chiral product are identical and thus do not need to be separated (Buschmann et al., 2000). The racemic mixture must have been affected by a weak perturbation in order that autocatalysis, which acts as an amplifier of enantioselectivity, could have led to only one of the two enantiomeric forms. This perturbation could have been due to the slight energy difference of the enantiomers referred to above, or to statistical fluctuations.

The question also arises as to where the chiral molecules came from. Were the L-amino acids or the D-sugars selected on the primeval Earth, or are extraterrestrial sources responsible for the homochirality? This second possibility is dealt with by hypotheses on the effect of circularly polarised light, of extraterrestrial origin, on chiral molecules in the molecular clouds from which the solar system was formed. One such hypothesis was proposed by Rubenstein et al. (1983) and developed further by others, particularly A. W. Bonner (Bonner and Rubenstein, 1987); both scientists worked at Stanford University. The authors believe that the actual radiation source was synchrotron radiation from supernovae. The excess of one enantiomeric form generated by this irradiation process would have needed to be transported to Earth by comets and meteorites, probably during the bombardment phase around 4.2–3.8 billion years ago.

This hypothesis has shared the same fate as many others on the biogenesis problem: it is still in dispute! Not only is the origin of the circularly polarised (CP)

UV radiation hypothetical, but so is the transport of molecules from outer space to Earth. Recent analyses of the Murchison meteorite by two scientists from the University of Arizona, Tucson (Cronin and Pizzarello, 1997; Cronin, 1998) have shown it to contain the four stereoisomeric amino acids DL-α-methylisoleucine and DL-α-methylalloisoleucine. In both cases, the L-enantiomer is present in a clear excess (7.0 and 9.1%). Similar results were obtained for two other α-methyl amino acids, isovaline and α-methylvaline. Contamination by terrestrial proteins can be ruled out, since these amino acids are either not found in nature or are present in only very small amounts. Since the carbonaceous chondrites are thought to have been formed around 4.5 billion years ago (see Sect. 3.3.2), the amino acids referred to above must have been subject to one or more asymmetric effects prior to biogenesis.

It is still unclear what kind of radiation sources can lead to asymmetric reactions. Jeremy Bailey from the Anglo-Australian Observatory in Epping, Australia, investigated which astronomical objects could be considered radiation sources (Bailey et al., 1998; Bailey, 2001). It was possible in laboratory experiments to generate a small enantiomeric excess of some amino acids by using circularly polarized UV light (Norden, 1977). This asymmetric photolysis involves photochemical decomposition of both D- and L- enantiomers, but at different rates, so the more stable form tends to survive. This process must be subject to autocatalytic multiplication.

It is still not clear which astronomical objects could serve as radiation sources. The best-known CPL sources are white dwarfs with a strong magnetic field. However, the probability that organic molecules could have been subjected to this radiation is very low. Other possible CPL sources are reflection nebulae in regions where very large numbers of stars are being formed; these cosmic regions show the presence of infrared CP radiation, and in fact, almost all the stars in such regions can serve as UV sources. The polarisation could be due to dust particles oriented by magnetic fields. These regions of space generally also contain organic molecules, and thus (probably) chiral biomolecules or their precursors. According to Bailey (2001), it is not possible to obtain exact estimates of the probability of the process involved, i.e., whether homochirality really had its origin in extraterrestrial processes such as the action of the universal weak interaction.

Another hypothesis on homochirality involves interaction of biomolecules with minerals, either at rock surfaces or at the sea bottom; thus, adsorption processes of biomolecules at chiral mineral surfaces have been studied. Klabunovskii and Thiemann (2000) used a large selection of analytical data, provided by other authors, to study whether natural, optically active quartz could have played a role in the emergence of optical activity on the primeval Earth. Some researchers consider it possible that enantioselective adsorption by one of the quartz species (L or D) could have led to the homochirality of biomolecules. Asymmetric adsorption at enantiomorphic quartz crystals has been detected: L-quartz preferentially adsorbs L-alanine. Asymmetrical hydrogenation using D- or L-quartz as active catalysts is also possible. However, if the information in a large number of publications is averaged out, as Klabunovskii and Thiemann could show, there is no clear preference in nature for one of the two enantiomorphic quartz structures. It is possible that rhomobohedral crystals of calcite, $CaCO_3$, can act selectively in adsorption of L- and D-amino acids; this property of calcite has been demonstrated for a racemic mixture of aspartic acid. Further work is, however, still required (Hazen et al., 2001).

The processes occurring at hydrothermal systems in prebiotic periods were without doubt highly complex, as was the chemistry of such systems; this is due to the different gradients, for example, of pH or temperature, present near hydrothermal vents. Studies of the behaviour of amino acids under simulated hydrothermal conditions showed that D- and L-alanine molecules were racemised at different rates; the process was clearly concentration-dependent. L-Alanine showed a low enantiomeric excess (ee) over D-alanine at increasing alanine concentrations. The same effect was observed with metal ions such as Zn^{2+} in the amino acid solution. Thus, homochiral enrichment of biomolecules in the primeval ocean could have resulted under the conditions present in hydrothermal systems (Nemoto et al., 2005).

It should be possible to use the special properties of chiral structures for particular separation problems. According to Belinski and Tencer, one possible way in which nature solved the ribose problem could have involved an enantioselective and diastereoselective purification process acting on a mixture of biomolecules, which left ribose as the only molecule available for further reactions. The authors propose a theoretical mechanism in which a type of chromatographic process occurs at chiral mineral surfaces. This paper is likely to stimulate new experiments as well as the quest for as yet unknown surfaces which can separate racemic carbohydrate mixtures. The question arises, however, as to whether there were minerals present on the young Earth which are now unknown, as they no longer exist on the Earth of today (Belinski and Tencer, 2007).

Szabo-Nagy and Keszethelyi (1999 and 2000) have carried out experiments which show a possible violation of parity in the crystallisation of racemates of tris(1,2-ethylenediamine-Co(III)) and the corresponding iridium compound.

Kondepudi (2000) from Wake Forest University in Winston-Salem obtained some results which appeared sensational: when crystallising a mixture of dissolved $NaClO_3$ and $NaBrO_3$ with stirring, an excess of one enantiomeric form (up to 80%) was formed. Prediction of which enantiomer will be in excess is not possible; it is a matter of chance. The process is probably due to chiral autocatalytic processes; similar phenomena were observed in melts.

It appears as if an axiom of stereochemistry, the absolute identity of the most important chemical and physical properties of chiral isomers, is no longer valid. Experiments using the amino acid tyrosine (Tyr) showed unexpected differences in the solubility of D-and L-Tyr in water: a supersaturated solution of 10 mM L-Tyr crystallised much more slowly than that of D-Tyr under the same conditions. The saturated solution of L-Tyr was more concentrated than that of D-Tyr. Supersaturated solutions of DL-Tyr in water formed precipitates containing mainly D-Tyr and DL-Tyr, so that there was an excess of L-Tyr in the saturated solution. The experiments were carried out with extremely great care in order to exclude the possibility of contamination. Further experiments will show whether this is a particular property of tyrosine, or whether other amino acids will show similar behaviour. Possible conclusions with respect to the homochirality problem of biomolecules cannot yet be drawn (Shinitzky et al., 2002).

These experimental results could not be confirmed by Lahav and co-workers; they suggest that impurities in the starting materials have a much greater effect on the crystallisation process than the PVED (Parity Violating Energy Difference). Extensive experimental studies indicate the importance of small quantities of impurities, particularly in early phases of crystallisation nucleus formation. Amino acids from various sources were used, and the analyses were carried out using the enantioselective gas chromatography technique (M. Lahav et al., 2006).

The Rosetta mission with its planned landing on a comet, with analysis of cometary material (see Sect. 3.2), should provide more information on the occurrence of chiral molecular species in the cosmos (Adam, 2002). The GC-MS apparatus installed in the robotic lander RoLand is also able to separate and analyse chiral organic molecules (Thiemann and Meierhenrich, 2001).

Laboratory data from two groups (see Sect. 3.2.4) indicate that chiral amino acid structures can be formed in simulations of the conditions present in interstellar space. The experimental results support the assumption that important asymmetrical reactions could have taken place on interstellar ice particles irradiated with circularly polarised UV light. The question as to whether such material was ever transported to the young Earth remains open. But the Rosetta mission may provide important answers on the problem of asymmetric syntheses of biomolecules under cosmic conditions (Meierhenrich and Thiemann, 2004).

B. DiGregorio (2006) presents a highly readable introduction to the homochirality problem in which he reports on work by Shinitzky and co-workers; they subjected synthetic D- and L-polyglutamic acid molecules, as well as the corresponding polylysine molecules, to an autocatalytic process in which left- or right-handed helical structures were formed. If these polymers are added to deuterated water, selective interactions with the asymmetrical isomers of water (ortho- and parawater) are observed; these differ only in their spin, all other properties being equal. These interactions led to a higher instability of polymer chains composed of D-amino acids. The postulate that the asymmetry of the weak interaction acts on molecular structures, thus favouring one enantiomeric form, is controversial.

Di Gregorio mentions another remarkable experimental result; Graham Cooks and co-workers found that the amino acid serine is able to form enantioselective cyclic octamers in two different ways, from aqueous solution or via sublimation from the solid state (Cooks et al., 2001).

According to R. M. Hazen, these results are in agreement with the postulate that some self-assembly processes of chiral molecules are highly enantioselective (DiGregorio, 2006).

The topical homochirality problem is presently being investigated in several research laboratories across the world. One new object of study is systems with eutectic mixtures. The addition of chiral dicarboxylic acids that co-crystallise with chiral amino acids to aqueous mixtures of D- and L-amino acids allows tuning of the eutectic composition of the amino acids; in several cases, these systems yield new eutectic compositions of 98% ee or higher. Thus, solid mixed crystals with a ratio valine:fumaric acid of 2:1 were formed (via hydrogen bonding) for both enantiopure and racemic valine.

Such systems also afford highly enriched aqueous solutions: if these were present on the primeval Earth for long enough periods, this may have sufficed to allow efficient asymmetric catalysts to become active (Klußmann et al., 2007).

Many compounds are less soluble as racemates than as their pure enantiomers. It thus appears probable that evaporation of an amino acid solution with a low ee should cause selective precipitation of the racemate crystals, which in turn should lead to an increase of the ee. Extremely simple manipulations, carried out in the chemistry department of Columbia University, led to a drastic increase in enantiomeric excess of phenylalanine: 500 mg phenylalanine (with a 1% ee of the L-component) was dissolved in water, and the resulting solution slowly evaporated until about 400 mg had crystallised out. The remaining solution contained a few mg of phenylalanine with 40% ee of the L-component (i.e., a 70:30 ratio of L to D). If 500 mg of such a solution (40% ee in water) is allowed to evaporate and is separated from the racemate, the result is about 100 mg, with 90% ee of the L-enantiomer (Breslow and Levine, 2006).

In conclusion, we can say that, in spite of the important results already obtained, the homochirality problem is still awaiting a solution!

References

Adam D (2002) Nature 420:723
Anfinsen C (I973) Angew Chem 85:1065
Bailey J, Chrysostomou A, Hough JH, Gledhill TM, McCall A, Clark S, Ménard F, Tamura M (1998) Science 281:672
Bailey J (2001) Orig Life Evol Biosphere 31:167
Bakasov A, Ha T-K, Quack M (1995) Ab initio Calculation of molecular energies including parity violating interactions. In: Chela-Flores J, Raulin F (eds) Chemical Evolution: Physics of the Origin of Life. Kluwer, Dordrecht Boston London, p 287
Ball P (1994) Designing the Molecular World, Chemistry at the Frontier, Princeton University Press
Belinski R, Tencer M (2007) Orig Life Evol Biosphere 37:167
Bong DT, Clark TD, Granja JR, Ghadiri MR (2001) Angew Chem 113:1017, Int Ed 40:988
Bonner WA, Rubenstein S (1987) Biosystems 20:99
Breslow R, Levine MS (2006) Proc Natl Acad Sci USA 103:12979
Buschmann H, Thede R, Heller D (2000) Angew Chem 112:4197, Int Ed 39:4033
Cronin JR Pizzarello S (1997) Science 275:951
Cronin JR (1998) Clues from the Origin of the Solar System: Meteorites. In: Brack A (Ed.) The Molecular Origins of Life. Cambridge University Press, pp 119–146
Davis P (2000) The Fifth Miracle. Simon and Schuster
DiGregorio B (2006) Microbe 1:471
Eddington AS (1932) The Nature of the Physical World. Cambridge University Press
Eigen M (1971a) Quart Rev Biophys 4:149
Eigen M (1971b) Naturwissenschaften 58:465
Goldanski VI (1986) Scientific American, February p 38
Goldanski VI, Kuzmin VV (1991) Nature 352:114

Hazen RN, Filley TR, Goodfriend GA (2001) Proc Natl Acad Sci USA 98:5487

Jantsch E (1980) The Self-Organizing Universe: Scientific and Human Implications of the Emerging Paradigm of Evolution. Pergamon Press, New York.

Jessen W (1978) Naturwissenschaften 65:449

Kauffman SA (1991) Scientific American, August p 64

Kauffman SA (1996) At Home in the Universe. Oxford University Press

Klabunovskii E, Thiemann W H-P (2000) Orig Life Evol Biosphere 30:431

Klußmann M, Izumi T, White AJP, Armstrong A, Blackmond DG (2007) J Am Chem Soc 129:7657

Kondepudi DK, Nelson GW (1985) Nature 314:438

Kondepudi DK (2000) Orig Life Evol Biosphere 30:214

Lahav M, Weissbuch I, Shavit E, Reiner C, Nicholson J, Schurig V (2006) Orig Life Evol Biosphere 36:151

Mason SF, Tranter GE (1983) J Chem Soc Chem Commun 1983:117

Meierhenrich UJ, Thiemann W H-P (2004) Orig of Life Evol Biosphere 34:111

Nemoto A, Horie M, Imai E-I, Honda H, Hatori K, Matsuno K (2005) Orig Life Evol Biosphere
35:167

Norden B (1977) Nature 266:567

Prigogine I (1980) From Being to Becoming – Time and Complexity in Natural Sciences. WH
Freeman and Company, New York

Prigogine I, Stengers I (1984) Order out of Chaos: Man's New Dialogue with Nature. Flamingo,
London

Rubenstein E, Bonner WA, Noves HP, Brown GS (1983) Nature 306:118

Schuster P, Sigmund K (1982) In: Hoppe W, Lohmann W, Markl H, Ziegler H (Eds.) Biophysics.
Springer, Berlin Heidelberg New York, p 907–947

Shinitzky M, Nudelman F, Barda Y, Halmowitz R, Chen E, Deamer DW (2002) Orig Life Evol
Biosphere 32:285

Snow CP (1993) The Two Cultures. Cambridge University Press

Szabo-Nagy A, Keszethelyi L (1999) Proc Natl Acad Sci USA 96:4252

Szabo-Nagy A, Keszethelyi L (2000) Orig Life Evol Biosphere 30:219

Thiemann W H-P, Meierhenrich U (2001) Orig Life Evol Biosphere 31:199

Chapter 10
Primeval Cells and Cell Models

10.1 Palaeontological Findings

The early period of the Earth's history, around 3.8–4 billion years ago, is completely shrouded in darkness; possible witnesses from this archaic period might help to cast some light. So we are looking for possible remains of the first primitive life forms on our planet—fossils, or to be exact, microfossils, which refers to the remains of living cells. What have survived are mainly only cell walls, which can be isolated from sedimentary rocks when the silicate-, sulphide- or carbonate-rich minerals are chemically dissolved away. The microfossils are found in the remaining carbon-containing residue, transparently thin platelets of which are prepared for microscopic studies.

Another important method for the detection of early life forms involves the ratio of the two carbon isotopes ^{12}C and ^{13}C. The lighter isotope ^{12}C ("normal carbon") is incorporated preferentially into biomolecules, so that the value of the ratio of ^{13}C to ^{12}C can indicate the presence of material which was formed in living things. It must of course be certain that other, non-biological processes can be excluded, though it is questionable whether this is always possible, in particular when we are unsure of the geological processes which occurred 3–4 billion years ago.

There are three types of evidence for the first life forms on Earth; they are independent of one another, but reinforce each other:

Stromatoliths
Cellular fossils
Biologically formed carbonaceous material

Stromatoliths are large columnar calcium carbonate structures produced by cyanobacteria and can occur in many forms.

Cellular fossils can be studied under the microscope, and more recently, laser Raman spectroscopy has been used (see below). The platelets often show cells with sizes, shapes, cell structure and colony shapes similar to those of today's microorganisms. These properties could not have survived if the rocks had been heated to temperatures above 420 K, and the fossils also would not have withstood high pressures.

H. Rauchfuss, *Chemical Evolution and the Origin of Life*,
© Springer-Verlag Berlin Heidelberg 2008

The *isotopic composition of carbon* in carbonaceous organic material (kerogen) from ancient sedimentary rocks gives information on whether photosynthetic organisms were present during rock formation or not. It can also provide information on biological activities if cellular structures had already been destroyed. Sulphur can be used in a similar way (Schopf, 1999).

The oldest rock formations on Earth are mainly found in three regions:

Southern Africa
Western Australia
Greenland

The historical development of palaeontology began slowly after Darwin's "search for the missing link", i.e., the missing fossils from the Precambrian which would have provided proof of his theory. In 1865, the Canadian scientist Sir John William Dawson discovered the "eozoon pseudofossils", and in 1883, James Hall discovered the "cryptozoon stromatoliths" (Schopf, 1999). But the intensive phase of palaeobiology of the Precambrian began only in the middle of the last century.

Of particular importance was the work of Barghoorn and Schopf (1966) on microfossils of the Fig Tree group, which are about 3.1 billion years old. The two palaeobiologists found twig-shaped microorganisms in the carbon-rich cherts and named them "*eubacterium isolatum*". They also discovered fossils from the Onverwacht group (Swaziland Supergroup), which are about 100 million years older. These microfossils probably derived from algae-like organisms. Gas-chromatographic analysis of carbon-containing compounds extracted from these sedimentary rocks shows the presence of straight-chain hydrocarbons with between 16 and 25 carbon atoms, with a maximum at 20. It is not clear whether these compounds are of biogenic origin.

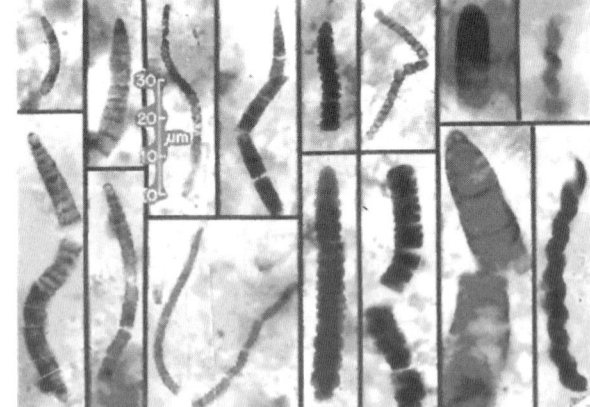

Fig. 10.1 Cellular, petrified, filamentous microfossils (cyanobacteria) from the Bitter Springs geological formation in central Australia; they are about 850 million years old. With kind permission of J. W. Schopf

The news of the oldest microfossils to be discovered was then a sensation (Schopf, 1993). They were found in western Australia in geological formations whose age had been determined some years before to be more than 3 billion years (Groves et al., 1981); interestingly, one of the formations is called "North Pole". The age of the fossils, which were mainly found in hornstone (a sedimentary rock), was determined as 3.465 billion years.

Fig. 10.2 Cyanobacteria-like, filamentous carbonaceous fossils from the 3.456-billion-year-old Apex chert in northwestern Australia; their origin and formation are still under discussion. The photographs are accompanied by the corresponding drawings. With kind permission of J. W. Schopf

MEDIUM DIAMETER (2-5μm) FILAMENTS, CYLINDRICAL CELLS

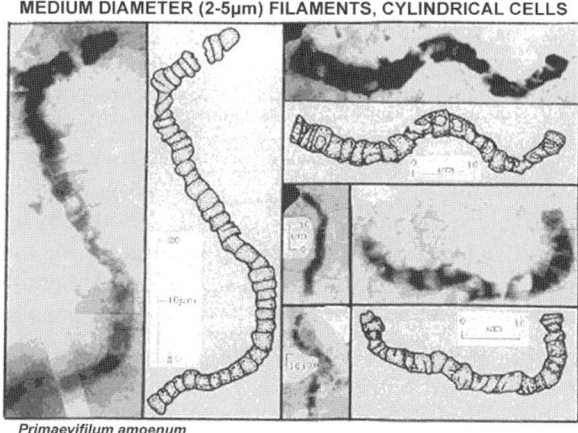

Primaevifilum amoenum

Archaic fossils can only be found in geological formations which have not suffered any dramatic changes or deformations since their formation. The area of Western Australia near Marble Bar and North Pole is part of the geological formation known as the Pilbara Shield (Buick et al., 1995). The hot, desolate North Pole region is about 40 km from the small town of Marble Bar. The sediments containing the fossils are in a former volcanic lagoon, which was formed in a complicated process (Groves et al., 1981). This area of western Australia was explored and described in detail years ago because of the many ore deposits it contains. In Swaziland, in the

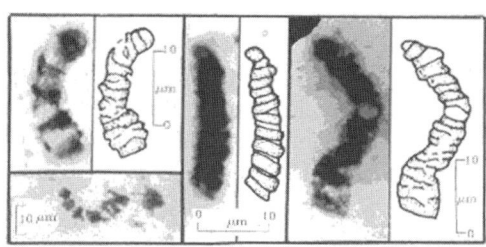

ROUNDED END CELLS - *Archaeoscillatoriopsis disciformis*

Fig. 10.3 Microfossils with differently formed end cells, from the same source as in Fig. 10.2 and thus of the same age. Again, the corresponding drawings are shown to make the structures clearer. With kind permission of J. W. Schopf

CONICAL END CELLS - *Primaevifilum conicoterminatum*

east of southern Africa, there are also rock formations which are 3–3.5 billion years old; even older rock has been found near Isua in Greenland, but it is almost impossible that the traces of life forms found there are genuine, because the Isua rock formations were subject to greater thermal variations than those in western Australia (we will return to this point below).

How can we determine geological ages of around 3.5 billion years? The method of choice is the well-known radiometric one, which is based on the decay of radioactive isotopes to give "daughter" isotopes. If the half-life of the isotope concerned is known, we can determine the age of the rock under study if we can obtain an exact value for the amount of the daughter isotope present. Thus, the method is based on two factors, one determined by nature, the other by technological progress, as highly sensitive mass spectrometers are required for the measurements. The nature of the rock material to be studied is also important: both the undecayed "parent isotope" and the daughter isotope formed by the decay must be encapsulated within the rock, with no chance of escape. If geological processes thwart the latter requirement, the resulting values are incorrect.

The "best sealed-in" minerals are zircons, zirconium silicate minerals which are formed when melted lava on the flanks of volcanoes solidifies. When the zircons crystallize out, they incorporate radioactive uranium (in particular ^{238}U), which decays in several steps, leading finally to the lead isotope ^{208}Pb. The rate of decay is very low, as the half-life of uranium-238 is 4.5×10^9 years. Thus, the U–Pb–zircon method for age determination of Precambrian rock is very important. The fossils studied by Schopf were sandwiched between two lava layers (Schopf, 1999). The volcanic layers were dated to $3.458 \pm 0.0019 \times 10^9$ years and $3.471 \pm 0.005 \times 10^9$ years; the age of the fossil layer (Apex chert) was thus determined to be about 3.465×10^9 years.

Three years after Bill Schopf's article had appeared in *Nature*, one from the group of Gustaf Arrhenius (Scripps Institute of Oceanography in La Jolla) was published (Mojzsis et al., 1996). They had studied minerals from Greenland using the most up-to-date technology and concluded that life might possibly be 300 million years older than had previously been assumed. Over 30 years ago, initial evidence was published on sedimentary rocks from near Isua in western Greenland, about 150 km northeast of Godthab (Nuuk) (Nagy, 1976; Walters et al., 1981). The Isua sediments had certainly been subjected to great stress; however, it seemed that traces of primitive life had been found. The $^{13}C/^{12}C$ ratio in the Isua rock indicated the carbon to be of biological origin (Schidlowski, 1988).

A $^{13}C/^{12}C$ ratio about 3% above that of a standard value (the so-called VPDB standard) had been found in western Australian rock samples from the Pilbara Formation: similar values were found for the Isua rock. However, since this no longer had its original morphology, the yeast-like relicts found by Pflug (1978) may not be real, and doubts have been cast, in particular by the American scientists J. W. Schopf and E. Roedder (Breuer, 1981, 1982). Neither the Isua microfossils nor their $^{13}C/^{12}C$ isotope ratio could convince Bill Schopf that 3.8-billion-year-old samples were really involved. The Isua graphite flakes could also be "a charred residuum from the primeval soup" or carbonaceous material brought to Earth by meteorites or comets (Schopf, 1999).

The analyses of the Isua rock were carried out in the Scripps Institute using the most modern technology: ion microprobe mass spectrometry. The material under study came from the island of Akilia, which is close to Isua. The graphite flakes were encapsulated in apatite [$Ca_3(PO_4)_3OH$, F] from the 3.8-billion-year-old rock; their volume was 10 μm^3 and they contained about 20 μg of carbon. However, the new analytical method had not passed all the necessary quality tests. Therefore, and for some geological reasons, the results did not appear entirely trustworthy, so that doubts as to the age of 3.8 billion years for the first traces of life forms arose (Holland, 1997). These doubts were reinforced when, five years after Holland's short article was published in *Science*, a "correction" from the Scripps Institute appeared (van Zuilen et al., 2002a, b). Here, the assumption that the Isua samples contained relics of the oldest living material was retracted. New geological studies of the rock formation led to the conclusion that the graphite flakes in the metacarbonate rock could have been formed by chemical disproportionation (at higher temperatures and pressures than had previously been assumed) via the following equation:

$$6\,FeCO_3 \rightarrow 2\,Fe_3O_4 + 5\,CO_2 + C$$

This correction shifts the point at which life emerged back to the value found by Schopf—3.5 billion years; thus, 300 million years more were available for the process of chemical and molecular evolution.

Three years before the correction, M. F. Rosing (1999) from the Geological Museum in Copenhagen provided confirmation of the work done by Schidlowski and Mojzsis; he studied pelagic sedimentary rocks (i.e., sediments from the abyssal plain of the deep ocean) from the Isua supracrustal belt and found a depletion in ^{13}C in the more than 3.7-billion-year-old sea floor sediments of around 19% (according to the VPDB standard). Rosing interprets the data as showing that the carbon found in the sediments could represent biogenic detritus.

Fig. 10.4 Fossilized cellular filamentous microorganisms (two examples of *Primaevifilum amoenum*). They are 3.456 billion years old and come from the Apex chert region in northwestern Australia. As well as the original images, drawings and the Raman spectra and Raman images, which indicate that the fossils have a carbonaceous (organic) composition, are shown. With kind permission of J. W. Schopf

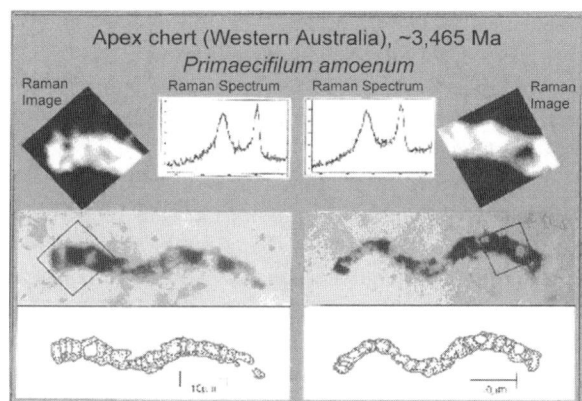

However, the question of when exactly the first life forms emerged is by no means settled: the date set by Schopf, 3.465×10^9 years, is now in doubt. Schopf has recently introduced a new, ultramodern analytical method, laser Raman spectroscopy, as a highly sensitive technique for the study of microscopic fossil material. With the help of this method, it is possible to determine the chemical composition and also the two-dimensional structure of fossils (Kudryavtsev et al., 2001).

Schopf used laser Raman spectroscopy to confirm his results on the oldest microfossils (Apex chert, western Australia) (Schopf, 2002a). However, Jill Dill Pasteris and Brigitte Wopenka from Washington University, St. Louis, do not agree with Schopf's highly positive conclusions. According to them, Schopf overinterpreted his Raman spectra, as they in fact show that the material studied consists of highly disordered carbonaceous material, its spectra resembling those of kerogens (Pasteris and Wopenka, 2002).

But the headwind is getting even stronger! A group consisting of eight scientists from four research institutes was not completely happy with the optimistic interpretation of Schopf's data (Brasier et al., 2002a). According to them, it is not clear that the Apex chert microfossils have their origin in primitive life forms. The British–Australian group headed by Brasier (Oxford University) points out two problems: the objects which Schopf interprets as fossilized bacterial remains show a variety of non-uniform structures, some of which could hardly come from protozoa. When studied under various conditions, thin polished slices of the fossil material show structures which are unusual for cells, and which are consistent with their formation as secondary artefacts of crystal growth. The second argument has to do with the geology of the site: the rocks from which the samples were taken are of hydrothermal origin, i.e., they were formed by the effect of hot water on minerals. The Brasier group considers that this process can lead to a carbon isotope ratio corresponding to that of biological material. It is thus possible that CO_2 was set free by volcanic activity at the site under study, and that isotope fractionation set in subsequent to reactions occurring at 520–620 K. This process, occurring mainly during recrystallisation from amorphous to spherulitic silica, was responsible for the simulation of a biological origin of the carbonaceous matter studied.

Further studies are thus required to determine whether the Earth's oldest putative "microfossil" assemblages from the Apex chert represent genuine primeval life forms or only a fatal delusion (Schopf et al., 2002b; Brasier et al., 2002b, 2004).

The great difficulties inherent in determining the origin of natural structures by studying morphological characteristics have recently been recognized and described (García-Ruiz, 2002).

At the end of 2003, new research results led to sensational headlines: "Minerals Cooked Up in the Laboratory Call Ancient Microfossils Into Question" was the title chosen by Richard A. Kerr for his article in *Science* dealing with synthetically prepared silicate carbonates. Their microstructures show morphologies which look exactly like those of filaments which had been assigned as cyanobacterial microfossils of the Precambrian Warrawoona chert formation in western Australia. The synthetic structures consist of silicate-encapsulated carbonate crystals, and in part have a helically twisted morphology reminiscent of biological objects. Simple hydrocarbons, the origin of which could also have been abiotic and inorganic, condense readily on these filaments and polymerise on mild heating to give kerogenic materials (García-Ruiz et al., 2003; Kerr, 2003).

New, improved analytical techniques can help to solve the problems referred to above. In a comprehensive article, William Schopf and co-workers introduced a new method, laser Raman imagery: this is a non-intrusive, non-destructive analytical technique which was used to analyse kerogenous microscopic fossils and associated carbonaceous sapropel. The material studied was between 400 and 2,100 million years old: much younger than the palaeontological-geological findings described above. According to the authors, Raman imagery can help to address questions about the biogenicity of ancient fossil-like objects, which is a long-standing problem in Precambrian palaeobiology (J.W. Schopf et al., 2005). The island of Akilia in western Greenland, about 22 km from Nuuk, is very well known in the scientific world because of the rock material discovered there which was found to be 3.85 billion years old. More than 20 years ago, M. Schidlowski had suggested that kerogen deposits found there with a low ^{13}C content might be the remains of the earliest organic life forms. Other geologists disagree, however, so the question is still open.

William Schopf studied supercrustal rock samples from Akilia; Raman and ion microscopic photographs showed the presence of carbon-containing inclusions in grains of apatite. The carbon isotope ratio was determined by secondary ion mass spectroscopy (SIMS): the $\delta^{13}C$ value was $-29\% \pm 4\%$, in agreement with earlier analyses. This in turn confirmed the values obtained by Mojzsis (1996), which had been questioned by Lepland et al. three years later. The final verdict on the oldest fossils in western Greenland may not be reached for several years yet (McKeegan et al., 2007; Eiler, 2007).

The scientific controversies presented above make clear the huge difficulties which confront much of biogenesis research. The coming years are likely to be quite exciting, and surprises can be expected!

10.2 The Problem of Model Cells

"The basic unit of life is the cell", according to Christian de Duve in his book "Blueprint for a Cell: The Nature and Origin of Life" (1991). The cells of contemporary organisms are highly complex systems containing a great number of very different types of molecules, from macromolecules to simple alkali ions. In spite of their tiny size (average volume around 10^{-8} ml), cells are equipped with all the functions necessary for life processes. A number of structures and characteristics are to be found in the cells of all life forms, and the assumption seems logical that the cells of all living systems derive from one single "primeval cell".

D. W. Deamer from the State University of California at Santa Cruz (1998) asks two important questions on the problem of the origin of the first cellular structures; these questions must be answered if we want to understand life's beginnings:

Did the simplest life forms develop a priori from already existent cellular structures, or
Did cellular life develop in a later phase of evolution?

Put simply, we can ask: what came first, the cell or an information-transmitting
system?

The research dealing with models for the first primitive cells has had one central
topic for many years: the "minimal cell". According to Luisi et al. (2006a), this is
defined as an artificial or semi-artificial cell which contains a minimal (but sufficient) number of components to keep the cell "alive". The cell is considered to be
living when three conditions are fulfilled:

Self maintenance
Reproduction
Evolvability

This definition does not identify one special structure; it is a descriptive term for a
variety of minimal cells. Although there is no consensus on this definition, scientists do agree on the main points. Thus, a minimal genome should have 200–300
genes. The question arises as to whether it is conceivable that this number can be
further reduced, or whether we need to devise other, as yet unknown, precursor
systems.

The minimal cell, as the simplest system which has all the required properties of life (metabolism, self-reproduction and the ability to evolve), is presently
studied as part of a new research discipline: synthetic biology. This includes sub-
jects such as synthesis in branches of biological systems, for example, of new
RNA species, new peptides and new nucleic acid analogues, as well as the synthesis of peptide nucleic acids. One example is the work of M. R. Ghadiri and
G. von Kiedrowski on self-replication of oligonucleotides and oligopeptides (Luisi,
2006b).

10.2.1 Some Introductory Remarks

The cells of all contemporary living organisms are surrounded by cell membranes,
which normally consist of a phospholipid bilayer, consisting of two layers of lipid
molecules, into which various amounts of proteins are incorporated. The basis for
the formation of mono- or bilayers is the physicochemical character of the molecules
involved; these are amphipathic (bifunctional) molecules, i.e., molecules which have
both a polar and also a non-polar group of atoms. Examples are the amino acid
phenylalanine (a) or the phospholipid phosphatidylcholine (b), which is important
in membrane formation. In each case, the polar group leads to hydrophilic, and the
non-polar group to hydrophobic character.

a) Phenylalanine

hydrophobic NH_3^+ hydrophilic

b) Phosphatidylcholine

$H_3C-(CH_2)_{15}-CH_2-\overset{\overset{O}{\|}}{C}-O-CH_2$

$H_3C-(CH_2)_{15}-CH_2-\overset{}{\underset{\overset{\|}{O}}{C}}-O-CH$ hydrophilic

$H_2C-O-\overset{\overset{-}{O}}{\underset{\overset{\|}{O}}{P}}-O-CH_2-CH_2$

$^+N(CH_3)_3$

hydrophobic

In amphiphilic molecules, the polar, hydrophilic part is known as the "head" and the non-polar, hydrophobic part the "tail" of the molecule. Hydrophilic molecules, or parts of molecules, try to interact with polar water molecules, while hydrophobic moieties try to avoid them.

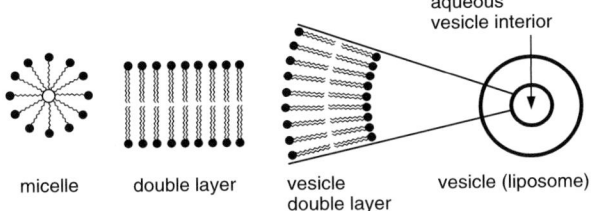

aqueous
vesicle interior

micelle   double layer   vesicle   vesicle (liposome)
double layer

Fig. 10.5 Schematic diagrams: a micelle consisting of ionized fatty acid molecules, a phospholipid bilayer and the vesicle bilayer of a liposome

If amphipathic molecules are mixed with water, three different types of lipid structure are possible; the type of aggregate formed depends on the physicochemical conditions and the lipid species involved. The thermodynamic parameter involved is the hydrophobic interaction.

> *Micelles* are the simplest possible structures. These are spherical entities in which the hydrophobic groups are directed towards the interior. Micelle formation occurs preferentially when the head groups are larger than the hydrophobic groups, as in fatty acids.
> A *bilayer* forms when two lipid layers come together; the hydrophobic groups in the two single layers interact and exclude water.

If the double layers are unstable, a closed vesicle (a *liposome*) is formed, the interior of which is filled with water.

10.2.2 The Historical Background

One of A. I. Oparin's great wishes was to be able to demonstrate primitive metabolism in a certain type of protocells, the coacervates. The physicochemical process of their formation can be compared to an unmixing, for example, a salting out effect; this process was first described by Bungenberg de Jong et al. (1930). Coacervate formation involves the aggregation of dehydrated colloidal particles; removal of water from the solvate envelope of the macromolecules involved is realized by a second, more hydrophilic, colloid in the system. The opposite charges of two or more types of macromolecule in the system are superimposed on this effect. The "cell-like balls" are often found in systems containing complex macromolecules. More than 200 such hydrophilic systems are known, e.g., gelatine/gum arabic, serum albumin/gum arabic etc. The coacervate drops are heterogeneous and polydispersive, i.e., of different sizes and diameters (1–500 µm).

Oparin and his school in the Biochemical Institute in Moscow worked for many years with coacervate systems. For Oparin, the origin of life was the moment of formation of the first cell. Coacervates are today of only historical importance; because of their low thermodynamic stability, they are considered dubious and too unstable.

A second historical model for protocells is provided by the "microspheres" (Fox, 1980; Nakashima, 1987; Lehninger, 1975). These are formed when hot saturated proteinoid solutions are allowed to cool (see Sect. 5.4.2). In recent years, the microspheres were also consigned to the limbo of unimportant scientific models. Perhaps there will come a time when coacervates or microspheres (in their original or in modified forms) find their way back into the scientific discussion.

10.2.3 New Developments

For several years, Pier Luigi Luisi and co-workers from the Polymer Institute of the ETH in Zürich have been working on the problem of the emergence of primitive membranes. The oldest fossils (see Sect. 10.1) indicate that the first life forms already had cell-like structures, i.e., life has been based on the following three important principles for 3.5 billion years (or maybe less?) (Luisi, 1996):

Metabolism
Reproduction (and as a consequence, evolution)
Compartmentalisation

Since at present, one of these three, reproduction (the "RNA world"), dominates the scientific discussion, the other two have automatically been neglected. It could, however, have been demarcation (cell formation) which was the precondition for the extremely sensitive chemistry of nucleotide synthesis and the subsequent polycondensation process to give the nucleic acids or similar precursors.

After more than 20 years, Walde et al. (1994) returned in a way to coacervate experiments, although using other methods. Walde (from the Luisi group) repeated nucleotide polymerisation of ADP to give polyadenylic acid, catalysed by polynucleotide phosphorylase (PNPase). But instead of Oparin's coacervates, the Zürich group used micelles and self-forming vesicles. They were able to demonstrate that enzyme-catalysed reactions can take place in these molecular structures, which can thus serve as protocell models. Two different supramolecular systems were used:

Reverse micelles of sodium bis(2-ethylhexyl)sulfosuccinate in isooctane
Oleic acid/oleate vesicles at pH 9

In the first case, the reactions occurred with quite good yields of polymers, for example, polyadenosine [poly(A)], which was precipitated from the micelles.

Poly(A) synthesis also occurred in the second system, but the product remained within the vesicles. Walde also determined the increase of the vesicle concentration, which corresponds to that expected for an autocatalytic process. In this experiment, the enzyme PNPase is first captured by the vesicle envelope, and in the second step, ADP and oleic anhydride are added; the anhydride is hydrolysed to the acid. ADP passes through the vesicle double layer and is polymerized in the interior of the vesicle by PNPase to give poly(A). Hydrolysis of the anhydride causes a constant additional delivery of vesicle-forming material, so that the amount of vesicle present increases during the poly(A) synthesis. These experiments demonstrated a model for a minimal cell. Autocatalytically synthesised giant vesicles could be prepared under similar conditions and observed under a microscope (Wik et al., 1995).

The favourable properties which mark out vesicles as protocell models were confirmed by computer simulation (Pohorill and Wilson, 1995). These researchers studied the molecular dynamics of simple membrane/water boundary layers; the bilayer surface fluctuated in time and space. The model membrane consisted of glycerine-1-monooleate; defects were present which allowed ion transport to occur, whereby negative ions passed through the bilayer more easily than positive ions. The membrane–water boundary layer should be particularly suited to reactions which are accelerated by heterogeneous catalysis. Thus, the authors believe that these vesicles fulfil almost all the conditions required for the first protocells on earth!

Several years earlier, Morowitz, Heinz and Deamer (1988) described their ideas on "the chemical logic of a minimal protocell"; Harold J. Morowitz from George Mason University in Fairfax, VA, is a proponent of the "metabolism first" hypothesis. The three scientists suggested that formation of replicating protocell vesicles was the first phase in the biogenesis process. The formation of the vesicle bilayer could have only occurred via amphiphilic molecules which were present in large amounts on the primeval Earth. If primitive pigment molecules were available for incorporation into the bilayer, they could have absorbed light energy and built up an electrochemical ion gradient. The energy (so the hypothesis goes) could have been used to convert simple molecules, such as hydrocarbons, into amphiphilic compounds such as monocarboxylic acids. The "principle of continuity" makes it possible to derive three assumptions on the logical formation of a protocell (Morowitz et al., 1988):

> The most plausible components are amphiphilic compounds. Growth occurs by means of chemical transformation of simple molecules into amphiphilic species.
>
> The conversion of light energy by primitive pigments affords energy for chemical reactions, making possible not only the growth of the protocells but also the generation of a chemiosmotic proton gradient which served for energy storage.
>
> Phosphates seem to be the most likely compounds for moderating the process between light absorption and chemical, energy-consuming molecular conversions.

Only a few years after Morowitz's article appeared, some of the models suggested were confirmed (or as good as confirmed).

It is important in studies on the problems of the formation of the first cells to put aside the ideas of present-day membrane chemistry and to remember that the boundaries between the first cell-like structures probably consisted of starting materials completely different from those in today's cell walls; there is no doubt that such materials were amphiphilic in character.

Thus, the question is whether such classes of molecules were present on the young Earth. The only witnesses capable of giving an answer to this question are meteorites (Deamer, 1988). The group of David Deamer studied Murchison material after extraction and hydropyrolysis (at 370–570 K, with reaction times of several hours or days). GC and MS analyses showed the presence of a series of organic compounds, including significant amounts of amphiphilic molecules such as octanoic (C_8) and nonanoic acids (C_9) as well as polar aromatic hydrocarbons.

The formation of relatively stable vesicles did not require the presence of pure compounds; mixtures of components could also have done the job. However, whether the concentrations of the compounds isolated from the Murchison meteorite would have been sufficient for the formation of prebiotic protocells or vesicles is unclear, even if concentration effects are assumed. Sequences in which the technical Fischer-Tropsch synthesis is the role model have been proposed as possible sources of amphiphilic building blocks.

The formation of lipid components in an aqueous phase at temperatures from 370 to 620 K was studied by Rushdie and Simoneit (2001), who heated aqueous solutions of oxalic acid in a steel vessel for 2 days; the yield of oxidized compounds reached a maximum (5.5% based on oxalic acid) between 420 and 520 K. A broad spectrum of compounds was obtained, from n-alkanes to the corresponding alcohols, aldehydes and ketones. At higher temperatures, i.e., above 520–570 K, cracking reactions competed with the synthetic reactions.

The manufacture of liposomes (which is, for example, carried out by the cosmetics industry) is a complex technical process. Thus, the question arises as to whether a simple mechanism for liposome formation exists which could have functioned under plausible prebiotic conditions. More than 20 years ago, Deamer and co-workers developed the so-called sandwich method; this made it possible to introduce large molecules into the interior of liposomes (Shew and Deamer, 1985) (Fig. 10.6).

Fig. 10.6 Macromolecules can also be encapsulated in liposomes. This is done by drying a mixture of liposomes and macromolecules; these form polylayer structures, the macromolecules being embedded in alternating layers. Almost half the macromolecules are incorporated in liposomes on rehydration (Deamer, 1998)

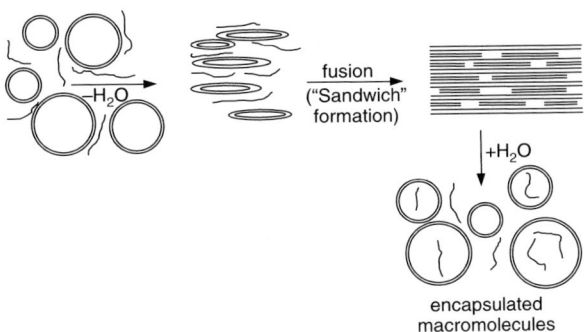

encapsulated
macromolecules

The mixture of liposomes and macromolecules was first dried under nitrogen; the two types of molecules formed a multilamellar film with sandwich structures. Larger liposomes, containing macromolecules (proteins or RNA) were formed on rehydration. This process could have occurred in hot regions of the young Earth with the help of the tidal rhythm of the oceans.

The experiments were carried out using C_{14}-phosphatidylcholine (PC) vesicles. The biochemical reaction which was planned to occur in the vesicles was the aforementioned RNA polymerisation reaction involving the enzyme polynucleotide phosphorylase (PNPase), which Oparin and co-workers had used many years ago in their work on coacervates. PNPase and added ADP then form oligonucleotides in the vesicles.

The dependence of the RNA synthesis in the vesicles on the length of the carbon chain in the two hydrophobic groups of the PC molecules is shown in Fig. 10.7.

Fig. 10.7 RNA synthesis in vesicles. Membrane permeability can be regulated by choosing the correct chain length of the fatty acids in the phospholipids. Short chains (**a**) make the bilayer so unstable that even large molecules such as proteases can enter the vesicle interior and damage the polymerase. Carbon chains which are too long (**b**) prevent the entry of substrate molecules such as ADP. RNA polymerisation in the vesicle occurs only with C_{14} fatty acids (**c**) (Deamer, 1998)

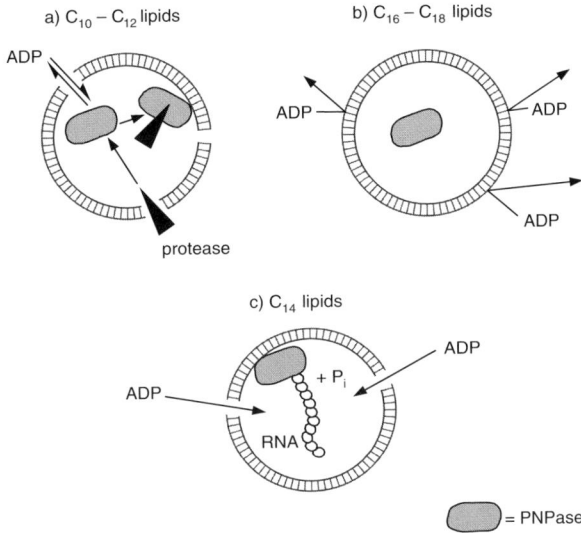

a) $C_{10} - C_{12}$ lipids

b) $C_{16} - C_{18}$ lipids

protease

c) C_{14} lipids

= PNPase

The permeability coefficient of 2.6×10^{-10} cm/s at 296 K measured by Deamer is sufficient to supply the enzyme in the liposomes with ADP. How could it be shown that RNA formation actually does take place in the vesicles? The increase in the RNA synthesis was detected by observing the fluorescence inside the vesicles. In the interior of the liposomes, the reaction rate is only about 20% of that found for the free enzyme, which shows that the liposome envelope does limit the efficiency of the process. The fluorescence measurements were carried out with the help of ethidium bromide, a fluorescence dye often used in nucleic acid chemistry.

Monnard and Deamer (2001) carried out further studies, using DMPC liposomes, to determine their properties under conditions of passive diffusion of dissolved molecules. The passage across the lipid bilayer is a precondition for the intake of "nutrient substances" via the vesicle envelope. The experiments showed that even polar molecules can enter the interior of the liposomes; oligonucleotides, however, cannot cross the lipid bilayer of DMPC vesicles.

These successes in the area of precellular structures led Daniel Segré et al. from the Weizmann Institute in Rehovot to publish the "lipid world" hypothesis together with Deamer. The authors are of the opinion that the (more probable) "lipid world" determined chemical and molecular evolution on the primeval Earth prior to the (less probable) "RNA world". Further development stages (or "worlds") finally led to biogenesis (Segré et al., 2001).

The concept of a lipid world is derived from the ability of amphiphilic molecules to form supramolecular structures such as micelles or vesicles/liposomes via spontaneous self-organisation processes. The critical judgement of many biogeneticists on the chances of an RNA world is well known. Apart from the weaknesses which have already been mentioned, RNA cannot take up energy from the environment. Thus, according to these critics, RNA cannot be involved in the formation of supramolecular structures. Amphiphilic molecules, however, have all the necessary properties. They are extremely versatile, since each individual molecule with a hydrophobic carbon chain and a polar group is amphiphilic. The utilization of external energy (for example, in the form of light) can take place by means of incorporation of pigment molecules. An increase in the size of micelles and vesicle populations has also been observed. Further experiments are required in order to come to a final evaluation of the lipid world hypothesis.

Attempts have recently been made to link the RNA world with the lipid world. Two groups involved in RNA and ribozyme research joined up with an expert on membrane biophysics (Szostak et al., 2001). They developed a model for the formation of the first protocells which takes into account both the most recent experimental results on replication systems and the self-organisation processes of amphiphilic substances to give supramolecular structures.

The question is how primitive a cell can be in order for it to still fulfil the needs of the definitions of the term "life". As already discussed (see Sect. 1.4), the definition of life is notoriously difficult (Luisi, 1998) and sometimes leads to misunderstandings in the community of biogeneticists.

Szostak et al. worked on the basis of a simple cellular system which can replicate itself autonomously and which is subject to Darwinian evolution. This simple protocell consists of an RNA replicase, which replicates in a self-replicating vesicle. If this system can take up small molecules from its environment (a type of "feeding"), i.e., precursors which are required for membrane construction and RNA synthesis, the protocells will grow and divide. The result should be the formation of improved replicases. Improved chances of survival are only likely if a sequence, coded by RNA, leads to better growth or replication of membrane components, e.g., by means of a ribozyme which catalyses the synthesis of amphiphilic lipids (Figs. 10.8 and 10.9). We can expect further important advances in the near future from this combination ("RNA + lipid world").

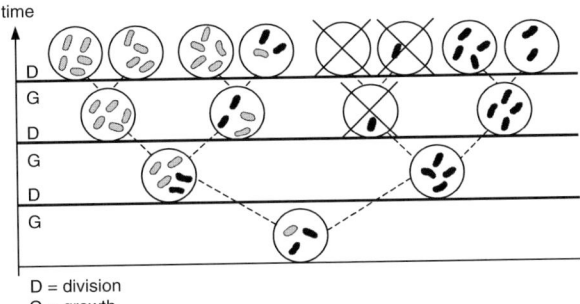

D = division
G = growth

Fig. 10.8 The importance of the vesicle for the Darwinian evolution of a replicase. Compartmentalisation ensures that "related" molecules tend to stay together. This permits superior mutant replicases (grey) to replicate more effectively than the parent (black) replicases. The evolutionary advantage spreads in the form of vesicles with superior replicase molecules, leading with a greater probability to vesicles with at least two replicase molecules (or a replicase and a matrix molecule). Vesicles with less than two replicase molecules are struck out: their progeny cannot continue the RNA self-replication. Thus, the vesicles with better replicases form the growing fraction of vesicles which carry forward the replicase activity (Szostak et al., 2001)

The clay mineral montmorillonite, which is often used in different prebiotic syntheses, is probably now the most important mineral for experiments on prebiotic chemistry. It has shown its abilities in the area of simulation experiments on the formation of primitive cellular compartments: montmorillonite accelerates the spontaneous conversion of fatty acid micelles to vesicles. Clay particles are often incorporated into the vesicle, just as is RNA, which is adsorbed at such clay particles. If the vesicles have been formed, they can continue to grow if fatty acids are "fed" to them via micelles. If the vesicles are pressed through 100 nm pore filters, they divide without dilution of their contents.

These successful experiments suggest possible prebiotic reaction pathways for the formation, growth and multiplication of the first cells (Hanczyc et al., 2003).

Fig. 10.9 Possible reaction
pathway for the formation of
a cell. The important
precursors are an RNA
replicase and a
self-replicating vesicle. The
combination of these two in a
protocell leads to a rapid,
evolutionary optimisation of
the replicase. The cellular
structure is completed if an
RNA-coded molecular
species, for example, a
lipid-synthesised ribozyme, is
added to the system (Szostak
et al., 2001)

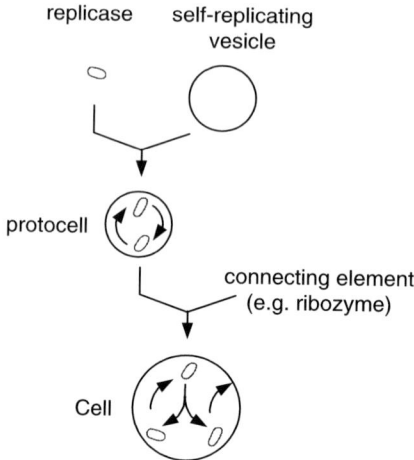

According to Hanczyc and Szostak, primitive cells had to rely on the self-organisation properties of their components, and on interactions with their environment, in
order to carry out basic functions such as growth and division; they did not have the
capacities of modern-day cells. Many types of vesicle with double-layer membranes
can exhibit very complex morphological changes, for example, growth, fusion, fis-
sion and budding, as well as interactions at the vesicle surfaces and internal vesicle
assembly. These dynamic properties of vesicles suggest models of how replication
processes in primitive cells could have taken place, not just because of chemical
and physical forces. Continuing studies of vesicle properties of systems containing
simple amphiphilic molecules may perhaps provide pointers to the identification of
the identity of actual prebiotic components of the first primitive cells (Hanczyc and
Szostak, 2004).

At the beginning of biogenesis, or during phases of the emergence of processes similar to life, organic compounds which were able to pass through the
membranes of primitive cells would have had a large selective advantage over
competing molecular species which could not do so. This problem was studied
by Sacerdote and Szostak, who concluded that semipermeable lipid double layers demonstrate stereoselectivity, so that ribose can pass preferentially through
membranes. This is true for both fatty acid and phospholipid membranes; ribose
is favoured over other aldopentoses, or hexoses. This kinetic advantage of ribose
could be one reason for the formation of an RNA world (Sacerdote and Szostak,
2005).

One of the preconditions for the beginning of primitive cellular life was the ability of dissolved polar molecules to reach the interior of protocells. On the other hand,
polycondensation of amino acids to give peptides was a precondition for metabolism
within the cell. D. Deamer's group was able to observe a non-enzymatic peptide
synthesis in vesicles: thioglutamic acid was able to pass through the lipid double
layer fast enough to allow polycondensation to give oligopeptides, with the reaction

mixture containing oligomers as high as 11mers. It is suggested that certain lipid surfaces can catalyse this oligomerisation reaction (Zepik et al., 2007).

The great importance of minerals in prebiotic chemical reactions is undisputed. Interactions between mineral surfaces and organic molecules, and their influence on self-organisation processes, have been the subject of much study. New results from Szostak and co-workers show that the formation of vesicles is not limited to one type of mineral, but can involve various types of surfaces. Different minerals were studied in order to find out how particle size, particle shape, composition and charge can influence vesicle formation. Thus, for example, montmorillonite (Na and K10), kaolinite, talc, aluminium silicates, quartz, perlite, pyrite, hydrotalcite and Teflon particles were studied. Vesicle formation was catalysed best by aluminium solicate, followed by hydrotalcite, kaolinite and talcum (Hanczyc et al., 2007).

D. W. Deamer and J. P. Dworkin have reported in detail on the contribution of chemistry and physics to the formation of the first primitive membranes during the emergence of precursors to life; the authors' discussion ranges from sources of amphiphilic compounds, growth processes in protocells, self-organisation mechanisms in mixtures of prebiotic organic compounds (e.g., from extracts of the Murchison meteorite) all the way to model systems for primitive cells (Deamer and Dworkin, 2005).

The subject of cells and cell-like structures is now a typical interdisciplinary area of research, which includes minerals, and thus rocks. It had previously been pointed out, for example, by H. Kuhn (see Sect. 8.4), that rock pores could have served in prebiotic reactions as primitive minicells for concentrating biomolecules. In the *Proceedings of the National Academy of Sciences*, Eugene V. Koonin begins his comments on an article dealing with the accumulation of nucleotides in the pores of a simulated hydrothermal system with the attractive title "An RNA-making Reactor for the Origin of Life". In their investigations, the authors had found that temperature gradients in hydrothermal vents converted the network of rock pores into an effective system for concentrating materials. Two hydrodynamic processes working concurrently, thermal convection and thermodiffusion at a temperature gradient, led to an enrichment of mononucleotides close to the bottom of a blocked pore by a factor of more than 1,000. Linking of the pores to form a network, and/or an increase in molecular size up to RNA or DNA containing 100 bases, led to a dramatic increase in the molar concentration. In nature, hydrothermal systems are surrounded by porous mineral deposits; the authors found that single nucleotides could undergo enrichment in such systems by a factor of more than 10^8 at the bottom of a blocked pore network (Baaske et al., 2007; Koonin, 2007).

10.3 The Tree of Life

The broad survey of the animal and plant world provided by Aristotle about 2,400 years ago included only a tiny fraction of all extant life forms. However, his compilation formed the basis of our knowledge of living things until the Middle Ages:

only 2,000 years later did Carl von Linné (1707–1778) try to develop a system of classification for all the plants and animals then known. The credits gained by Linné for his "system of nature" are not diminished by the modern conclusion that the great man was mistaken when he considered all types of animals and plants to be changeless, i.e., that they look today exactly as they did when they first appeared on Earth. Fossilized life forms were considered in Linné's time to be some of nature's little games, which should not be taken seriously.

The rethinking process set in with Charles Darwin: according to him, all life forms originate from only a few antecedents, and these in turn from an even smaller number of forebears. Thus, it is possible to construct a type of family tree for all living things, showing how they are related.

Ernst Haeckl (1834–1919), professor in Jena, also developed such a family tree (plantae, protista and animalia); these were all derived from a single common root, the Radix Moneres. For many years, there were fierce arguments on the structure of such family trees, until finally, most biologists agreed that all life must have had its origin in a single cell, the "last universal common ancestor". What possibilities did scientists have to try to verify their hypotheses and theories until about 40 years ago? All they had for coming to conclusions on the phylogenic relationships between organisms, and kingdoms of organisms, was comparisons of anatomical and physiological character. Although these methods were successful for the recognition of relationships in higher life forms, they broke down for protozoa. But since microorganisms have inhabited our planet for much longer than multicell life forms, scientists were in a blind alley.

A way out of this situation was found by the genius Linus Pauling, who (with Emile Zuckerkandl) suggested in the middle of the 1960s that comparisons should be based not on external characteristics of life forms but on molecular criteria—the molecular structures of proteins or genes, i.e., the sequence of monomers in these macromolecules, the "molecular palaeontological record". At that time, only protein analysis, i.e., the determination of the sequence of amino acids in a protein, was possible. Thus, hemoglobins and cytochromes were mainly used for clearing up phylogenetic relationships.

Carl Woese, microbiologist and geneticist from the University of Illinois, found that this "protein clock" was not reliable enough. He wanted to use RNA as the material for comparison, not gene products. The chosen source of RNA was a type of cell organelle which is of vital importance in protein biosynthesis: the ribosome. To be more exact, he chose only a certain region of it—the RNA component of the smaller of the two ribosome halves, the so-called 16S rRNA (in bacteria). This polynucleotide consists of 1,540 nucleotides. The determination of the nucleotide sequences was not possible with the same degree of accuracy as today, so Woese decided not to compare the exact base sequences of the 16S rRNA of different organisms; instead, he used a shortened procedure to achieve his goal. He determined the degree of difference between fractions of the 16S rRNA formed in cleavage reactions with RNase T1.

Woese was able to confirm the previously existing view of the world in which life forms are divided into two kingdoms:

The *eukaryotes*: these include animals, plants, fungi and protozoa, the DNA of which is enclosed in a membrane-enclosed organelle (the cell nucleus). They have a cytoskeleton (a fine membrane-like network in the interior of the cell, which provides stability) and contain mitochondria. Higher plants, as well as algae, are equipped with chloroplasts for photosynthesis.

The *prokaryotes* are simple cells without a nucleus. Until about 25 years ago, this term was used for all bacteria.

However, Carl Woese proposed the following surprising hypothesis: apart from these two known kingdoms of life, there must be a third. The organisms in this third group resemble bacteria externally but have other completely different and unusual properties. They are "eccentrics", as they live in environments which, prior to their discovery, were considered to be hostile to life: in geysers, black smokers, in water with a high salt content, and at unusual pH values. The members of this third kingdom are called archaebacteria (or archaea), and they differ greatly from normal bacteria. There is a large group known as methanogens, which can exist only in the absence of oxygen; they produce methane by reduction of CO_2 and other carbon-containing compounds. Other groups are the halophiles and thermophiles.

Many biologists were initially very critical and sceptical about Woese's ideas, but this changed when he provided more evidence to support his hypothesis. Thus, the cell membranes of the archaea are made up of very unusual lipids. While the lipids found in eubacteria and eukaryotes contain straight-chain fatty acids, those from archaebacteria consist of a glycerine derivative in which phytanol residues replace the fatty acids (phytanols are long-chain fat alcohols with a methyl group at every fourth carbon atom). These are joined to the glycerine component via ether linkages.

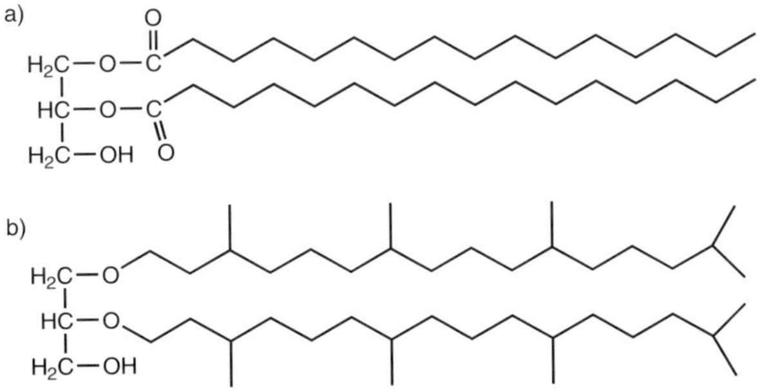

Fig. 10.10 Differences in the membrane lipids between eukaryotes (including eubacteria) and the archaea. The main components of the membrane in the eukaryotes are fatty acid glycerine esters, while in archaea, the membranes contain mainly di- (or other) ethers of glycerine with phytanol residues

A second example of the differences between important biomolecules in archaebacteria and eubacteria is their DNA-dependent RNA polymerase. The enzyme found in archaea resembles that in eukaryotes more than it does those in bacteria;

this is true both for the molecular construction and for reactions involved in the translation process. In contrast to the polymerases of eubacteria, those of the archaea are insensitive to the antibiotics rifampicin and streptolydigin (Kandler, 1981). Detailed information on archaebacteria can be found in the book by M. Groß (2001).

Woese chose the name archaebacteria because these microorganisms grow best under conditions which were probably found on the primeval Earth between 3.5 and 4 billion years ago: hot boiling water and thermal vents, highly acidic environment, oxygen-free atmosphere and high salt concentrations.

After the acceptance of the third kingdom by the experts, the question arose as to whether or not eukaryotes had developed from archaebacteria. Information on certain similarities in transcription and translation mechanisms suggested that the two groups were related in some way. The situation was clarified by Peter Gogarten et al. (1989) and a Japanese group, who studied the amino acid sequence of two important proteins, one being a V-type proton-pump ATPase. Both groups concluded that archaea and bacteria must descend from the last universal common ancestor, with separation of the two occurring later. This generally agreed picture of the family tree of life can be summed up as follows:

> Two original kingdoms first developed from the last universal common ancestor: bacteria and archaea.
> The eukaryotes developed from archaea-like precursors.
> According to the endosymbiotic hypothesis, the eukaryotes used genes from both bacteria (alpha-protobacteria) and cyanobacteria. The first led to the development of mitochondria, the second to that of chloroplasts, i.e., cell organelles which are highly important for energy production (ATP synthesis) and photosynthesis.

The number of organisms whose genomes have been completely determined has increased considerably in recent years. Information on more than half a dozen archaea, and more than two dozen bacteria, was available at the turn of the century (Doolittle, 2000); since then, this number has increased considerably. An exact analysis of the sequence data shows that there are discrepancies in the generally agreed family tree of life. There are a number of bacterial genes in archaea; for example, *Archaeoglobus fulgidus* has a bacterial form of the enzyme HMG (hydroxymethylglutaryl)-CoA reductase, an important enzyme for the synthesis of membrane lipids. This microorganism, which lives in hydrothermal systems, also uses other bacterial genes in order to survive.

These facts do not agree with the assumption that organisms evolved in straight lines. Since the Darwinian era, it has been usual to use the term "genetic transfer" to refer to the transmission of information *within* a species. However, for more than 30 years, phenomena have been observed which show that gene transfer *between* species of life forms is also possible.

The basis of the traditional tree of life was the "vertical" passing on of genetic information, i.e., from generation to generation. The discrepancies visible in the previous model of the phylogenetic tree of life can only be explained if a second process was involved, the so-called "horizontal gene transfer" (Maier et al., 1996): by this, we mean the transfer of genetic material from one type of organism to another.

Horizontal gene transfer (HGT) could also explain some unsolved riddles, such as the fact that genes which normally occur in bacteria are present in some archaea. HGT has been recognised for some time, but no one realised that the phenomenon was an important principle of the development of living systems, even at the dawn of life.

According to Ford Doolittle from Dalhousie University in Halifax (Doolittle, 2000), the transition from archaea to eukaryotes had been described in too simple a manner, or even wrongly interpreted. He believed that the original eukaryotes evolved from a type of precursor cell, which was itself the product of a complex series of horizontal gene transfer processes. A modified phylogenetic genealogical tree also includes three kingdoms, but the relationships between them now become more complex. It appears that there were many interconnections between the developing species, particularly in the first few million years after the emergence of life. The question arises as to whether the exceptional position of the 16S rRNA sequences previously assumed should remain unchallenged. Many researchers are of the opinion that the genes for 16S rRNA, as well as others responsible for transcription and translation, changed very little, so the original tree of life remains valid. But according to Doolittle, this "non-exchangeability" is an assumption which has hardly been checked. 16S rRNA has an important property: it consists of progressive and conservative elements, so that its analysis (sequence determination) provided information on very early, and also much later, events in phylogenetic development (Kandler, 1981).

An initial suggestion made by Ford Doolittle shows a jumble of interconnections between the lines of development, rather than simple branches in the phylogenetic tree. These interconnections resemble a mycelium and have almost nothing in common with the original model, except for the termini of the three kingdoms. In a review article in *Science*, Elizabeth Pennisi (2001) chose the colourful metaphor of a tangled bramble bush to describe the new model.

Carl Woese also regards horizontal gene transfer as one of the main driving forces for the evolution of cellular organisation. In 2002, he proposed a model based on the conjecture (supported by data) that the dynamic of horizontal gene transfer is primarily determined by the organization of the recipient cell (Woese, 2002). HGT thus becomes one of the most important keys for understanding cell evolution. Woese had already pointed out the central importance of HGT for the universal phylogenetic tree of life (Woese, 2000).

As already mentioned, the HGT principle has been known for many years; however, until about 15 years ago, it was regarded as only a weak force in the evolutionary events involved in the development of primeval cells, whose organisation must have been very simple. They probably had a "loose" structure which could easily be modified by HGT processes. An important characteristic of Woese's model is the nature of the evolutionary process: Woese assumes that cell design could be achieved only by the overall performance of HGT. It was the community as a whole, the ecosystem, which evolved. Individual cell designs were fundamentally distinct, because in each case, the initial conditions were somewhat different.

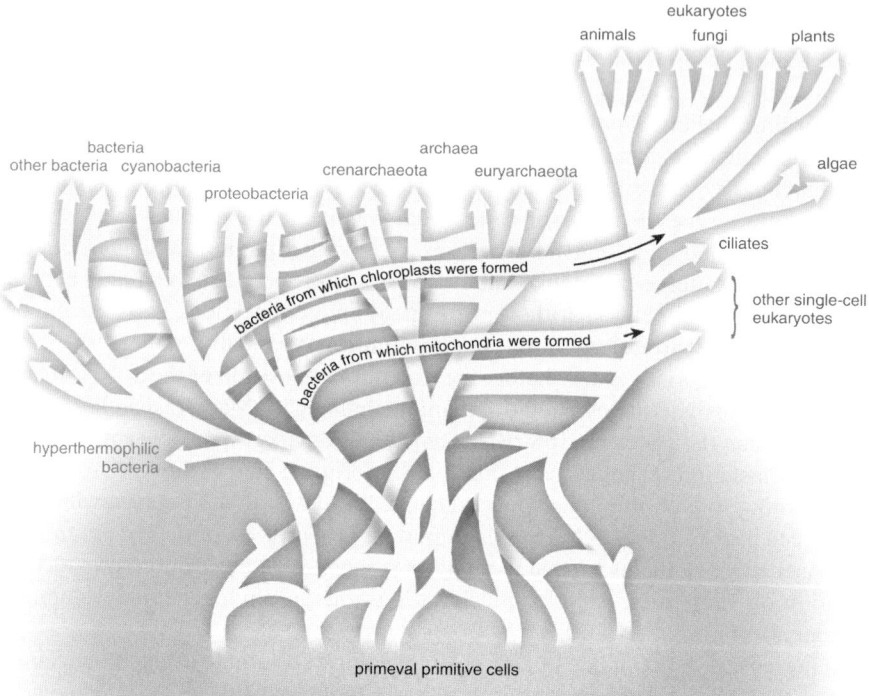

Fig. 10.11 The "modified tree of life" still has the usual tree-like structure and also confirms that the eukaryotes originally took over mitochondria and chloroplasts from bacteria. It does, however, also show a network of links between the branches. The many interconnections indicate a frequent transfer of genes between unicellular organisms. The modified tree of life is not derived, as had previously been assumed, from a single cell (the hypothetical "primeval cell"). Instead, the three main kingdoms are more likely to have developed from a community of primitive cells with different genomes (Doolittle, 2000)

This model appears to do away with some previous ambiguities; however, when looked at in a less optimistic manner, it in fact poses a lot of new questions!

A study published around ten years ago caused animated discussions among the experts: Russell F. Doolittle (1996) from the Institute of Molecular Genetics in San Diego investigated the amino acid sequences of 57 different enzymes in the 15 main phylogenetic groups. A total of 531 sequences were used to examine the tree of life via protein sequences, just as Woese did with ribosome RNA. Doolittle found more or less the same types of branching, as did Woese: however, to the astonishment of the San Diego group and the experts worldwide, the date of divergence between prokaryotes and eukaryotes was not around 3.5×10^9 years ago, but about 1.8 billion years later! Thus, there is a huge discrepancy between the findings (considered as verified) on the oldest fossils, the cyanobacteria-like filaments in the Apex chert in western Australia (see Sect. 10.1) and the timescale for the formation of the last common universal ancestor determined by Doolittle and co-workers using the "molecular clock". Wills and Bada (2000) describe the problem as the "Doolittle event", since it could suggest that the Earth was subjected to a cataclysmic catastrophe, which destroyed all existent life forms. It is completely unclear (and very questionable) whether the process of the evolution of life could have recommenced after this hypothetical "event", or have been regenerated, starting from some organisms which *had* survived.

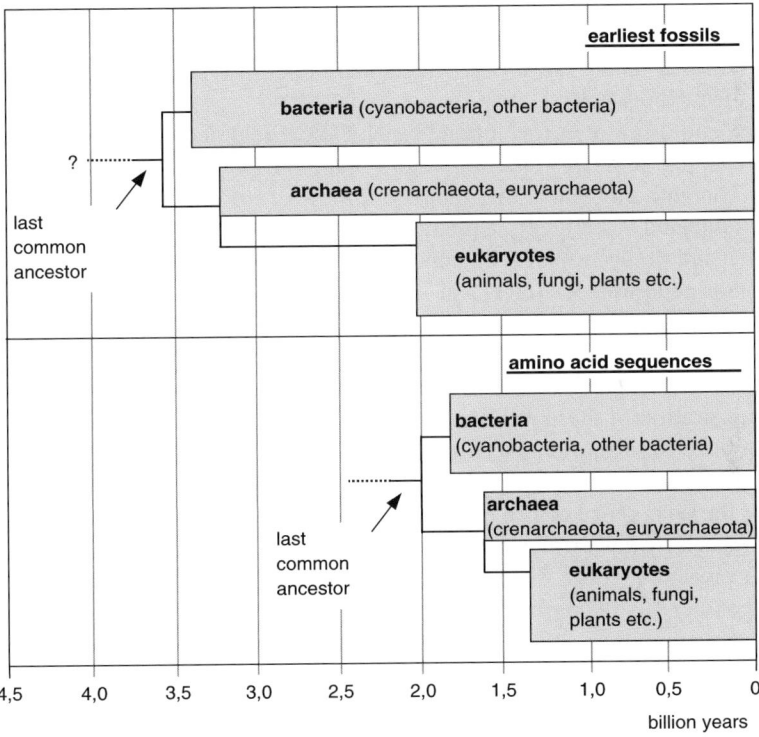

Fig. 10.12 Two quite different methods of studying the history of life on our planet, and thus of determining the starting point of the tree of life:

(a) Using the earliest fossils, with the help of Precambrian, palaeobiological studies
(b) Using the amino acid sequences of a large number of proteins

The difference between the two timescales is about 1.8 billion years!

There are a number of other possibilities for the explanation of the "Doolittle event" (Mooers and Redfield, 1996). One of these could be due to the analytical method used by Doolittle, as he assumed a relatively constant rate of amino acid substitution. This assumption may not be justified and should be checked.

It is not only the vast difference of about 1.8 billion years which is so surprising: the timescale for the separation point of the main branches of the tree of life is clearly shifted. The catastrophe hypothesis put forward to explain this difference appears unlikely, since there are no signs of such a phenomenal "obliteration of all life" on Earth. Another explanation could be that the data from the amino acid sequences provide only information on the way in which life forms diverged, but not on the timescale (Schopf, 1998). This interpretation of the "Doolittle event" by Schopf was provided at a time when doubts had not yet been cast on the dating of the first fossils at 3.45 billion years, published by him in 1993.

If, in the near future, results were to appear which show that the first traces of life are not in fact 3.45 billion years old, the Doolittle event would return to the discussion and become more probable than it is at present. Thus, the tree of life is still decorated with many questions, which will hopefully decrease, and finally disappear, in the next few years.

The article "Biology's next revolution" by N. Goldenfeld and C. Woese, which is well worth reading, indicates that our picture of the biosphere is likely to change dramatically. The authors believe that the huge inflow of information to which we are presently subjected is leading to a picture of microbes as gene-swapping collectives which demands a revision of such concepts as organism, species and evolution itself. Horizontal gene transfer (HGT) in particular allows us to explain phenomena which could not previously be understood. The molecular reductionism of the twentieth century must now be replaced by a modern, interdisciplinary approach that embraces collective phenomena.

The interpenetration of these phenomena, mediated by HGT, is also likely to be of great importance for evolution. According to the authors, the genetic code, which plays a key role in all forms of life, leads to the prediction that the first steps of the process of the emergence of life evolved in a Lamarckian manner, with vertical descent driven back by powerful early forms of HGT.

Biology is presently going through an important phase, as the importance of collective phenomena is becoming very clear (as is the case in other areas). The above-mentioned important and highly complex questions will perhaps only be solved using statistical mechanics and/or dynamic system theory (N. Goldenfeld and C. Woese, 2007).

References

Baaske P, Weinert FM, Duhr S, Lemke KH, Ussel MJ, Braun D (2007) Proc Natl Acad Sci USA 104:9346
Barghoorn ES, Schopf JW (1966) Science 152:758
Brasier MD, Green OR, Jephcoat AP, Kleppe AK, VanKranendonk MJ, Lindsay JF, Steele A, Grassineau NV (2002a) Nature 416:76
Brasier MD, Green OR, Jephcoat AP, Kleppe AK, VanKranendonk MJ, Lindsay JF, Steele A, Grassineau NV (2002b) Book of Program and Abstracts, ISSOL-02, Oaxaca, p 51
Brasier MD, Green OR, Lindsay JF, Steele A (2004) Orig Life Evol Biosphere 34:257

Buick R, Thornett JR, McNaughton NJ, Smith JB, Barley ME, Savage M (1995) Nature 375:574

Bungenberg de Jong H, Decker WA, Swan OS (1930) Biochem Z 221:392

Deamer DW (1998) Membrane Compartments in Prebiotic Evolution. In: Brack A (Ed.) The Molecular Origins of Life. Cambridge University Press, p 189

Deamer DW, Dworkin JP (2005) Chemistry and Physics of Primitive Membranes. In: Walde P (Ed.) Prebiotic Chemistry, Topics in Current Chemistry 259:1, Springer Berlin Heidelberg New York

de Duve C (1991) Blueprint for a Cell: The Nature and Origin of Life. Patterson Burlington NC

Doolittle RF, Feng D-F, Tsang S, Cho G, Little E (1996) Science 291:470

Doolittle WF (2000) Scientific American, February

Eiler JM (2007) Science 317:1046

Fox SW (1980) Nature 67:378

García-Ruiz JM (2002) Astrobiology 2:353

García-Ruiz JM, Hyde ST, Carnerup AN, Christy AG, Van Karnendonk MJ, Welham NJ (2003) Science 302:1194

Gogarten JP, Kibak H, Dittrich P, Taiz L, Bowman EJ, Manolson MF, Poole RJ, Date T, Oshima T, Konishi J, Denda K, Yoshida M (1989) Proc Natl Acad Sci USA 86:6661

Goldenfeld N, Woese C (2007) Nature 445:369

Groß M (2001) Life on the Edge. Amazing Creatures Thriving in Extreme Environments Perseus Books

Groves DI, Dunlopp JSR, Buick R (1981) Scientific American, October p 56

Hanczyc MM, Fujikawa SM, Szostak JW (2003) Science 302:618

Hanczyc MM, Szostak YW (2004) Curr Opin Chem Biol 8:660

Hanczyc MM, Mansy SS, Szostak YW (2007) Orig Life Evol Biosphere 37:67

Holland HD (1997) Science 275:38

Kandler O (1981) Naturwissenschaften 68:183

Kerr RA (2003) Science 302:1134

Kudryavtsev AB, Schopf JW, Agresti DG, Wdowiak TJ (2001) Proc Natl Acad Sci USA 98:823

Lehninger AL (1975) Biochemistry. Worht Publ Inc, New York

Luisi PL (1996) Orig Life Evol Biosphere 26:272

Luisi PL (1998) Orig Life Evol Biosphere 28:613

Luisi PL, Ferri F, Stano P (2006a) Naturwissenschaften 93:1

Luisi PL (2006b) Orig Life Evol Biosphere 36:605

Maier U-G, Hofmann CJB, Sitte P (1996) Naturwissenschaften 83:103

McKeegan KD, Kudryavtsev AB, Schopf JW (2007) Geology 35:591

Mojzsis SJ, Arrhenius G, McKeegan KD, Harrison TM, Nutman AP, Friend GRL (1996) Nature 384:55

Mooers AD, Redfield RD (1996) Nature 379:587

Morowitz HJ, Heinz B, Deamer DW (1988) Orig Life Evol Biosphere 18:281

Monnard PA, Deamer DW (2001) Orig Life Evol Biosphere 31:147

Nagy B (1976) Naturwissenschaften 63:499

Nakashima T (1987) Topics in Current Chemistry 139:57

Pasteris D, Wopenka B (2002) Nature 420:476

Pennisi E (2001) Science 293:197

Pflug HD (1978) Naturwissenschaften 65:611

Pohorille A, Wilson MA (1995) Orig Life Evol Biosphere 25:21

Rosing MT (1999) Science 283:674

Rushdi AL, Simoneit BRT (2001) Orig Life Evol Biosphere 31:103

Sacerdote MG, Szostak YW (2005) Proc Natl Acad Sci, USA, 102:6004

Schidlowski M (1988) Nature 333:313

Schopf JW (1993) Science 260:640

Schopf JW (1998) Tracing the Roots of the Universal Tree of Life. In: Brack A (Ed) The Molecular Origins of Life. Cambridge University Press, pp 336-362

Schopf JW (1999) Cradle of Life. Princeton University Press, Princeton

Schopf JW, Kudryavtsev AB, Agresti DG, Wdowiak TJ, Czaja AD (2002a) Nature 416:73

Schopf JW, Czaja AD, Kudryavtsev AB, Agressti DG, Wdowiak TJ, Kempe A, AltermanW, Heckl WM (2002b), Book of Program and Abstracts ISSOL-02, Oaxaca, p 51

Schopf JW, Kudryavtsev AB, Agresti DG, Czaja AD, Wdowiak TJ (2005) Astrobiology 5:333

Segré D, Ben-Eli D, Deamer DW, Lancet D (2001) Orig Life Evol Biosphere 31:119

Shew RL, Deamer DW (1985) Biochim Biophys Acta 816:1

Szostak JW, Bartel DP, Luisi PL (2001) Nature 409:387

van Zuilen MA, Lepland A, Arrhenius G (2002a) Book of Program and Abstracts ISSOL-02, Oaxaca, p 46

van Zuilen MA, Lepland A, Arrhenius G (2002b) Nature 418:627

Walde P, Goto A, Monnard PA, Wessiken M, Luisi PL (1994) J Amer Chem Soc 116:7541

Walters C, Shimoyama A, Ponnamperuma C (1981) Origin of Life 11:473

Wick R, Walde P, Luisi PL (1995) J Amer Chem Soc 117:1435

Wills C, Bada J (2000) The Spark of Life. Perseus Publishers, Cambridge, Mass

Woese CR (1981) Scientific American, June p 94

Woese CR (2000) Proc Natl Acad Sci USA 97:8392

Woese CR (2002) Proc Natl Acad Sci USA 99:874

Zepik HH, Rajamani S, Laurel M-C, Deamer DW (2007) Orig Life Evol Biosphere DOI 10.1007/s11084-007-9070-9

Chapter 11
Exo/Astrobiology and Other Related Subjects

Although the terms exobiology and astrobiology really mean the same thing, "astrobiology", introduced by NASA in 1995, has become the one of choice. This branch of science reaches from cosmochemistry via biogenesis to all the other themes involving research on traces of life (of whatever sort): on planets and on moons, both within and outside our solar system.

In 1996, NASA set up the Astrobiology Institute at the Ames Center in Mountain View (California). Several well-known institutes joined up to form this centre, including the Carnegie Institution, Washington, DC, (evolution of hydrothermal systems), Harvard University (terrestrial geochemistry and palaeontology), Scripps Institute, University of California at San Diego (prebiotic chemistry, self-replicating systems) and the University of Colorado at Boulder (RNA catalysis, habitability of planets). The 2000 budget was $20 million, which rose to $65 million by 2005; huge budget cuts now seem likely. There are special doctoral and postdoctoral programs.

Europe has a comparable institution, the European Exo/Astrobiology Network Association (EANA), which brings together national research activities, so that the coordination with other organisations (particularly in the USA) is more effective. The organisation has various specific objectives, including the following:

Bringing together active European researchers to link up their research programs

Funding exchange visits between laboratories and sharing facilities for research

Attracting young scientists to participate practically in this evolving interdisciplinary field

Promoting this important field of research to European funding agencies and politicians

Interfacing the Research Network with European bodies (e.g., ESA and the European Commission)

EANA was founded in May 2001, during the First European Workshop on Exo/Astrobiology in Frascati, Italy, which was attended by 200 scientists (ESA, 2001).

Astrobiological research is, however, not only concerned with the search for life on other heavenly bodies; it also includes work on life forms in extreme habitats, in hydrothermal systems and, for example, in deep rock formations (Pedersen, 1997) (see Sect. 7.2.1).

H. Rauchfuss, *Chemical Evolution and the Origin of Life*,
© Springer-Verlag Berlin Heidelberg 2008

11.1 Extraterrestrial Life

Life which exists, or could exist, outside our Earth is generally known as extraterrestrial life. A distinction is also made between life (or possible life) *within* or *outside* the solar system. In spite of what is claimed in many science fiction books and films, there is no single piece of evidence for a living system outside Earth. The coming years and decades will hopefully provide clarity on the question as to whether we are really alone in the universe or not.

There are three objects within the solar system which are the subject of research on possible extraterrestrial life, traces of life, biomolecules or their precursors:

> The planet Mars
> The Jovian moon Europa
> The Saturnian moon Titan

11.1.1 Life in Our Solar System

11.1.1.1 Extraterrestrial Life on Mars?

As far as life on other planets is concerned, Mars occupies a special place:

> Mars is a direct neighbour of Earth.
> The climate on Mars is not so inimical to life as that on Earth's other direct neighbour, Venus.
> Historical events, such as the discovery of the (hypothetical) canals on Mars, led to strong general interest in Mars in the mid-nineteenth century.

The phase of science fiction stories about little green men from Mars is hopefully over, and the question can only be: are there, or were there ever, real living systems there, or can we find traces of precursors of life in the form of biomolecules?

Free water, the essential precondition for life as we know it, has recently been detected on Mars: images taken by the high resolution stereo camera (HRSC) on board ESA's Mars Express spacecraft show a patch of water ice on the floor of an unnamed crater near the Martian north pole. Geomorphic studies indicate that the surface of Mars can be divided into two types (Jaumann et al., 2002):

> A southern highland hemisphere, several thousand metres above the normal level, covered with craters and similar to the Moon's highlands
> The lowlands in the north, which could correspond to ocean floors on Earth

At the Martian equator there is a system of gorges much larger than those known on Earth, although the diameter of Mars is only about half that of Earth. The huge shield volcano Olympus Mons is 26 km high, and the gorges are probably 8 km deep. Large erosion channels and river beds suggest that there must have been large amounts of liquid water on Mars at some time. In the 1970s, the Mariner 9 spacecraft took photographs of valleys resembling river valleys on Earth. More recent satellite photographs suggest other interpretations, for example, that the valleys could have been formed by the collapse of surface structures which were washed away by flowing water. It appears that large water masses once flowed through the valleys into the northern lowlands, and it is possible that large areas of alluvial soil were formed.

However, the origin of the water on Mars is still unknown. Since the Earth and Mars have some common features in their history, the water on Mars could have come both from its interior and from comets and asteroids. The huge size of the Martian shield volcanoes, one class of which resembles the shield volcanoes Kilauea and Mauna Kea on Hawaii, suggests that a large proportion of the water was of volcanic origin.

But where could the water have gone to? Two possibilities are under discussion:

A fraction of the water is hidden in deeper layers of the planet's interior.
Another fraction was decomposed photolytically, the hydrogen set free, escaping into space.

A total of 33 missions to Mars had been carried out up to 2002; however, only eight US missions were successful.

Three further important projects were initiated in 2003:

On June 2, the ESA space vehicle Mars Express with its lander, Beagle, was launched and landed on Mars on December 23; however, contact with Beagle was not possible.
A few days later, on June 10, the US vehicle Spirit (MER-A) started its journey; on board was a Mars rover which was intended to investigate rock samples and to look for water. The landing in the crater Gusev took place on January 4, 2004.
Finally, the sister spacecraft Opportunity (MER-B) was launched on July 8, 2003, and landed on January 25, 2004; it also had a Mars rover on board.

Two more missions started in 2005 and 2007:

August 12, 2005, saw the launch of the US spacecraft Mars Reconnaissance Orbiter, which entered orbit around Mars on March 10, 2006. This craft has high-resolution cameras on board to permit a more exact mapping of the Martian surface (as a precondition for the search for suitable landing grounds).
Almost exactly two years later, on August 4, 2007, the NASA mission Phoenix started, launched by a Delta II rocket. The landing in the Martian polar area is planned for May 25, 2008. The Phoenix lander will start work in the as yet unexplored "Green Valley" area at around 69° north latitude. The most important goal of this mission is to find out whether water ice is present in polar areas less than one metre below the surface; it will also look for organic molecules. The Phoenix lander will be able to dig down a metre, using its robot arm. The soil samples will then be investigated by the laboratory instruments MECA and TEGA. MECA (Microscopy, Electrochemistry and Conductivity Analyser) will allow four experiments to be carried out in order to characterise the Martian soil.

The geological history of Mars can be divided into three epochs:

The Noachian epoch, between 3.8 and 3.5 billion years ago, characterised by massive bombardment.

The Hesperian epoch, between 3.5 and 1.8 billion years ago, in which the lava flatlands were formed.

The Amazonian epoch, which includes the last 1.8 billion years; in this period, there were only a small number of meteorite impacts, and Mons Olympus was formed.

The extensive layered sediments at the south pole, which contain water ice, will provide information on climatic variations. The subsurface sounding radar instrument SHARAD (Shallow Radar) on board the Mars Reconnaissance Orbiter carried out a detailed cartographic study of the subsurface at the Martian south pole. The data indicate that the sediments there have been subjected to considerable erosion (R. Seu et al., 2007). The density of the material deposited at the Martian south pole was calculated by M. T. Zuber and co-workers; by combining data from the gravitational field with those from various instruments on board the Mars Orbiter, they obtained a value of $1,200\,kg/m^3$. This value corresponds to that calculated for water ice containing about 15% dust (Zuber et al., 2007).

The layers of sediment at the Martian south pole do not consist of pure ice: they are interspersed by layers of dust. The latest data were obtained by the Mars Advanced Radar for Subsurface and Ionospheric Sounding apparatus (MARSIS) on board the Mars Express Orbiter. The radar waves from the instrument pass through the ice layers until they reach the base layer, which can be at a depth of up to 3.7 km. The distribution of the ice at the south pole is asymmetric, and its total volume has been estimated to be $1.6 \times 10^6\,km^3$: this corresponds to an amount of water which would cover the whole planet with a layer 11 metres deep (Plaut et al., 2007).

The north polar region of Mars consists of variously layered sediments, the upper layer consisting of water ice and thus having a high albedo. More detailed pictures ($\sim$ 30 cm per pixel) were obtained from the High Resolution Imaging Science Experiment (HiRISE) on board the Mars Reconnaissance Orbiter (MRO). The sediment layers in this region are only about 10 cm thick but seem to be covered with a layer of dust. A detailed analysis of the HiRISE pictures of the north pole deposits indicate that complex, multi-step processes must have occurred at the polar icecaps (Herkenhoff et al., 2007).

As described previously, it is absolutely certain that water ice is present at both poles; it is, however, completely uncertain whether water is present in other regions, and the historic processes occurring there are also presently unknown.

The Mars exploration rover Opportunity found sandy sediments in the plain called Meridiani Planum; these consist of sulphates and hematite globules. The sulphate sediments were probably not deposited in surface water, but have their origin in sulphate-containing ground water, which rose to the surface and evaporated there: this process must have taken place at the end of the Noachium. The OMEGA (Observatoire pour la Minéralogie, l'Eau, les Glaces et l'Activité) data indicate that the combination of iron with sulphates can also be found in other regions of the Martian surface (Bibring et al., 2007).

Using the HiRISE on board the MRO with a picture resolution of up to 25–32 cm per pixel, a large research group consisting of NASA employees, US universities and two Swiss institutes has carried out a more detailed study of some key regions of Mars: thus, stone blocks up to 2 m in size, which are found everywhere in the centre of Mars up to high latitudes, could be clearly identified. In the case of some objects (bright deposits), it can no longer be confirmed that these are relicts formed by flowing water. The planet Mars seems to have been wet for a shorter time than had previously been assumed (McEven et al., 2007).

The most recent research results show that it is difficult to detect formerly liquid water on Mars. Pictures of what are assumed to be seafloors and riverbeds are no longer considered to be certain evidence that flowing water was always present in all the regions of Mars. Critical examination of some landscapes indicates that they were carved by lava flows and not by water. Only in the case of the edges of craters, and some river gullies, does it appear quite clear that they were formed by liquid water. However, at the time of writing, the Mars Reconnaissance Orbiter mission is only a few weeks old, so further important results (and perhaps surprises) on the surface and mineralogy of Mars can be expected (Baker, 2007).

In spite of many new, and in some cases sensational, results concerning the Red Planet, we are still no nearer to answering the question of life on Mars. Four alternatives appear possible for Mars:

Living systems will be found.
Formerly living systems will be discovered.
Biomolecules or their precursors will be detected.
None of these three will occur.

Could it be the case that microorganisms, like the suspected fossils in the Mars meteorite ALH 84001, exist in the Martian soil? This question leads to the counterquestion as to whether it has previously been possible to detect and study life (primitive life forms) under highly extreme conditions. Are there such conditions on Earth? We now know quite a lot about extremophiles such as the thermophilic, halophilic and hyperthermophilic microorganisms.

Only recently has it been possible to demonstrate clearly that microorganisms are present in plutonic rock. Karsten Pedersen, a microbiologist from Göteborg University, found microorganisms (anaerobic sulphur bacteria) in water samples taken at a depth of 500 m, and later (during drilling done in Gravberg in Dalarna), at a depth of 3,500 m; these bacteria use only the water present in hairline cracks in the rock, and substances dissolved in that water, for their metabolism (Pedersen, 1997). Whether biogenesis is possible under such conditions, as discussed by Pedersen, is naturally only pure speculation. It appears feasible that small molecules such as CO_2, CO, CH_4, H_2, H_2S, NH_3 and water vapour could have permeated through the rock. We can also not exclude the possibility that some layers of rock could be catalytically active, and that different temperature zones can occur. The question immediately arises as to whether biomolecules might be formed under such extreme conditions.

Stevens and McKinley (1995) found similarly undemanding microorganisms in the groundwater from deep rock layers (basalts) north of the Columbia river in the state of Washington (USA). The water samples collected at a depth of 1,500 m were shown to contain hydrogen as well as autotrophic microorganisms: methanogens (methane-forming bacteria), which occur in such high concentrations that they could hardly survive on the small amounts of carbon-containing material in the water. So where does the hydrogen, which could reduce the CO_2 dissolved in the water to organic biomass, come from? The solution to this puzzle came from an explosion! The latter was initiated by a welding flame in a heap of basalt rock. The iron (II) contained in the basalt reduces water to free hydrogen; the bacteria use the hydrogen to reduce CO_2, forming biomass and methane. The iron (III) produced in this process is converted to relatively insoluble substances such as magnetite $[Fe_2O_3][FeO]$, and the reaction equilibrium is shifted towards the formation of H_2. This hypothesis was confirmed experimentally: a reaction mixture containing only pure, oxygen-free water and milled basalt rock supports microbe growth. It thus seems certain that certain microorganisms (chemolithotrophs) can survive under extreme conditions (only water and rock). It does, however, appear too rash to take this as the foundation of a biogenesis process (Kaiser, 1995; Groß, 1997). Such new information on life forms existing in environments which are "hostile to life" does, however, open the door to further speculation on the phenomenon of life and its origin (see also Fredrickson and Onstott, 1996). The search for life on Mars will be much more difficult and drawn out than has often been suggested.

During the Viking mission in 1976, it was not possible to go deeper than 6–8 cm when collecting samples from the Martian surface. The limits of this mission have now been demonstrated by Navarro-González and co-workers: the TV-GC-MS method (TV = thermal volatilization) used in the Viking mission was unable to detect traces of organic compounds, as shown by experiments involving Mars-like soils from the Antarctic Dry Valleys, the Atacama desert and the Libyan desert. The oxidation of organic material to CO_2 is greatly retarded by iron oxides and/or their salts in iron-containing soils (such as jarosite from Rio Tinto) and simulated Mars soils (containing palagonite). The CO_2 values found in the Viking mission (50–700 ppm) indicate the presence of organic material, but the real carbon concentration my have been much higher (Wu, 2007; Navarro-González et al., 2006).

The Phoenix landing planned for 2008 should allow the collection of soil samples up to a depth of a metre. Even if no unequivocally positive results are obtained, i.e., if no biomolecules (or even living systems) are detected, a clear demonstration of the presence of carbon compounds would still be a success. A negative result, however, would tell us nothing. It must be remembered that, in spite of many decades of intensive searching (under conditions which are much more favourable than on faraway Mars), it has only been possible in recent years to find the many types of microorganisms which can live under extreme conditions on Earth. Thus, a recent report deals with thermophilic, anaerobic microorganisms which were found in Virginia 2,700 m under the Earth's surface (Kounaves, 2007).

This makes it clear that a vast amount of work must still be done on Mars in order to obtain a final answer as to whether life exists, or ever existed, on the Red Planet.

11.1.1.2 Traces of Life on Europa?

The most important information available on the Jovian moon Europa has been presented in Sect. 3.1.5. Europa is an important target for many studies involving satellite missions, but also an object of much speculation: it has become of great interest for scientists since the discovery (or assumption) that there is an ocean with liquid water under its ice shield. In comparison with Mars, however, the probability of finding life or its precursors on Europa is much lower. There is, however, still the possibility that biomolecules were synthesised there, even under Europa's drastic climatic conditions. Hydrogen peroxide, sulphuric acid and carbon dioxide have been detected on the moon's surface, so that the conditions can be considered as oxidising. This could be a problem for the synthesis of biomolecules, although low temperatures provide a stabilising factor. Brown-coloured regions of the surface could indicate the presence of salts (perhaps $MgSO_4$) (Pappalardo et al., 1999). However, the presence of salts does not mean that carbon chemistry occurs, so any discussions regarding signs of life on Europa are purely speculative.

Earth and the moon Europa do have one thing in common: they are the only bodies in the solar system to have free oxygen in their atmospheres, although the concentration on Europa is much lower. However, Europa's oxygen is probably abiotic in origin, probably coming from photolytic processes, i.e., decomposition of water ice due to solar UV irradiation and the influence of charged particles on the moon's surface. In this process, the liberated hydrogen diffused into space while the oxygen remained (Lammer et al., 2002a; Sandstrand, 2002). The presence of several substances which are "hostile to life", such as H_2O_2, O_2, H_2SO_4 etc., greatly reduces the probability of finding life or precursors of it on Europa.

The change of emphasis to other projects has unfortunately caused NASA to postpone all plans for an extensive study of the moon Europa to the next decade.

11.1.1.3 Organic Synthesis on Titan?

Of the three extraterrestrial targets in our solar system, the Saturnian moon Titan is the least likely to provide signs of life. To quote Christopher McKay from the NASA Ames Research Center, "Titan is an interesting world. For example, its organic haze layer could be an example of the prebiotic chemistry which led to life on Earth". Direct links to extraterrestrial life have not, however, yet been found, as water (one of the main preconditions for life) has not been detected on Titan, apart from traces of water vapour in the higher layers of the Titanian atmosphere (Brack, 2002).

More than 10 years ago, Smith et al. (1994) used the Hubble space telescope to show the presence of bright surface structures on Titan. Astronomers from several institutes suggested that water might well be present on the Titanian surface, basing their arguments on spectra of the light reflected from the surface. These spectra use wavelengths in a small "window" which makes it possible to look through the haze layers in the atmosphere. Thus, it was suggested that the surface of Titan could contain areas of water ice and others consisting of hydrocarbons (Griffith et al., 2003).

The start of the Cassini-Huygens double mission in 1997, already mentioned in Sect. 3.1.6, came to a successful conclusion with the landing of the Huygens probe on Titan (January 4, 2005) after a journey of around three billion kilometres! This technical and scientific accomplishment is distinguished by several important features. The huge distance from Titan to Earth (about 10 AU) means that radio signals from Earth to the Saturnian system take about 80 minutes. The spacecraft reached its goal, Titan, only 14 minutes late after a flight lasting seven years (we should not compare this with commercial aircraft delays on Earth!). Measurements on its physical and chemical profile were made during the 21/2 hour parachute-assisted descent of Huygens through the Titanian atmosphere. The data were transmitted to the Cassini orbiter, which passed them on to Earth. Wind velocities of up to 450 km/h were measured at a height of 120 km. Instruments from several European countries for the analysis of the atmospheric layers, the aerosol layers, cloud structures and the surface of Titan were on board the Huygens probe. The surface phase of the mission lasted for 70 minutes—much longer than expected; at the landing point of the probe, the surface appeared to have the consistency of moist, loose sand.

One of the six scientific instruments aboard this spectacular mission was HASI (Huygens Atmosphere Structure Instrument), which determined and characterised the physical properties of the Titanian atmosphere during the long (170 km) glide down to the surface. One section of HASI investigated the electrical characteristics of Titan's atmosphere and its surface, including the registration of lightning flashes. Acoustic measurements were also carried out in order to try to detect distant electrical discharges (Jerney et al., 2002; Fisher et al., 2002).

Successful attempts to reveal the secrets of the heterogeneous Titanian surface made use of radar astronomy. Donald B. Campbell and co-workers (in cooperation with other institutes) used the 300-m Arecibo radar telescope to send hundreds of kilowatts of microwave radiation towards Titan. The radar echoes recorded after about 2 h indicated the presence of liquid hydrocarbons (Campbell et al., 2003; Lorenz, 2003).

Now, after many years of uncertainty, we have more exact information on the formation of the orange aerosol layers which are found in the lower layers of the Titanian atmosphere. These hydrocarbon nitrile aerosols are known as "tholins", a name suggested by Carl Sagan, which in Greek means "muddy". Although the solar flux on Titan is only about 1% of that incident on Earth, the Titanian atmosphere (about 95% N_2 and 5% CH_4) is also subject to bombardment by particles from the Saturnian magnetosphere. This leads to a chemistry which is unique in our solar system: tholin formation begins about 1,000 km above Titan's surface: not only dissociation processes, leading to products such as C_2H_2, C_2H_4, C_2H_8 or HCN, occur in this region (ionosphere/thermosphere), but ions such as $C_2H_5^+$, $HCNH^+$ or $C_4H_3^+$ are generated by the influence of high-energy particles.

The dissociation and ionisation products react with one another in lower layers of the atmosphere to give more complex compounds such as benzene and similar organic molecules with masses between 100 and 350 Da, which are converted to organic anions (20–8,000 Da), which then in turn lead to tholins.

It is assumed that tholins also played an important role in the "PAH world" hypothesis. The PAH world could have existed as a "pre-RNA world" on the primeval Earth. It is suggested that PAHs can undergo stacking, thus forming structures to which nucleobases are chemically bound (to OH functions which are formed by photochemical derivatisation of the PAHs). The PAH stacks replace the more complex phosphate–D-ribose–phosphate chains in the nucleic acids (Ehrenfreund et al., 2006).

These new results only became possible when the groups of H. Waite and B. Magee at the South West Research Institute (SWRI) in San Antonio, Texas, combined analytical data obtained from two different instruments. These are the ion and neutral mass spectrometer (INMS), which obtained data on small to medium-sized molecules in the upper atmosphere of Titan (950–1,150 km) and the Cassini plasma spectrometer (CAPS), which also registered particles from tholins, both positive (100–350 Da) and negative ions (20–8,000 Da) (Waite et al., 2007; Atrey, 2007).

The search for methane oceans or lakes, the existence of which had been proposed a long time back, began soon after the landing of the Huygens probe on Titan and the arrival of the first pictures of Titan's surface from the Cassini orbiter; however, it was initially unsuccessful. Dendritically branched drainage-like structures, as well as methane clouds and a high moisture content (around 50%) were detected in the vicinity of the point where the Huygens probe landed. Continually regenerated sources of methane which could counter methane loss due to decomposition were also not detected. On the other hand, the Cassini orbiter showed the presence of dry, almost desert-like regions with dune-like formations, while some areas seemed to have been subjected to fluvial erosion.

G. Mitri and co-workers calculated the minimum area of hydrocarbon lakes which would be necessary to preserve the relative methane humidity in the lower regions of the atmosphere. The result was surprising: the calculations indicated that only between 0.002 and 0.2% of the total surface area of Titan would be required (Mitri et al., 2007).

After a long period of intensive searching, lakes containing liquid methane or ethane have been observed. On July 22, 2006, Cassini radar measurements of Titan (T16) showed the presence of dark patches, which resemble terrestrial lakes, over terrain north of $70°$ latitude. Some are black, indicating that the surfaces are extremely smooth, either liquid surfaces or smooth rock areas. Their shapes, and particularly those of their edges, show a morphological resemblance to lakes. Some of the patches have channels (perhaps riverbeds?) leading into them; some lakes are circular, others irregularly shaped. They are between 3 and 70 km across, and in some cases, they have steep rims, like those seen in volcanic craters. The depressions are apparently not always filled with liquid. There may be a precipitation cycle, in which the atmosphere, particularly the methane, condenses out in the Titanian winter and evaporates again; in the summer, the lakes might shrink or even dry out. Future flybys should bring more clarity with respect to the present assumptions. It should, however, be remembered that until now, only a few percent of the total surface area of Titan has been studied (Stofan et al., 2007; Sotin, 2007).

Studies on the haze layers in the Titan atmosphere become more interesting when they are seen in conjunction with possible processes occurring in the primeval Earth's atmosphere. A group involving several US institutes carried out experiments with simulated atmospheres, both for Titan and the primordial Earth. The analysis of an aerosol MS showed that laboratory aerosols could be seen as analogues of the haze layers of Titan. The composition of the aerosol and the shapes and sizes of the particles were determined as a function of the gas mixtures. The results showed that about 4 billion years ago, the young Earth was probably surrounded by an aerosol layer which played a large role in determining the prebiotic chemistry. The authors estimate an aerosol production of about 10^{14} g per year, and thus consider the aerosols to have been the primary source of the organic molecules which led to the formation of biomolecules. (Trainer et al., 2006).

Important weather details are not only provided by the newest information from the Cassini orbiter: the Very Large Telescope in the Atacama desert and the W. M. Keck Observatory on Hawaii are also involved. Near-IR spectra show increased cloudiness in the Titanian troposphere on the morning side, i.e., there are methane clouds at a height of about 30 km and methane drizzle at the surface (Ádámkovics et al., 2007).

Eighteen months after the Huygens landing, there was still no sign of the expected hydrocarbon oceans, so several scientists have been searching for explanations. As already described (Sect. 3.1.6), the photodissociation of methane leads to ethane. D. Hunten from the University of Arizona, Tucson, has proposed that ethane is bonded to smog particles. The "smust" particles (from "smog" and "dust") sink to the surface of Titan and form dunes and the dark coating on the ground. It has been estimated that the amount of "smust" material formed since the emergence of the solar system is such that Titan must be covered with a layer which is many metres thick (Hunten, 2006).

G. Tobie et al. (2006) have suggested that methane gas emissions from methane clathrates may possibly be a source of the methane in Titan's atmosphere. According to this group's model, the Titanian atmosphere contained about 0.1% methane shortly after the moon's formation, with the heat from the interior of the young satellite driving the gas to the surface. Some of this gas entered the atmosphere, while the remainder was fixed in a kilometre-thick layer of methane clathrate, which swam on a sea consisting of water containing dissolved ammonia. After the interior of the moon had cooled, the water froze to give a solid layer. According to this hypothesis, methane gas broke through this layer occasionally (perhaps via cryovolcanism) and entered the atmosphere.

As the workshops on exo/astrobiology show, many groups in various countries are working together, using simulation experiments and theoretical techniques, to forecast and interpret the processes occurring on the Saturnian moon Titan.

Fig. 11.1 Pseudo-colour radar picture of the north polar region of Titan (NASA/JPL, 2007)

11.1.2 Extrasolar Life

11.1.2.1 The Search for Extrasolar Planets

The search for extrasolar life requires us to find planets outside the solar system. But since stars are only visible as small dots of light, it will be very difficult to find planets in outer space, since these do not emit light. This is true even for the nearest star, Alpha Centauri, which is about 4.3 light years away (roughly 7,000 times the distance from the sun to Pluto).

How could extrasolar planets be detected? In recent years, astronomers have developed four different methods of finding non-emitting objects (planets) and even, to some extent, characterizing them. These are:

The Doppler shift or radial velocity method
The transit method
The microgravitation lens (microlens) method
Astronomical measurements

The *Doppler shift* method works on the principle that the presence of heavy planets causes the central object in the system, the star, to orbit around a common centre of mass. If the star is moving towards the observer on Earth, the light waves are shifted to shorter wavelengths; if the star is moving away, the shift is towards longer wavelengths (a red shift). The rate at which the star is moving towards or away from the Earth is called the radial velocity; this method has been very successful in the search for distant planets. The exact determination of the velocity and position change of the star provides information on the magnitude of the gravitational effect of the invisible planet, from which its mass and orbit can be calculated. Many of the planets initially discovered have masses similar to that of Jupiter. Their orbits lie near the central star, so the Doppler effect is particularly easy to observe. Later refinement of the method has made it possible to detect and measure smaller planets. Such interactions between the central star and the planets can also be observed in our solar system: Jupiter forces the sun to "totter" a little, so that orbital oscillations of about 12 m/s occur.

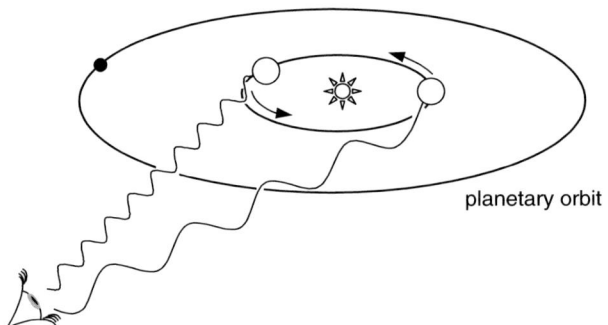

planetary orbit

Fig. 11.2 The invisible planet causes the central star (the "sun") to "stagger" a little; in doing so, it moves either towards the Earth or away from it. In the "approaching" phase, the light waves are shifted to shorter wavelengths, while they are shifted to longer wavelengths in the "leaving" phase. These small spectral changes can be registered very exactly and analyzed quantitatively

The *transit method* requires that the central star, the planet and the observer are connected by a line of sight. The "dark" planet passes across the light source and thus diminishes its light intensity to some extent. Observation is only possible when observer, star and planet are in a favourable position, i.e., the planet lies between the star and the observer. In spite of this requirement, the method permits the discovery of planets of about the size of the Earth; information is also available on the size, mass and density of the planet as well as on its orbit. Because of its limits of applicability, this method is not often used. In the case of the star OGLE-TR-56, it was possible to detect an extrasolar planet, the orbit of which is very close to its sun: only a twentieth of the distance of Mercury away from it. The temperature of the planet was determined to be around 1,900 K; its diameter is about 1.3 times larger than that of Jupiter, its density about 500 kg/m^3 (Brown, 2003; Konacki, 2003).

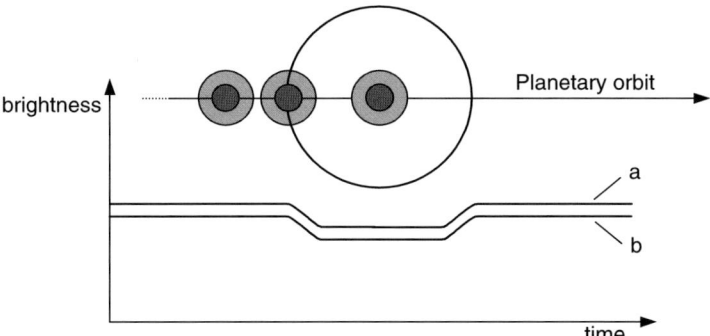

Fig. 11.3 The transit method is based on the unusual constellation of there being a straight line from the Earth via the invisible planet to its sun. As the planet crosses the face of the sun, there is a slight (but measurable) decrease in its apparent brightness (**a**). If the planet has an atmosphere, the line is displaced (**b**)

The third method for detecting the presence of distant planets and planetary systems is the *microgravitation lens method*. Here, the light rays from a star which is behind a planet are deflected, as a consequence of Einstein's general theory of gravitation. Light rays are bent if they pass through a region which is subject to a large mass. The mass of the planet acts like an optical lens: it focuses the light rays, leading to a small increase in the star's brightness, and to an apparent change in its position.

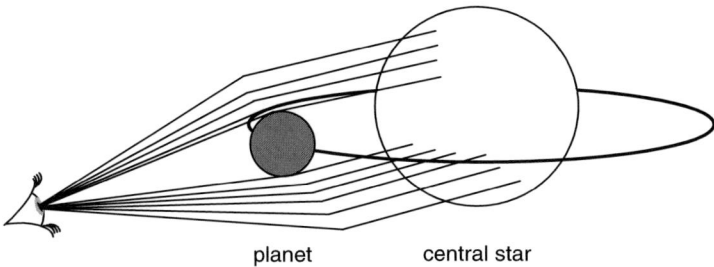

Fig. 11.4 The microgravitation lens method, which will probably be the method of choice in the future. When the invisible planet appears to pass in front of the central star, its gravitation acts like a lens, which focuses the star's light rays. There is a temporary increase in the star's light intensity and an apparent shift in its position

The fourth method, astronomical measurement, is (like the Doppler method) based on the observation and measurement of tiny motions of a star which are due to the mass(es) of the orbiting planet(s). A primary goal of the Space Interferometry Mission (SIM) is the discovery of planets similar to Earth, which orbit around stars like the sun. The SIM is due to start in 2009, and the measurements carried out will increase the accuracy of determinations of distances of stars in our galaxy several hundred times. Exact information can be found in the catalogue which forms part of the Extrasolar Planets Encyclopaedia (www.exoplanet.eu); most discoveries have been made using the radial velocity method. Various parameters of the new planets are listed in the catalogue. As of July 9, 2008, the following had been discovered:

263 planetary systems
307 planets
31 systems with several planets

The masses of the planets so far discovered vary between about 0.02 and 18 Jupiter masses. There are also very large variations in the values of the semi-major axis of the planetary orbits. If the first two methods for the discovery of extrasolar planets are compared (Doppler and transit methods), Doyle et al. (2000) point out the following facts: around 40,000 photons are required to determine the transit of an extrasolar planet across the star HD 209548 using a photometer. But detection of the same system using variations in radial velocity requires 10 million photons.

The number of known extrasolar planets grows monthly; the first discovery was made on October 1, 1995, by Michael Mayor and Didier Queloz from the University of Geneva, who found a planet orbiting around the sun-like star 51 Pegasus.

Using the Hubble telescope, the extrasolar planet HD 209458b (a gas planet with 0.7 times the mass of Jupiter) has been shown to have an extensive external atmosphere consisting of atomic hydrogen; it is possible that the hydrogen is escaping from the planet (Vidal-Madjar et al., 2003).

One of the projects planned for the next decade is Darwin, to be organised by ESA. Darwin will be a flotilla of four or five spacecraft that will search for Earth-like planets around other stars and analyse their atmospheres for the chemical signature of life. Three of the spacecraft will carry 3–4 m "space telescopes", which will form the Infrared Space Interferometer IRSI; they will be stationed 1.5 million kilometres from Earth, in the opposite direction from the Sun, at the Lagrangian Point L2 (a libration point at which the gravitational forces of the Earth and the sun cancel out).

As has already been done on Earth, the telescopes will form an interferometer which will redirect light to the central hub spacecraft, simulating the existence of a much larger telescope; the individual telescopes and the hub must stay in formation with millimetre precision.

The detection of small extrasolar planets (of around the size of the Earth) will be done by registering the infrared light which they emit. Interference filters will blot out the light emitted by the star in question. Because of the huge distance from the Earth, effects due to its atmosphere and its IR radiation will not interfere. Darwin is intended not only to discover planets but to analyse their atmospheres for possible signs of life.

The Darwin mission will not be carried out until the middle of the next decade. However, the COROT (Convection, Rotation and Planetary Transits) telescope was launched from the Baikonur cosmodrome in December 2006. The satellite, which weighs 630 kg, circles the Earth at a height of about 900 km in a polar orbit. The mission is planned to last 2 1/2 years, and more than 120,000 stars are to be observed.

Fig. 11.5 One of the telescopes in the Darwin flotilla. With kind permission of ESA

The 27-cm telescope is equipped with four CCD detectors and measures the light curves of bright stars in the wavelength range 370–950 nm. The scientific goals of the mission are:

> The discovery of planets outside our solar system, using the transit method. For the first time, the search will be directed particularly towards planets which are only slightly larger than the Earth. The first, Corot-Exo-1b, was discovered in April 2007.
> The detection of star vibrations (astroseismology).

The COROT mission is French-led and involves several European partners as well as one in Brazil.

The start of NASA's Kepler mission is planned for February 2009 and has goals similar to those of the COROT project, though rather more ambitious:it is intended to determine the percentage of terrestrial and larger planets there are in or near the habitable zone of a wide variety of stars, and also to determine the distribution of sizes and shapes of the orbits of these planets.

The Kepler mission also supports the objectives of the future NASA Origins missions Space Interferometry Mission (SIM) and Terrestrial Planet Finder (TPF).

11.1.2.2 Habitable Regions of the Universe

It is assumed that the greatest part of our solar system, and indeed of the Milky Way, is hostile to life. The term "habitable zone" (Franck et al., 2002) takes into account parameters such as the size of the planet, its distance from its sun, the radiation intensity of that sun and the presence of water and carbon on the planet's surface. The habitable zone in our solar system comprises a belt which encompasses the orbits of Earth and Mars. Although there is discussion as to whether life may perhaps be present on the Jovian moon Europa, such life forms would have properties which are as yet unknown.

Can we assume that there is a habitable zone somewhere in our galaxy? This seems feasible: astronomers divide the Milky Way into four regions, which cannot always be exactly separated:

The thin disk
The thick disk
The galactic bulge
The stellar halo

It is very unlikely that Earth-like planets will be found in the halo region, since the stars in the halo were formed in the early phase of the development of the galaxy and thus contain very little of the heavy elements. Our solar system is in the region of the thin galactic disk, around 28,000 light years from the central point of the galaxy (which has an overall diameter of 100,000 light years). The content of "metals" (i.e., elements with a higher atomic number than helium) decreases steadily with increasing distance from the centre of the galaxy. In recent years, astronomers have been able to determine the gradient of this phenomenon, which is about a 5% decrease in metal content per 1,000 light years (Gonzales et al., 2001).

In the "galactic lifebelt", the amount of rock-forming elements is high enough for Earth-like planets to have been formed. However, two conditions must be fulfilled:

The intensity of incident radiation must be as low as possible.
Heavenly bodies with large masses, which can attract asteroids, large planetesimals and comets, must be present in the vicinity of the life-sustaining suns, so that impact catastrophes can be avoided.

If Earth-like planets are present in the regions of the thick disk and the solar halo, they will be subject to much more intense cosmic radiation from neutron stars and supernovae than is the moderate zone in which our solar system is situated. According to Gonzalez et al. (2001), the chance of finding higher life forms is in fact much lower than assumed by many experts. The ecological niches are small and few. The galactic lifebelt cannot yet be exactly delineated and described; however, it seems that the probability that intelligent life could have evolved (during hundreds of millions of years) in the globular clusters, the central region of the galaxy or the regions at its edges is very low.

Australian astronomers have attempted to characterise the habitable zone of the Milky Way more exactly. The models which they used were based on the following assumptions for complex life in our galaxy:

There must be enough heavy elements present to form terrestrial planets.
There must have been enough time for biological evolution.
The environment must be free from life-threatening supernovae.

The group identified a circular belt between 7 and 9 kpc[1] from the centre of the galaxy. This zone (Fig. 11.6) consists of a population of stars which developed between 8 and 4 billion years ago; it contains about 10% of all the stars in the galaxy, and around 57% of the stars in the habitable zone are older than our sun (Lineweaver et al., 2004).

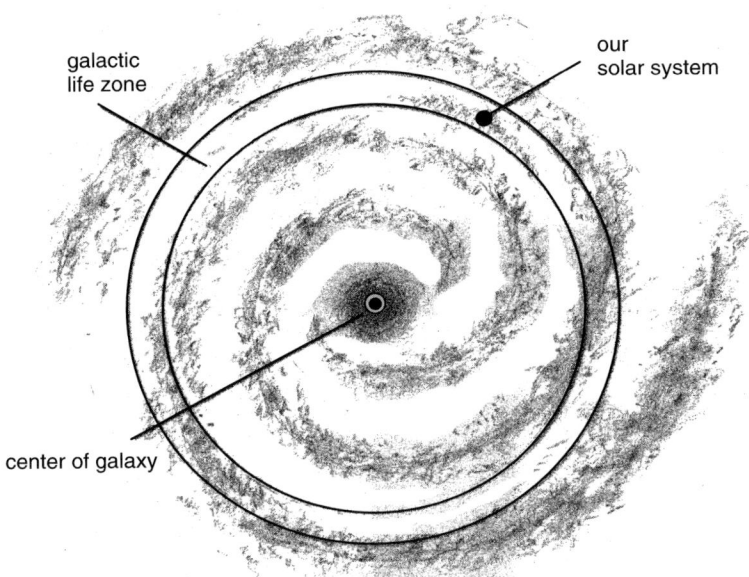

Fig. 11.6 Schematic representation of the Milky Way with the galactic "life zone", in which life should be possible. The centre of the galaxy is kept practically sterile by extreme radiation, while areas in which stars are formed are localized in the spiral arms

When, many, many million years in the future, our sun expands in its final phase to become a red giant, the "habitable" zone of our solar system will shift by 1–2 AU, to the region where Triton, Pluto/Charon and the Kuiper Belt are found. This zone is referred to as the "delayed gratification habitable zone". All the heavenly bodies in this zone contain water and organic material, so that chemical and molecular

---

[1] A kiloparsec (kpc) is 3,260 light years.

evolution processes seem feasible. In the Kuiper belt, with its more than 10^5 objects (with a radius of 50 km or more), the total surface area is probably three to four times as great as that of the four terrestrial planets. Speculations in this area appear utopian: but we must not forget that the Milky Way contains around 10^9 red giants! Thus, a large number of star/planet systems with delayed habitable zones could exist in our galaxy. Of course, the question as to whether chemical, molecular or even biological evolution could occur in such systems is a matter of pure conjecture (Stern, 2003).

11.1.2.3 ETI and SETI

"So where are they then?" This question, now famous, is attributed to Enrico Fermi: he was reacting to an observation, made by a member of the audience at a physics conference in Los Alamos in 1950, that there must be extraterrestrial intelligence (ETI). (The search for ETI is known as SETI, search for extraterrestrial intelligence.) The question is now known as the "Fermi paradox"; initial investigations with the goal of solving it took place a few years after Fermi's question was posed.

At the end of the day, the question of extraterrestrial life is based on two principles: the Copernican principle and the mediocrity principle. The first states that the Earth is not in a central, specially favoured position. The mediocrity principle states that life on Earth is nothing special, so that life (in whatever form) could evolve in many places in the universe.

As to the scientific studies dealing with Fermi's question, as early as 1960, the astronomer Frank Drake directed the 26-m radio telescope of the National Radio Astronomy Observatory in Green Bank, West Virginia, towards the constellation Cetus, the whale, to try to receive signals which that system might emit. The experiment was very speculative, because (according to our Earth-fixated ideas) it required the presence of a planet-like system. In addition, the putative intelligent beings must be capable of transmitting radio signals. The star chosen was Tau Ceti, which is only 11 light years from our solar system, and thus a neighbour! This first "eavesdropping" in outer space (project Ozma) brought no results, but F. Drake and his group organised a conference to discuss the possibilities and limitations of solving the problem. Drake divided the question of ETI into a series of single factors, each of which should be easier to deal with. Three variables were seen as particularly relevant:

(a) Physical conditions: ETI would need to have its own environment, which would resemble the Earth in many ways.
(b) Biological conditions: life should be able to emerge anywhere in the universe where physical and chemical conditions make it feasible. The development of intelligence would need to be driven by external forces.
(c) Socio-cultural conditions: intelligent life must develop to give a civilisation, which will at some stage develop a technique for interstellar communication.

The result of all the discussions was the Drake equation, which in spite of its mathematical formalism, provides no final answers, but at best estimates of the number of technically higher developed systems in the Milky Way. The Drake equation is as follows:

$$N = R^* \times f_s \times f_p \times n_e \times f_l \times f_i \times f_c \times L$$

where

$N =$ the number of intelligent civilizations within our own galaxy which are able to communicate

$R^* =$ the average rate of star formation in our galaxy

$f_s =$ the fraction of stars that are suitable suns for planetary systems

$f_p =$ the fraction of suitable suns with planetary systems

$n_e =$ the mean number of planets that are capable of supporting life

$f_l =$ the fraction of such planets on which life actually originates

$f_i =$ the fraction of such planets on which some form of intelligence arises

$f_c =$ the fraction of such intelligent species that develop the ability and desire to communicate with other civilizations

$L =$ the mean lifetime (in years) of a communicative civilization

If we simplify by assuming that the various factors are independent of one another, we can obtain an estimate of the number of communicating civilisations just by multiplying the values of the factors.

Apart from the factor R^*, all the values need to be estimated. Thus, the values of N vary by several orders of magnitude (from 10^8 to 10^{-3})! The value of R^* is assumed to be 20 per year. Since estimates of all the other factors vary so much, some researchers consider that attempts to use the Drake equation to shed more light on the ETI question are useless.

Taking a value of 10^7 for N would mean that in our galaxy (with its perhaps 100 billion stars), there could be several million planets with life forms capable of interstellar communication. However, if these were distributed statistically, the nearest would still be 200 light years away from Earth. One point is important: the term "probability" used in the Drake equation is interpreted in the sense of "subjective probability" (a term from the nomenclature used by statisticians and probability theorists), as the numerical value of this probability is determined only by the experience of the scientist concerned (Casti, 1989). Casti also provides more information on the Drake factors (apart from the factor f_s) in the chapter "Where are they then?" In summary, we can say that the Drake equation is a first attempt to quantify the ETI problem in order to move from the area of science fiction and pure speculation to that of serious scientific debate.

The Fermi paradox relates to the question as to whether intelligent life exists somewhere in space. But of course ETI species would need to be able to make themselves known by means of technical signals for us to detect their existence. Let us, however, pose the question as to whether life itself, including the most primitive life forms, is really to be found somewhere else in the universe: such systems, however small, would have to satisfy the conditions so far defined for the term "life" (see Sect. 1.4).

The search for extraterrestrial intelligence was originally coordinated by the SETI Institute in Mountain View, California, founded in 1984; however, in 1992, some of its supporters decided to stop financing the institute, as they were no longer happy with the "crazy hunt for extraterrestrials". From 1992–2004, the Center for SETI Research at the SETI Institute was funded entirely by donations from individuals and grants from private foundations. In 2005, a NASA grant of $5 million was awarded for work on signal detection by the Allen Telescope Array.

11.1.2.4 Panspermia

At the beginning of the twentieth century, the panspermia hypothesis was very popular, having support from such well-known scientists as Lord Kelvin, von Helmholtz and, in particular, Svante Arrhenius. This bold hypothesis, according to which "seeds" of life can be transported across the universe by means of radiation pressure, later dropped out of sight. Around 1980, the well-known British astronomer Fred Hoyle, together with Chandra Wickramasinghe, provided new support for the theory (Hoyle, 1985). Francis Crick (1982) proposed the modified form of "directed panspermia": according to this, life evolved in a region of the universe in which the conditions were perhaps more favourable than those on Earth. According to Crick and Orgel, the "seeds of life" came to Earth on an unmanned space vehicle and started to multiply in a primeval ocean. Crick and Orgel are, of course, aware that their hypothesis is very vague, as there is no evidence of any type for such a process.

The panspermia hypothesis does not provide any explanation for the origin of life, as it presupposes the existence of functioning life forms. Two types of panspermia can be discussed, depending on the distances to be covered:

Interplanetary panspermia: the transport of life forms within the solar system
Interstellar panspermia: the transport from one star to another

As far as the way in which transport could take place is concerned, the following possibilities can be discussed (von Bloh et al., 2003):

Lithopanspermia: the transport of the seeds of life by meteorites
Radiopanspermia: microbes are driven through space by radiation pressure
The "directed" panspermia referred to above

In which development phase of the universe could there have been the greatest chance of panspermia processes taking place? A research group from the Potsdam Institute for Climate Research has tried to provide answers to this difficult question on the basis of research results from astronomy and astrophysics. Using mathematical models, they concluded that the maximum number of habitable planets in our galaxy must have been present at the time when our solar system and the young Earth were evolving (von Bloh et al., 2003).

Much more important is the problem of the survival of microorganisms for longer periods of time under the extreme conditions present in outer space; there is absolutely no comparison between these conditions and those in laboratories on Earth! The first question is in what form such microorganisms would be subjected to cosmic conditions:

Unprotected, as "naked" cells, i.e., spores
Protected by material such as ice or rock

Which drastic influences are brought to bear on microorganisms in outer space? The following factors must be taken into consideration:

Temperature
Ultra-high vacuum
Radiation (of differing types and with differing energies)

Temperatures in outer space vary from close to absolute zero in the depths of space to extremely high temperatures in the neighbourhood of suns.

There is no doubt that the most important parameter in the organisms familiar to us is water content. The lapidary sentence "no life without water" is valid for all aspects of biogenesis, whether on the primeval Earth or on another heavenly body. The life processes in all living species known to Man are based on liquid water, which has a number of special properties (Brack, 1993). The dehydrating effect of a high vacuum is assumed to be the most important limiting factor in the transport of microbes between heavenly bodies. This effect would naturally depend on the time required for such a transfer, since some spores can survive for what are, in cosmic dimensions, short periods.

To this point, two types of experiments have been carried out to study the survival rates of microorganisms:

Laboratory experiments
Experiments in space (using either balloons or satellites in orbit)

Naturally, experiments carried out in space are considered to be more authentic than laboratory experiments; this must not necessarily be the case, however.

The optimum water content of most cells is around 80%. Liquid water is absolutely necessary for the stability of the lipid membrane and the hydrophobic regions in proteins. The hydrophilic fractions of the nucleic acids and the proteins require liquid water for maintaining their three-dimensional structures and thus their functionality.

When water is removed, the hydrophobic structures of proteins and membranes are the first to suffer, as these are stabilised by the water. Experiments were carried out mainly with *Bacillus subtilis* spores, but also with other microorganisms, such as *Aspergillus* species. If the cells lose water, practically all the metabolic processes which occur in the aqueous phase cease. Only a few operations, such as electron transfers in protein complexes involving metal ions, are affected either not at all or only slightly. Injuries to living cells occur in a complex manner. Thus, it was found in laboratory experiments that, at under 92% relative humidity, double helix DNA changes from the B conformation (with eleven nucleotide pairs per rotation) to the A conformation (ten pairs per rotation). A further drop in the relative humidity, to below 80%, leads to formation of the C conformation with 9.3 pairs per rotation. Thus, it is possible that drying out, together with high mechanical strain, may cause the chains to break.

However, some microorganisms, the so-called anhydrobiotes, are able to deal better with a lack of water. One of their strategies for survival is the accumulation of non-reducing sugars (disaccharides) which replace the lost water during dry periods and can, for example, stabilise membrane structures, even when water is not present.

The survival of microorganisms in an extremely dry region, the Atacama desert in Chile, was investigated by Dose and co-workers (2001). They found that about 15% of the *Bacillus subtilis* spores and about 30% of the *Aspergillus niger* conidia survived for 15 months in the dark. If, however, the microorganisms were subjected to sunlight (at an energy flux of about $300 \, kJ/m^2$ between 280 and 320 nm), their survival chances dropped to zero after around 100 h. These studies showed that microorganisms in protected dormant forms, such as spores, could survive for longer periods in shady areas (vessels with a lid under the conditions mentioned above), but certainly not for years or centuries.

But how do microorganisms behave in outer space? Answers to this question require experiments to be carried out in space, as (however well they may try to simulate conditions in outer space) laboratory experiments are often considered to be "artificial" and "unrealistic". Thus, microbes have been put on board a number of space vehicles and subjected to outer space conditions to probe the effect of various variables on the survival probability of *Bacillus subtilis* spores.

It has been known for some time that certain types of bacteria spend a certain amount of their lifespan in a dormant state. The bacteria are then known as "endospores", or just "spores". In this state, they appear not to undergo any metabolic processes and are important particularly because of their heat resistance. The formation of spores is a highly complex process of bacterial cell differentiation.

The behaviour of spores under space conditions has been studied for nearly 30 years both in the laboratory (under simulated space conditions) and outside the Earth's atmosphere. More than 20 years ago, a group from Cologne, Mainz and Frankfurt/Main began experiments with *Bacillus subtilis* spores in orbit (Spacelab I) in order to determine the influence of solar UV irradiation at different wavelengths. This work was later continued as part of various space missions, including a long-term experiment which lasted almost 6 years, the Long Duration Exposure Facility (LDEF) (Horneck et al., 1984; Horneck, 1993).

More recent experiments on the ability of *Bacillus subtilis* spores to survive in space were carried out on behalf of NASA by the Russian FOTON satellite. The NASA appliance, BIOPAN, allowed various experiments to be carried out on the spores; three flights were carried out, in 1994, 1997 and 1999. An orbit took 90 minutes; FOTON rotated, so that BIOPAN passed into and out of sunlight both during rotation and during each orbit. More exact details are given by Horneck et al., 2001.

The spores were subjected to various conditions in space:

Completely unprotected
Protected by a thin film of clay or by aluminium foil
Protected by a quartz disk
Mixed with various protective substances

The dry protective materials were:

Clay
Ground red sandstone
Meteorite material
Simulated Mars soil
Sand from the Martian meteorite Zagami

The experiments were intended to clarify the question as to whether transmission of spores from Mars to Earth could be feasible. To do this, a "Mars meteorite" was simulated, i.e., the spores were mixed with powdered rock and the mixture pressed together to give a small cube about $1\,cm^3$ in size. The spore concentration was about the same as in normal soil on Earth. The samples were in orbit for around 2 weeks, and their survival ability was determined on their return to Earth, compared with the corresponding samples which had been left on Earth (control experiment).

The result was that only a small fraction (10^{-6}) of the spores which had been subjected to space conditions without protection survived. The same was true for those which were protected by a quartz disk or covered by a thin layer of clay. However, the survival rate of the spores mixed with the protective materials listed above was about five orders of magnitude higher, while almost 100% of those in the small cube survived (Horneck et al., 2002a). This experiment suggests that small pieces of rock, only a few centimetres in diameter, could act as a transport medium between certain heavenly bodies. However, the "classical" panspermia hypothesis, involving seeds of life on grains of dust, must be completely unreal.

Table 11.1 Survival of *Bacillus subtilis* spores subjected in different missions to the ultra-high vacuum conditions (10^{-6}–10^{-4} Pa) prevalent in outer space (Horneck et al., 2002b)

Mission	Time spent in vacuum	Rate of survival in % at the end of the influence of space conditions			
		Spores in thin layers		Spores in thick layers with sugar	
	Days	In space	Control	In space	Control
Spacelab1	10	69.3 ± 15.8	85.3 ± 2.6		
EURECA	377	32.1 ± 16.3	32.7 ± 5.6	45.5 ± 0.01	62.7 ± 8.2
LDEF	2,107	1.4 ± 0.8	5.2 ± 2.9	67.2 ± 10.2	77.0 ± 6.0

It has been known for a long time that solar UV irradiation is the greatest danger for unprotected spores. A multilayer system with an aluminium covering, and containing added protective materials such as glucose (up to 10^{-4} M), had clear protective properties. The high resistance of the *Bacillus subtilis* spores is probably due to two effects:

The dehydrated, mineralised (particularly with Ca^{2+}) core, which is surrounded by a thick protective layer (followed by two or more coatings of the spore cell)

Table 11.2 Water partial pressure and relative humidity under various conditions

	Water vapour partial pressure (mb)	Relative humidity (%)
Earth's atmosphere, temperate zones	12–23	50–90
Dry deserts	1–8	5–30
Laboratory (silica gel)	1	5
High vacuum	Approx. 10^{-8}	Approx. 0
Orbit close to Earth	Approx. 10^{-10}	Approx. 0
Surface of Mars	Approx. 10^{-3}	Approx. 0
Deep space	Approx. 0	Approx. 0

Saturation of DNA with small, acid-soluble proteins, which fulfil complex functions

The damaging effect of the UV irradiation is amplified in space by the additional effect on the cells of the ultrahigh vacuum; Nicholson et al. (2000) provide additional information on the resistance of the *Bacillus subtilis* spores.

Criticisms of laboratory simulation experiments (which generally last between several hours and a few days) are often based on statements made on the basis of extrapolation of the experimental results to cover processes in outer space, i.e., judgements based on experiments taking 2–3 days are made on processes occurring in outer space which can take many million years. In this case, the *Bacillus subtilis* research can offer one very positive experiment, as the space mission LDEF took almost six years to complete.

Lichens were also subjected to deep-space conditions: the species *Rhizocarpon geographicon* and *Xanthoria elegans* (from the Spanish highlands and an Antarctic dry valley, respectively) survived a 16-day flight with the Russian FOTON M2 satellite surprisingly well. All the lichens exhibited almost the same photosynthetic activity before and after their excursion into space, regardless of the optical filters used. Various microscopic studies also showed no detectable ultrastructural changes in most of the algal and fungal cells of the lichen thalli. However, compromised membranes were observed in a large fraction of the cells: it is questionable whether these results indicate a survival half-life of the lichens (Sancho et al., 2007).

11.2 Artificial Life (AL or ALife)

Although the research area of artificial life (AL) does not really fall within the scope of this book, it will be introduced briefly below. The dream that humankind can act as the creator of life can be found in various works of literature, for example Goethe's *Faust*, Strindberg's *By the Open Sea* or Mary Shelley's *Frankenstein*. There were also attempts to solve the problem of life from purely technical and mechanical points of view, for example the famous "Canard Digérateur" (Digesting Duck) created by Jacques de Vaucanson in the middle of the eighteenth century; this caused a great sensation in Paris, as it could drink, eat, digest food and swim in water.

This was only a first primitive step towards the phenomenon of life. In the last 20 years or so, due to the enormous developments in computers and computer science, it has become possible to work seriously on AL, the roots of which lie in computer simulation. Three factors are required in order to study the phenomenon of artificial life:

A freely programmable computer

The insight that it is not possible to start with the most complicated system (Man)

The abstraction that life must not of necessity be something material, but that it can, for example, consist only of pure information

The real beginning of scientific work on AL was in Fall 1987, when a group of scientists from the fields of computer science, biology, chemistry and philosophy met for the "First International Conference on Artificial Life" at the Los Alamos National Laboratories in New Mexico. When we hear the name Los Alamos, we generally think of the development of the first atomic bomb, but this meeting was purely peaceful in intent, as it dealt with computer models for the simulation of biological processes, such as plant growth or protein synthesis; the conference was organized by Christopher Langton.

The systems classified as AL have some common features, which are, however, not always all found in individual AL systems:

An information-processing system which is capable of assimilating, storing and transmitting information.

Such a system is assigned to an "environment" and interacts with this environment via incoming and outgoing information.

The system obeys rules of behaviour which determine the interactions.

These rules of behaviour cause the system to behave dynamically in its environment.

According to Langton, the creed of investigators of AL is that their studies are intended to look for "the spirit in the machine; an essence which emanates from it but is independent of it" (Casti, 1989). Langton asked himself how cellular machines (for example J. H. Conway's "Game of Life") (Eigen and Winkler, 1993) can be used for studies on real life. For Langton, the important functional roles are (as in real life) played by proteins (for catalysis, regulation, defence mechanisms and the conservation of structures) and nucleic acids as information carriers. Langton considers that these functional activities can be linked to the rules for a cell machine (as in the Conway system). Thus, each of the functional activities of a living organism can formally be represented by a logical "machine". The result of linking up individual machines is an object which can be described as "living", even if the object is purely artificial (Casti, 1989).

The mathematician John von Neumann was one of the pioneers in AL; he developed an analogy between the functions of a living organism and those of a machine. The latter consisted of two important parts which, when the computer industry developed, were referred to as software and hardware. Hardware processes information, while software embodies information (Dyson, 1985).

John von Neumann constructed several models (originally known as "kinematic models") with the goal of incorporating the logical core of self-production into the system. Others studied von Neumann's model and modified it, for example, Walter Fontana, who devised the "AlChemy" (Algorithmic Chemistry) system. This can be described as artificial chemistry, an evolution reactor in which the objects which react are not molecules but mathematical functions.

In recent years, the "transitions from nonliving to living matter" have been the subject of three seminars, bringing together theoreticians and experimentalists in the Los Alamos National Laboratory, in the Santa Fe Institute and in Dortmund. The biogenesis problem was expanded to the question, how can simple life forms be synthesised in the laboratory? Artificial cells (sometimes called protocells) could be quite different from the cell types known today, or from primeval cells; they might, for example, be orders of magnitude smaller than a bacterium. The seminars posed three questions for further work:

Where is the boundary between physical and biological phenomena?
What are the main obstacles for the integration of genes and energetic processes in an encasing unit?
How can the flow of information from theory and simulation to artificial cell experiments be improved? (Rasmussen, 2004)

It has still not been possible to obtain complete clarity and consensus on a definition of the modern terms "artificial life" and "synthetic life".

We shall have to wait and see whether the patent application made by Craig Venter for a unicellular synthetic organism with 381 synthetic genes ("Synthia") will become reality. Optimists estimate that it will be possible within the next 3–10 years to create "wet artificial life". The next few years promise to be interesting!

11.3 The "When" Problem

The phenomenon of time has become an important subject in theoretical physics and cosmophysics in recent years: previously, however, philosophers were almost alone in concerning themselves with the secrets of time. Time plays an important role in biogenesis: the two time-related questions which appear most important are: when did biogenesis take place, and how much time did the process require?

According to what we now know, life must have originated sometime between the Earth's formation, about 4.5 billion years ago, and the appearance of the first fossils (3.465 billion years ago; see Sect. 10.1). The intervening period of about a billion years is greatly limited by the indisputable fact that the Earth was subject to heavy bombardment by asteroids and planetesimals.

Leslie Orgel (1998) describes the difficulties of obtaining a clear answer to the "when question" using a simple graphic example. Suppose we ask the question, how long does it take to get from A to B? This question cannot be answered unless we know where A and B are, and what the road from A to B is like [Note: from A (abiotic) to B (biotic)]. According to Orgel, we cannot say anything about the time span necessary for the transition from lifeless matter to the first living system, as we know nothing about the distance from start to finish or about the terrain separating the two.

Many investigations, confirmed by theoretical studies (Maher and Stevenson, 1988), indicate that the heavy bombardment of the primeval Earth about 3.8 billion years ago led to its sterilisation, so it was practically "uninhabitable". The time stamp for the first stromatolith-forming prokaryotes, about 3.5 billion years ago (Schopf, 1993), does not in fact mark the beginning of the biogenesis process, but the appearance of quite complex life forms. Thus, the point at which the first primitive living systems appeared must have been considerably earlier.

Lazcano and Miller (1994) suggest that the time necessary for the first simple life forms to evolve to give cyanobacteria was probably not longer than 10 million years. One of their reasons for suggesting this value is the assumption that the biomolecules dissolved in the primeval ocean (primordial soup hypothesis!) could not have withstood passage through the Earth's hydrothermal systems (see Sect. 7.2). The authors do, however, consider it possible that self-replicating systems capable of Darwinian evolution could have developed much more quickly, prior to destruction or modification. There is clear evidence of gene duplication processes which must have taken place in the early phases of evolution.

Lazcano, an evolutionary biologist, and Miller, an experienced prebiotic chemist, believe that the most important bottlenecks in the biogenesis process leading from the primeval soup to the RNA world, and thence to cyanobacteria, are the following:

The origin and emergence of replicating systems
The development of protein biosynthesis
The evolutionary development of starter types, from which proteins later evolved
    via gene duplication and mutation

In 1998, about three years after the Lazcano/Miller publication appeared, Leslie Orgel published the article which we have already mentioned above. He subjected their ideas to a critical examination and came to the rather pessimistic conclusion that some of their speculations were much too bold.

According to Orgel, the period during which bacteria evolved, after the "great bombardment" of the young Earth had come to an end, can be arbitrarily divided into various phases:

An accumulation period (synthesis of organic building blocks),
A pre-organisation period (in which the first replicating systems evolved) and
The maturation period (in which the first "bacterium" evolved from the first
    replication system).

Orgel calls attention to one logical point in the "time frame" discussion. Points of general knowledge are regrettably often not taken into account: thus, the main uncertainty in the sum of a small number of terms always comes from the term or terms which is or are the least well-defined. If nothing is known about *one* of the terms, virtually nothing is known about the *sum* of the terms, even if all the others are relatively well-defined.

Orgel suggests that we know almost nothing about the length of time which was required for the RNA world to be formed from the prebiotic world (primordial soup?). It is impossible to estimate how long a process will take if the nature of the process is not known. These simple logical conclusions are often forgotten: Leslie Orgel is a laudable exception!

The passage of the whole of the water in the oceans through hydrothermal systems (in about 10 million years), as discussed by Miller and Lazcano, is also not a convincing argument for a possible thermal destruction of all the biomolecules dissolved in the primeval ocean, as there could have been other smaller bodies of water on the primordial Earth which were not subject to such a passage.

Estimates of the lengths of time in which the process of the evolution of life could have occurred are also provided by geologists and palaeontologists. These are, however, less interested in the time required for replication processes to evolve than in the length and intensity of the bombardment of the young Earth by asteroids and large planetesimals. Oberbeck and Fogleman (1989) assume that the geological condition of the primeval Earth played a decisive role in the length of time required for biogenesis. In other words, the number of gargantuan impacts which led to the sterilisation of the Earth also determined the time in which life could have developed. Thus, around 3.8 billion years ago, there was only a short time available for life to evolve (2.5–11 million years), whereas in the quieter phase, around 3.5 billion years ago, a much longer period for biogenesis can be envisioned (between 67 and 133 million years). If, however, a more conservative model is chosen for the bombardment, the length of time available around 3.8 billion years ago increases to about 25 million years. These model calculations were carried out on the assumption that the kinetic energy of the impacts was 2×10^{34}–2×10^{35} ergs; this would have sufficed to evaporate the primeval oceans. The authors deduce from their models that life could even have evolved more than once on the young Earth!

More questions than answers; and so we can still only conjecture as to how long it took for life to emerge and when the first living systems formed, replicated, evolved and even perhaps disintegrated again into their component molecules.

References

Ádámkovics M, Wong MH, Laver C, de Pater I (2007) Science Online DOI:10:1126/ science.1146244
Atrey S (2007) Science 316:843
Baker J (2007) Science 317:1705

Bibring JP, Arvidson RE, Gendrin A, Gondet B, Langevin Y, Le Mouelic S, Mangold N, Morris RV, Mustard JF, Poulet F, Quantin C, Sotin C (2007) Science 317:1206

Brack A (1993) Orig Life Evol Biosphere 23:3

Brack A (2002) Water, the Spring of Life. In: Horneck G, Baumstark-Khan C (Eds.) Astrobiology. Springer, Berlin Heidelberg New York, p 79

Brown T (2003) Nature 421:488

Campbell DB, Black GJ, Carter LM, Ostro SJ (2003) Science 302:431

Casti J (1989) Paradigms Lost: Images of Man in the Mirror of Science. William Morrow & Co

Crick F (1982) Life Itself. Simon and Schuster, New York

Dose K, Bieger-Dose A, Ernst B, Feister U, Gómez-Silva B, Klein A, Risi S, Stridde C (2001) Orig Life Evol Biosphere 31:287

Doyle L, Deeg H-J, Brown TM (2000) Scientific American, September

Dyson F (1985) Origins of Life, Cambridge University Press

Ehrenfreund P, Rasmussen S, Cleaves J, Chen L (2006) Astrobiology 6:490

Eigen M, Winkler R (1993) The Laws of the Game: How The Principles of Nature Govern Chance, Princeton University Press

ESA – Exo-Astro-Biology (2001) European Space Agency, Publications Division

Extrasolar Planets Encyclopedia http://www.obspm.fr/planets

Finster KW, Jensen A-M, Hansen A, Lomstein BA, Nørnberg P, Merrison JP (2002) Second European Workshop on Exo/Astrobiology. Graz, Abstracts p 187

Fischer G, Dokano T, Macher D, Lammer H, Rucker HO (2002) Second European Workshop on Exo/Astrobiology, Graz, Abstracts p 211

Foing B (2002) Space Activities in Exo-Astrobiology. In: Horneck G, Baumstark-Khan C (Eds.) Astrobiology. Springer, Berlin Heidelberg New York, p 389

Franck S, von Bloh W, Bounama C, Steffen M, Schönberner D, Schnellnhuber H-J (2002) Habitable Zones in Extrasolar Planetary Systems. In: Horneck G, Baumstark-Khan C (Eds.) Astrobiology. Springer, Berlin Heidelberg New York, p 47

Fredrickson JK, Onstott TC (1996) Scientific American, October

Gonzales G, Brownlee D, Ward PD (2001) Scientific American, October

Greenberg R, Tufts BR, Geissler P, Hoppa GV (2002) Europa's Crust and Ocean: How Tides Create a Potentially Habitable Physical Setting. In: Horneck G, Baumstark-Khan C (Eds.) Astrobiology. Springer, Berlin Heidelberg New York, p 111

Griffith PA, Owen T, Gebable TR, Rayner J, Rannou P (2003) Science 300:628

Groß M (2001) Life on the Edge. Amazing Creatures Thriving in Extreme Environments. Perseus Books

Herkenhoff KE, Byrne S, Russell PS, Fischbaugh KE, McEwen AS (2007) Science 317:1711

Horneck G, Bücker H, Dose K, Martens KD, Mennigman HD, Reitz G, Requardt H, Weber P (1984) Origins of Life 14:825

Horneck G (1993) Orig Life Evol Biosphere 23:37

Horneck G, Rettberg P, Reitz G, Wehner J, Eschweiler U, Strauch K, Panitz C, Starke V, Baumstark-Khan C (2001) Orig Life Evol Biosphere 31:527

Horneck G, Rettberg P, Reitz G, Panitz C, Rabbow E (2002a) Orig Life Evol Biosphere 32:542

Horneck G, Mileikowski C, Melosh HJ, Wilson JW, Cucinotta FA, Gladman B (2002b) Viable Transfer of Microorganisms in a Solar System and Beyond. In: Horneck G, Baumstark-Khan C (Eds.) Astrobiology. Springer, Berlin Heidelberg New York, pp 57–76

Hoyle F (1985) The Intelligent Universe: A New View of Creation and Evolution. Mermaid Books

Hunten DM (2006) Nature 443:669

Jaumann R, Hauber E, Lanz J, Hoffmann H, Neukum G (2002) In: Horneck G, Baumstark-Khan C (Eds.) Astrobiology. Springer, Berlin Heidelberg New York, p 89

Jensen J, Merrison JP, Kirch KM, Nørnberg P (2002) Second European Workshop on Exo/Astrobiology. Graz, Abstracts p 187

Jerney I, Aydogor Ö, Besser BP, Falkner P, Grard R, Molina-Cuberoz GJ, Hamelin M, Lopez-Moreno JJ, Schwingenschuh K, Fulchignoni M (2002) Second European Workshop on Exo/Astrobiology. Graz, Abstracts p 210

Kaiser J (1995) Science 270:377

Kereszturi A (2002) Second European Workshop on Exo/Astrobiology. Graz, Abstracts p 210

Konacki M, Torres G, Iha S, Sasselov DD (2003) Nature 421:507

Kounaves S (2007) Nature 449:281

Lammer H, Stumptner W, Molina Cuberos GJ (2002a) Martian Atmospheric Evolution: Implications of an Ancient Intrinsic Magnetic Field. In: Horneck G, Baumstark-Khan C (Eds.) Astrobiology. Springer, Berlin Heidelberg New York, p 203

Lammer H, Wurz P, ten Kate IL, Ruiterkamp R (2002b) Second European Workshop on Exo/Astrobiology. Graz, Abstracts p 209

Lazcano A, Miller SL (1994) J Mol Biol 39:546

Lindgren K, Nordahl M (1997) In: Forskningsrådets årsbok 1997, Swedish Science Press, Uppsala, p 105

Lineweaver CH, Fenner Y, Gibson BK (2004) Science 303:59

Lomstein BA, Hansen A, Merrison JP, Nørnberg P, Finster KW (2002) Second European Workshop on Exo/Astrobiology. Graz, Abstracts p 60

Lorenz R (2003) Science 302:403

Maher KA, Stevenson DJ (1988) Nature 331:612

Mayor M, Queloz D (1995) Nature 378:355

McEwen AS, Hansen CJ, De la Mere WA, Eliason EM, Kerkenhoff KE, Keszthelyi L, Gulick VC, Kirk RL, Mellon MT, Grant JA, Thomas N, Weitz CM, Squyrez SW, Bridges NT, Murchie SL, Seelos F, Seelos K, Okubo CH, Milazzo MP, Tornabene LL, Jaeger WL, Byrne S, Russell TS, Griffes JL, Martínez Alonso S, Davatzes A, Chuang FC, Thomson WJ, Fischbaugh KE, Dundas CM, Kolb KJ, Banks ME, Wray JJ (2007) Science 317: 1706

Merrison JP, Finster KW, Gunnlaugsson HP, Jensen J, Kinch K, Lomstein BA, Mugford R, Nørnberg P (2002) Second European Workshop on Exo/Astrobiology. Graz, Abstracts p 187

Mitri G, Showman AP, Lunine JI, Lorenz RD (2007) Icarus 186:385

Navarro-González R, Navarro KF, de la Rosa J, Iñiguez E, Molina P, Miranda LD, Morales P, Cienfuegos E, Coll P, Raulin F, Amils R, McKay CP (2006) Proc Natl Acad Sci USA 103:16089

Nicholson WL, Munakata N, Horneck G, Melosh HJ, Setlow P (2000) Microb Mol Biol Rev 64:548

Nørnberg P, Finster KW, Folkmann F, Gunnlaugsson HP, Hansen A, Kinch K, Lomstein BA, Merrison JP (2002) Second European Workshop on Exo/Astrobiology. Graz, Abstracts p 59

Oberbeck VE, Fogleman G (1989) Orig Life Evol Biosphere 19:549

Orgel LE (1998) Orig Life Evol Biosphere 28:91

Pedersen K (1997) FEMS Microbiology Reviews 20:399

Plaut JJ, Picardi G, Safaeinili A, Ivanov AB, Milkovich SM, Cicchetti A, Kofman W, Mouginot J, Farrell WM, Phillips RJ, Clifford SM, Frigeri A, Orosei R, Federico C, Williams IP, Gurnet DA, Nielsen L, Hagfors T, Heggy E, Stofan ER, Plettemeier D, Watters TR, Leuschen CJ, Edenhofer P (2007) Science 316:92

Rasmussen S, Chen L, Deamer D, Krakauer DC, Packard NH, Stadler PF, Bedau MA (2004) Science 303:963

Sancho LG, De la Torre R, Horneck G, Ascaso C, De los Rios A, Pintado A, Wirzchos J, Schuster M (2007) Astrobiology 7:443

Schopf JW (1993) Science 260:640

Seu R, Phillips RJ, Alberti G, Biccari D, Bonaventura F, Bortone M, Calabrese D, Campell BA, Cartacci M, Carter LM, Catello C, Croce A, Croci R, Cutigni M, Di Placido A, Dinardo S, Frederico C, Flamini E, Fois F, Frigeri A, Fuga O, Giacomoni E, Gim Y, Guelfi M, Holt JW, Kofman W, Leuschen CJ, Marinangeli L, Marras P, Masedea A, Mattei S, Mecozzi R, Milkovich SM, Morlupi A, Mouginot J, Orosei R, Papa C, Paternò T, Persi del Marmo P, Pettinelli E, Pica G, Picardi G, Plaut JJ, Provenziani M, Putzig NE, Russo F, Safaeinili A, Salzillo E, Santovito MR, Smrekar SE, Tattarletti B, Vicari B (2007) Science 317:1715

Smith PH, Lemmon MT, Lorenz RD, Sromosky LA, Caldwell JJ, Allison MD (1994) Icarus 119:336

Sotin C (2007) Nature 445:29

Stern SA (2003) Astrobiology 3:317

Stevens TO, McKinley JP (1995) Science 270:450

Stofan ER, Elachi C, Lunine JI, Lorenz RD, Stiles B, Mitchell KL, Ostro S, Soderblom L, Wood C, Zebker H, Wall S, Janssen M, Kirk R, Lopes R, Paganelli F, Radebaugh J, Wye L, Anderson Y, Allison M, Boehmer R, Callahan P, Encrenaz P, Flamini E, Francescetti G, Gim Y, Hamilton G, Hensley S, Johnson WTK, Kelleher K, Muhleman D, Paillou P, Picardi G, Posa F, Roth L, Seu R, Shaffer S, Vetrella S, West R (2007) Nature 445:61

Sundstrand LJ (2002) Second European Workshop on Exo/Astrobiology. Graz, Abstracts p 209

Tobie G, Lunine JI, Sotin C (2006) Nature 440:61

Trainer MG, Pavlov AA, DeVitt HL, Jimenez JL, McKay CP, Toon OB, Tolbert MA (2006) Proc Natl Acad Sci, USA 103:18035

Vidal-Madjar A, Lecavelier des Etangs A, Désert J-M, Ballester GE, Ferlet R, Hébrard G, Major M (2003) Nature 422:143

von Bloh W, Franck S, Bounama C, Schellnhuber H-J (2003) Orig Life Evol Biospherre 33:219

Waite Jr JH, Young DT, Cravens TE, Coates AJ, Crary FJ, Magee B, Westlake J (2007) Science 316:870

Wu C (2007) Nature 448:742

Zuber MT, Phillips RJ, Andrews-Hannah JC, Asmar SW, Konopliv AS, Lemoine FG, Plaut JJ, Smith DE, Smrekar SE (2007) Science 317:1718

Epilogue

If, after reading the eleven chapters of this book, we look back on the various scientific endeavours involved in studying the "great problem of life", we realize that there are probably more open questions than there are successfully answered ones.

Thus, the approaches, models, hypotheses, theories, experiments and computer simulations presented in the various chapters, as well as the speculations and lively controversies, have not yet led to a clear picture of the origins of life. While the individual research results discussed above provide important contributions, and individual pieces of the jigsaw puzzle, vital parts of the complete picture are still missing: the process of life's emergence from inanimate precursors can still not be described satisfactorily.

Pessimists are of the opinion that this area of science is in a hopeless state. Optimists, however, point to the great successes achieved in the last 50 years and are sure that these will form the basis for further progress. The more pessimistically inclined observer of the biogenesis scene is made aware of the large number of often conflicting models and "worlds". There are certainly no satisfactory answers to the following questions:

Were the first biomolecules formed in the primeval atmosphere or
At hydrothermal vents in the depths of the primeval oceans or
On the surface of the young Earth, at clay mineral surfaces or
Via thioesters?
In a "primeval soup" or
On pyrite crystals?
Was there a "pre-RNA world" before the hypothetical "RNA world" or
Did other molecular species act as information carriers?
Was there in fact an "HCN world" or
Were biomolecules "delivered" from outer space?
Could living systems from space have reached Earth without suffering great damage?

This series of questions could easily be continued!

All the models, hypotheses and theories on the origins of life are based on two assumptions:

> The unchanging validity and applicability of the laws of physics and of the fundamental natural constants across a period of four billion years
> The applicability of the theory of evolution to molecular systems

Most biogenesis researchers agree with these postulates, and also with the theory (formulated for the first time by Manfred Eigen nearly 40 years ago) that autocatalysis was an important principle for evolving systems. Very recently, there have been attempts to apply concepts from the area of "complex systems" to look for solutions (or partial solutions) to the problem of biogenesis. Four researchers from different scientific disciplines appear to have devised an important approach.[1] They wish to use "functional information" as a quantitative measure of the complexity (and thus of the biocomplexity) of a system. The concept of functional information is demonstrated using three examples [letter sequences, Avida artificial life genomes and biopolymers (RNA aptamers)]. The next few years will certainly make it clearer which of the biogenesis models is most likely to reveal the process of how the important steps from inanimate to living systems could have occurred. No one can predict whether the exact reaction pathways will ever be discovered.

Thus, the next years and decades will lead to many important results: from research laboratories where hypotheses and theories are forged, from geological and palaeontological institutes, and from all the scientific institutions which are working in the broad field of astrophysics.

One thing seems clear. As R. Shapiro put it so well in his book *Origins: A Skeptic's Guide to the Creation of Life on Earth*: "In the origin of life, however, if no surprises were forthcoming, that would be the most surprising result of all."

---

[1] Hazen RM, Griffin PL, Carothers SM, Szostak JW (2007) Proc Natl Acad Sci USA 104:8574 suppl.1.

List of Abbreviations

Å	Å ngström unit (10^{-10} m)
ADP	adenosine diphosphate
AEG	N-(2-aminoethyl)-glycine
AIBS	α-aminoisobutyric acid residues
AMP	adenosine monophosphate
AmTP	amidotriphosphate
ATP	adenosine triphosphate
AU	astronomical unit (the distance from the Earth to the Sun)
CAPS	Cassini plasma spectrometer
CDI	carbonyldiimidazole
CM	carbonaceous chondrites, Mighei-like group (M from Mighei in the Ukraine)
CoA	coenzyme A
CPL	circularly polarised light
DLR	Deutsches Zentrum für Luft- und Raumfahrt
DMPC	1,2-dimyristoyl-sn-glycero-3-phosphocholine
EANA	Exo/Astrobiology Network Association
EC	Enzyme Commission
EDC	1-ethyl-3-(3-dimethylaminopropyl)-carbodiimide
EDMA	ethylenediaminomonoacetic acid
ee	enantiomeric excess
ESA	European Space Agency
ESO	European Space Observatory
ETH	Eidgenössische Technische Hochschule (Zürich, Switzerland)
ETI	extraterrestrial intelligence
EUV	extreme ultraviolet
FMQ	fayalite-magnetite-quartz
FTT	Fischer-Tropsch type
GAP	glycolaldehyde phosphate
GC-MS	gas chromatography coupled with mass spectrometry
GLPC	gas–liquid partition chromatography

GRS	gamma ray spectrometer
HASI	Huygens atmospheric structure instrument
HC	hydrocarbon
HGT	horizontal gene transfer
HMG	hydroxymethylglutaryl
HPLC	high pressure liquid chromatography
HRSC	high resolution stereo camera
INMS	ion and neutral mass spectrometer
IR	infrared
IRSI	infrared space interferometer
ISM	interstellar matter
ISO	infrared space observatory
ISSOL	International Society for the Study of the Origin of Life
KG	ketoglutarate
KMQ	potash feldspar-muscovite-quartz
λ	wavelength
LDEF	Long Duration Exposure Facility
LDH	layered double hydroxide
MARSIS	Mars Advanced Radar for Subsurface and Ionosphere Sounding
mL	millilitre (10^{-3} litre)
μm	micrometre (10^{-6} m)
mRNA	messenger RNA
MRO	Mars Reconnaisance Orbiter
MS	mass spectrometry
NASA	National Aeronautics and Space Administration
NIMS	Near-Infrared Mapping Spectrometer
NDP	nucleoside diphosphate
nm	nanometre (10^{-9} m)
NMP	nucleoside monophosphate
NTP	nucleoside triphosphate
OAA	oxalacetate
OPA	olefinic peptide nucleic acid
PAH	polycyclic aromatic hydrocarbons
PC	phosphatidylcholine
PNA	peptide nucleic acids
PNPase	polynucleotide phosphorylase
P_i	inorganic phosphate
PP_i	inorganic diphosphate
PPM	pyrite-pyrrhotite-magnetite
PRPP	5-phosphoribosyl-1-pyrophosphate
PVED	parity violating energy difference
RCC	reductive citrate cycle
rRNA	ribosomal RNA
SDI	special differential image
SETI	Search for Extra-Terrestrial Intelligence

SIM	Space Interferometry Mission
SIMS	secondary ion mass spectroscopy
SIPS	salt-induced peptide synthesis
SPREAD	surface promoted replication and exponential amplification of DNA analogues
SSI	solid-state imaging
TBDMS	*tert*-butyldimethylsilyl
TCA	tricarboxylic acid
TMS	trimethylsilyl
TNA	(L)-α-threofuranosyl-($3' \rightarrow 2'$)-oligonucleotide
TPF	Terrestrial Planet Finder
tRNA	transfer RNA
UV	ultraviolet
VLT	Very Large Telescope
WMAP	Wilkinson Microwave Anisotropy Probe

Glossary of Terms

abiotic	not present in living organisms or linked with them.
achiral	achiral molecules are not "handed", i.e., they do not exist as non-superimposable image and mirror image.
acyl group	the group of atoms –CO–R (R = alkyl).
adaptive optics	technology to improve the performance of optical systems by reducing the effects of rapidly changing optical distortion.
aetiology	the study of causes or origins (for example of a disease).
accretion	process in which the size of something gradually increases by steady addition of smaller parts, e.g., the increase in size of planetesimals by gravitationally attracting more matter.
activation	process whereby something (e.g., a molecule) is prepared or excited for a subsequent reaction.
activation energy	the energy barrier that must be overcome in order for a chemical reaction to occur. The less reactive a reaction mixture, the greater the activation energy. Activation energy can be lowered by use of a catalyst.
aliphatics	organic compounds in which the chains of carbon atoms are either straight or branched, but not cyclic.
allotropes	different forms (modifications) in which an element can occur.
aminolysis	chemical reaction in which a molecule reacts with one (or more) molecules of ammonia or an amine, e.g., reaction of alcohols or alkyl halides.
amphiphathic	property of molecules with a hydrophilic and a hydrophobic functionality.
anhydride	acid anhydrides are formed from organic and inorganic acids via elimination of water.
animism	belief in souls; more generally the belief that souls inhabit all or most objects.
antiparticle	antimatter particle. Each particle has an antiparticle with the same mass and spin. The two species differ with respect to their electrical (and other) charges, the signs of which are opposite.
Apex chert	fossil-containing geological formation in western Australia.
aphelion	the point in the orbit of a planet or comet that is at the greatest distance from the sun.

asthenosphere	weak or "soft" zone of the upper mantle of the Earth which lies just beneath the lithosphere and in which seismic waves are greatly slowed down.
autocatalysis	form of catalysis in which the product of the reaction influences it by itself acting as a catalyst.
AVIDA	Computer software in the area of "artificial life" (AL), i.e., digital organisms.
Beilstein	the most extensive handbook of organic chemistry. It was founded by Friedrich Konrad Beilstein (1828–1901), professor in Göttingen and St. Petersburg.
Biuret test	used to show the presence of peptides (tripeptides and upwards) and proteins. A complex is formed with Cu^{2+} ions when at least two – CO–NH- groups are present.
caldera	large volcanic feature formed by the collapse of land following a volcanic eruption, often formed by the collapse of the roof of the magma chamber.
carbonyl group	name of the functional group $=C=O$ (found, for example, in aldehydes and ketones).
catalysis	increase of the rate of a chemical reaction caused by the presence of a catalyst, which remains unchanged at the end of the reaction. A catalyst increases the rate of reaction but does not alter the equilibrium state of the system.
collapse	process occurring in diffuse interstellar gas clouds which can lead to the formation of stars. The gravitational forces directed towards the centre of the gas cloud must be greater than the pressure forces directed towards its exterior.
conservation of angular momentum	fundamental principle of mechanics. The angular momentum remains constant if a system is only subject to internal forces (i.e., no external torque is present).
cosmology	study of the universe in its totality.
chert	type of rock consisting of microcrystalline quartz (SiO_2).
clathrate	chemical compound consisting of a lattice of one type of molecule which has trapped a second type of molecule.
cyanobacteria	microorganisms which are capable of photosynthesis (and thus of setting oxygen free).
derivatisation	the transformation of a chemical compound into a derivative.
deamination	cleavage of the amino group from amines, amino acids and amides.
diagenesis	physical and chemical processes which cause changes in a sediment after deposition. At higher pressures, the pores become smaller and the packing closer. There is a clear transition between diagenesis and metamorphosis, since temperature and pressure changes are also important in diagenesis.
dissipative structures	term coined by the Nobel Prize winner I. Prigogine for open systems which still have definite structures (non-equilibrium structures).
Doppler effect	physical phenomenon, occurring at all wavelengths (e.g., sound or light), when the source and the observer move towards or away from one another. Named after the Austrian physicist C. J. Doppler (1803–1853).
1,1-electrolyte	particle with one positive and one negative charge, e.g., Na^+Cl^- in a crystal lattice.
electron capture	uptake of an electron from the atomic shell by the nucleus. The following reaction occurs:

$$p + e^- \rightarrow n + \upsilon_e$$

i.e., proton and electron give a neutron and an electron neutrino.

endospores	small group of bacteria which are capable of forming particularly resistant cell forms known as spores.
entity	that which is perceived or known or inferred to have its own distinct existence.
entropy	quantity derived from the Second Law of thermodynamics. In a closed system, it is a measure of the irreversibility of a process. According to Clausius, it is a measure of the state of order of a thermodynamic system.
evolution	development process in which systems change clearly over long periods or many generations.
exhalations	gases emitted from the Earth's interior.
exogenous	something coming from outside (an organism or a system). Its opposite is endogenous
extra-terrestrial	existing outside the Earth.
extremophilic microorganisms	microorganisms which live under extreme conditions, e.g., in salt solutions or at extreme temperatures.
ferredoxin	water-soluble 12 kD protein with a [2Fe-2S] cluster.
Fischer-Tropsch synthesis	technical process for the preparation of liquid hydrocarbons (coal liquefaction) according to the general equation:

$$n\,CO + 2n\,H_2 \rightarrow (CH_2)_n + n\,H_2O.$$

fluxes	amount (of substance, energy etc.) that flows through a unit area per unit time.
fossils	mineralized or otherwise preserved remains or traces (such as footprints) of animals, plants and other organisms. Fossils are not found in volcanic rock or in primitive rocks.
fullerenes	can be considered as the third (or fourth) allotropic form of carbon. They consist of hollow spheres or ellipsoids with, for example, 60 or 70 carbon atoms.
genome	totality of all genes present in an organism.
glycolysis	highly important metabolic pathway in the living cell, in which carbohydrates (such as starch) are converted in many steps to low-molecular weight products such as pyruvate or acetyl-CoA. This catabolic process is used for the synthesis of substances, for example, ATP.
guncotton	form of cellulose nitrate; in its dry form, it is sensitive to shock and friction. Guncotton can easily be caused to explode under the influence of sparks or fire.
halophilic microorganisms	microorganisms which live in saline solutions of high concentration.
heterocycles	cyclic organic compounds in which the ring contains at least one other type of atom apart from carbon (mostly O, S, N).
homochirality	predominance of one of the two enantiomeric forms of a molecule (e.g., the L-amino acids in proteins).
horizontal gene transfer	exchange of genetic material between different species.
hybridization	in molecular biology: nucleic acid hybridization is the process of joining two complementary strands of DNA.
hydrolysis	reaction in which a molecule is cleaved by reaction with water.
hydrophilic	character of molecules which interact preferentially with water. The opposite is hydrophobic.
hydrophobic interaction	phenomenon occurring when non-polar (hydrophobic) molecules or parts of molecules interact, as, for example, in the hydrophobic amino acid side chains in proteins, which help to stabilise the protein structure.
hyperthermophilic microorganisms	microorganisms living at temperatures above 353 K.

interference	addition (superposition) of two or more waves, resulting in a new wave pattern.
interferometer	optical instrument using interference phenomena for precision measurements. Such instruments are used for measuring distances and angles, but in particular in spectroscopic measurements.
isotropic	isotropic substances have the same properties in all directions.
^{40}K–^{40}Ar method	method for the determination of the age of rocks. The amount of argon gas (^{40}Ar) formed in the radioactive decay of the potassium isotope ^{40}K is a measure of the time elapsed during which the rocks and minerals remained undisturbed.
Kalevala	legends collected in Finnish Karelia between 1827 and 1849 by the doctor Elias Lönnrot which he turned into an epic poem with 50 cantos (runes).
libration points	five points in space at which the attractive forces on a small body, due to two large bodies, cancel out.
ligases	enzymes which catalyse the joining up of two molecules, requiring energetic support from nucleoside triphosphates (generally ATP) to do so.
ligation	formation of a linkage between two atoms or molecules.
lithosphere	solid outermost layer of the Earth, which includes the upper layer of the mantle and the crust.
mass spectrometer	instrument for determining the mass of molecules, molecule fragments, and ions; one of the most important analytical instruments. Consists of 3 elements: ion source, mass analyzer and detector.
melanoids	dark yellow to brown-coloured pigments of differing constitution. They are formed in the Maillard reaction between sugars and amino acids.
mole fraction	now also called amount fraction. Amount of a constituent divided by the total amount of all constituents in the mixture. Amount fraction is equal to the number fraction: the number of entities of one constituent divided by the total number of entities in the mixture.
monotheism	belief in one deity or God.
OH sources	molecular clouds in which OH radicals have been detected.
palindrome	sequence of letters or words which reads the same in either direction.
partial pressure	in a gas mixture, each gas has a partial pressure which is the pressure which the gas would have if it alone occupied the volume. The total pressure of a gas mixture is equal to the sum of the partial pressures of each individual gas in the mixture.
permeability coefficient (p_x)	measure of the ability of ions to pass through a membrane. The value of p_x is obtained by dividing the diffusion constant of the ion in question (e.g., x) for the membrane passage by the thickness of the membrane.
plasma	term with various meanings. Introduced in physics by I. Langmuir (1930): a very hot gas, the properties of which are determined by cleavage of molecules to give ions, radicals, electrons and excited neutral particles. A majority of the material in outer space exists in the form of plasma.
plate tectonics	geological theory according to which the lithosphere consists of a number of plates which float on the asthenosphere and move very slowly (up to about 10 cm/year).
polyols	polyhydric alcohols, i.e., alcohols with two or more OH groups.
polytheism	belief in several or many gods.
prebiotic	occurring before biotic processes set in.

primary structure	in proteins: the sequence of the amino acids.
primer	oligonucleotide required as starter for a replication process.
prions	infectious protein particle, about 4–6 μm in diameter, which can cause fatal nerve diseases (scrapie in sheep, BSE in cows, and Creutzfeld-Jacob disease in humans). The proteins are folded abnormally (i.e., they have abnormal tertiary structures).
proteinogenic	the 20 naturally occurring amino acids occurring in proteins are referred to as proteinogenic.
proteinoids	protein-like polymers synthesised in the laboratory.
pyroglutamic acid	formed from glutamic acid by cyclisation (with elimination of water), e.g., on heating.
pyrolysis	thermal decomposition of substances.
quarks	fundamental particles of six types (up, down, charm, strange, top and bottom). Antiparticles of quarks are called antiquarks. Quarks have charges of 2/3 or $-1/3$ of the elementary charge.
racemisation	process in which an optically active compound is converted into a racemate, i.e., a mixture of equal amounts of the two optically active forms.
radial velocity (of a star)	velocity of the motion of a star in the line of sight, i.e., either towards or away from the observer.
Raman spectroscopy	spectroscopic technique based on the so-called Raman effect. Light incident on a molecule excites an electron, which then relaxes, generating Raman scattering. The weak Raman lines give information complementary to infrared (IR) absorptions. Raman spectroscopy normally uses monochromatic laser light in the visible range.
replication	process of identical duplication of a nucleic acid molecule.
retrograde	backward.
reverse micelles	micelles in which the hydrophilic parts of the micelle-forming molecules form the interior.
ribosomes	complex cell organelles in which protein biosynthesis occurs.
semipermeable membrane	membrane which will allow only certain particles to pass through it by diffusion.
solar wind	stream of charged particles ejected from the sun's corona, consisting mostly of high-energy electrons and protons (about 1 keV) able to escape the sun's gravitational field. Near Earth, the velocity of the solar wind is on average 400 km/s.
spontaneous generation	formation of living things from inanimate material without a long series of evolutionary steps.
stacking interaction	effects due to an ordered arrangement of molecules, such as the parallel stacking of aromatic ring systems.
radiochemical reactions	chemical reactions induced by high-energy radiation.
stromatoliths	large columnar calcium carbonate structures produced by cyanobacteria.
subduction	process in which one tectonic plate moves under another.
system	in thermodynamics, an *open system* is one which exchanges matter and energy with its environment. It never reaches real thermodynamic equilibrium, but only a "flow equilibrium". In contrast, a *closed system* is one which exchanges neither matter nor energy with its environment.
thermophilic microorganisms	microorganisms which live at high temperatures (up to 353 K).
T_m **value**	melting point at which half of the helix structure has been destroyed. In DNA, there is a linear relationship between the number of guanine-cytosine base pairs and the T_m value: the larger the G-C content of the DNA, the higher the T_m value.

tholins   a not clearly defined group of heteropolymeric organic compounds, formed on Titan under the influence of solar UV irradiation from hydrocarbons such as methane, ethane and other compounds, as well as nitrogen. Tholins have also been discovered in the proto-planetary discs of young stars. They are a subject of discussion in relation to the "PAH world" hypothesis.

topology   branch of mathematics; an extension of geometry. It defines and studies properties of spaces and maps, such as connectedness, compactness and continuity. Quantities such as lengths and angles are not taken into account.

T Tauri stars   stars which are in the T Tauri phase, i.e., a stage of development often characterised by extremely powerful stellar winds with velocities of 200–300 km/s.

tunnel effect   term from quantum mechanics referring to the fact (which classical physics cannot explain) that a relatively low-energy particle can overcome an energy barrier by appearing to "tunnel through" this "mountain". Consequence of particle/wave duality.

vacuoles   compartments in the interiors of cells that contain cell liquid and serve various purposes.

vital force   hypothetical force supposedly present in all living organisms.

wobble   inexact pairing of the third base of a codon with a non-complementary base of the anticodon which may occur when codon and anticodon interact; the first and second codon bases obey the rules of base pairing exactly.

Wöhler's synthesis of urea   synthesis of urea by heating ammonium cyanate, first carried out by Wöhler in 1828. This was the first time that an "organic" substance had been prepared starting from one which was "inorganic" without the help of a living organism.

Index

Printing: Krips bv, Meppel, The Netherlands
Binding: Stürtz, Würzburg, Germany